CAERPHILLY COUNTY BOROUGH
3 8030 08331
KT-194-693

THE SECRET POISONER

THE SECRET POISONER

A CENTURY OF MURDER

LINDA STRATMANN

YALE UNIVERSITY PRESS
NEW HAVEN AND LONDON

For information about this and other Yale University Press publications, please contact:

U.S. Office: sales.press@yale.edu yalebooks.com
Europe Office: sales@yaleup.co.uk yalebooks.co.uk

Typeset in Adobe Caslon Regular by IDSUK (DataConnection) Ltd
Printed in Great Britain by Gomer Press Ltd, Llandysul, Ceredigion, Wales

Library of Congress Cataloging-in-Publication Data

Names: Stratmann, Linda, author.
Title: The secret poisoner : a century of murder / Linda Stratmann.
Description: New Haven : Yale University Press, 2016.
LCCN 2015037508 | ISBN 9780300204735 (hardback)
LCSH: Poisoners—History—19th century. | Murderers—History—19th century. | BISAC: HISTORY / Modern / 19th Century. | TRUE CRIME / Murder / General. | SCIENCE / History. | SOCIAL SCIENCE / Criminology.
Classification: LCC HV6553 .S775 2016 | DDC 364.152/30922—dc23
LC record available at http://lccn.loc.gov/2015037508

A catalogue record for this book is available from the British Library.

10 9 8 7 6 5 4 3 2 1

This book is dedicated to the Crime Writers' Association, whose members include the very best of our fiction and non-fiction authors. The friendship and encouragement of fellow writers is beyond value.

Poisoning has always been a crime of the deepest dye, because of its treacherous and secret character.[1]

De toutes les armes que le génie de l'homme a inventées pour nuire à son semblable, le poison est la plus lâche; l'empoisonneur est le plus méprisable des criminels.[2]
(Of all the weapons that the genius of man has invented to harm his neighbour, poison is the most cowardly; the poisoner is the most contemptible of criminals.)

Contents

List of Illustrations

Author's Note

This book is not and was never intended to be a simple compendium of poison murders, a collection of unconnected famous and infamous cases, neither does it claim to be an academic study; rather it tells a single unified story of a duel of wits and resources, the main thrust of which lasted approximately a century. Starting with the dawn of modern forensic toxicology and the first serious demands for control of poison sales in the second decade of the nineteenth century, I have chronicled the efforts of science and the law to combat what was then seen as a grave threat to society: the secret homicidal poisoner. It was a campaign in which progress was often fraught with contradictions. Even as chemists devised increasingly sophisticated methods to detect toxins in human remains, their researches isolated and refined new and powerful methods of murder that posed fresh challenges to the analyst. Repeated attempts by parliament to restrict the availability of poisons to the public were confounded and sometimes obstructed both by the need of the poor to have easy access to cheap medicine and the means of destroying vermin, and by the professional demands of doctors and pharmacists.

The nineteenth century also saw the rise of the specialist – toxicologists and analytical chemists who were authorities on poisons, and whose antipathies and jealousies sometimes led to bitter and undignified wrangles that dented public confidence in medicine. When experts were called upon to testify at murder trials, there arose an uneasy interaction between the medical and legal professions, since the reputation of a witness could be made or destroyed in a single cross-examination.

I have illustrated the story by selecting those cases that show how poison murders stimulated both medical research and legal reform and influenced public opinion, and how increased restrictions on the availability of poison and improvements in methods of detection modified the methods of would-be murderers. I must apologise in advance, therefore, to anyone who was hoping that I would include a specific case and does not find it in these pages, also to those who feel that I have given a case that interests them far too little space. I, too, feel the loss of individual cases and detail, and I greatly regret that I was obliged to omit ones that, while fascinating in themselves, did not meet the brief or duplicated a point already made.

Those of you who have read my fiction will know that my lady detective does not spend her time wading boot-deep in unpleasant body fluids. I make no excuses for this – it just doesn't feel right in those books. Here, however, in the world of the real, nothing will be delicately hidden. I would be doing the reader of medical and crime history a terrible disservice if I drew a discreet veil over the suffering of poison victims, both human and animal, or the unsavoury details of the subsequent post-mortem examinations. Consequently, many scenes were uncomfortable to write about, but, in a truthful book, this is necessary. In order to understand where we are now, we need to know where we have come from.

* * *

I would like to thank everyone who has helped me in the preparation of this book, most especially my editor at Yale University Press, Heather McCallum, who suggested an idea to which it was impossible for me to say 'no', the enormously helpful staff at the British Library and The National Archives, and author Kate Clarke, who has done heroic work digging out some obscure references. During my research, I enjoyed recreating some of the historical dishes that were the last meals of poison victims, and must thank my husband Gary, who courageously agreed to taste them.

Introduction
The Awful Sentence of Death[1]

Wet and gloomy weather on 26 July 1815 did not prevent an immense crowd from gathering at an early hour to cluster expectantly about the portable gallows that had been drawn up by a team of horses before the Debtors' Door of Newgate Prison. Three convicts were due to suffer the extreme penalty. William Oldfield had raped a nine-year-old girl, and Abraham Adams had committed sodomy, but the real object of interest was twenty-two-year-old Eliza Fenning, a cook convicted of attempting to murder her master and his family by poisoning their dumplings with arsenic.[2] She had become a cause célèbre, not only because a young and attractive woman about to die inevitably excited sympathy, but also because of her steadfast refusal to confess to any crime. There was a strong feeling amongst the public at large that her guilt was far from certain.

Following her conviction on 11 April,[3] a petition begging for mercy had been sent to the Prince Regent, and Eliza had written both to the Earl of Eldon, Lord Chancellor, and to Viscount Sidmouth, the Home Secretary. She received many visitors to whom she protested her innocence in a manner that was passionate, open and convincing. Her supporters held meetings at which the facts of the case were analysed in minute detail, conducted experiments with arsenic and dumplings, and made allegations concerning the fragile sanity of her employer. So powerful was the movement in her favour that few believed that she would not be reprieved; she herself thought so until the morning of her execution. On the preceding day, Sir John Silvester, senior judge and Recorder of London, who had presided at the trial, and Mr Beckett, Under Secretary of State for the Home Office, responded to public concern by holding a meeting for the express purpose of discussing the fate of Eliza Fenning.[4] The whole case was gone over minutely, and every fact put forward in Eliza's favour examined. A second meeting came to the same conclusion as the first: that nothing should interrupt the course of justice.

At eight o'clock, the Sheriffs[5] walked along the grim subterranean passage that led from the Justice Hall of the Central Criminal Court (Old Bailey) to the press-yard, an open space that separated the prison from the law court,[6] and where the scaffold was kept when not in use.

The prisoners, emerging from their darkly forbidding condemned cells into the grey light of morning, were prepared for execution. The blacksmith had just struck off the men's shackles when Eliza, fortified with wine, water and a smelling bottle, freshly washed, and carefully attired in virginal white, joined them, leaning on the arm of a young doctor with whom she had been praying.[7] Oldfield the child rapist, who had once derided religion but had found faith while awaiting execution, had attached himself to Eliza's cause, perhaps in an effort to reinvent himself as a defender of injured youth. He urged her to pray, and they clasped hands, assuring each other that they would all be happy in the afterlife. The hangman John Langley then bound Eliza's elbows to her body with a cord, tied her hands together in front, and wound the rope with which she was to be hanged around her waist.

A procession was formed, the Sheriffs leading the way, taking the prisoners across the yard and out through the Debtors' Door to be confronted with the fearsome sight of the gallows. This was the waiting multitude's first view of Eliza, and to many she was a breathtaking image of innocence, even martyrdom, in a gown of pure white muslin, a muslin cap bound with white satin ribbon, and pale lilac boots, laced in front. Before she could mount the scaffold, the prison chaplain, Reverend Cotton, paused and asked Eliza if she had anything to say. Despite her terrible predicament, she maintained her composure and spoke firmly. 'Before the just and Almighty God, and by the faith of the Holy Sacrament I have taken, I am innocent of the offence with which I am charged.'[8]

Eliza was the first of the three to mount the steps to the platform, which she did with the steady demeanour she had shown throughout. The hangman, finding the white cotton caps he had brought too small to draw over her face, had to tie a piece of muslin over her eyes, and then, when this was insufficient, he added, despite her pleas, a dirty pocket handkerchief. He then placed the noose about her neck and secured the other end to the beam above. During all these proceedings, Eliza was repeatedly exhorted to confess, but continued to maintain her innocence, hinting mysteriously that God would make the truth known during the course of the day.

At half past eight, the signal was given and the hangman pulled the lever releasing the trapdoor. It was a short drop, only about eighteen inches, so Eliza did not die of a broken neck but by strangulation, the crowds looking on with an awful fascination. She was said to have struggled only briefly, and it is possible that, hidden by the platform from public gaze, Langley swung his weight from her legs to grant her a speedier death. The three bodies were left to hang for an hour, and the crowds dispersed, some to the Eagle Street lodgings of the Fenning family to show their support, some to create a disturbance outside the Chancery Lane home of Eliza's former employers, the poisoned Turner family, whose suspicions had led to her death. For three days, the mob, carefully watched by the police, milled around the Turners' premises, placing straw against it and muttering threats of destruction, until two of the most active members were arrested and the others dispersed.

Had Eliza been executed for murder, her body would have been sent to the Royal College of Surgeons for dissection, but, for a fee of fourteen shillings and sixpence, which her distraught family was obliged to borrow, it was returned to her parents for burial. In the back room of 14 Eagle Street,[9] the body was exhibited for five days and the narrow lane was crowded with sympathetic or simply inquisitive visitors. Donations were freely made and received and no doubt paid for the extraordinarily elaborate funeral that followed.

On 31 July, a procession proceeded up Lamb's Conduit Street to the burial ground of St George the Martyr, near Brunswick Square, Eliza's pall being borne by six young women robed in angelic white. Windows and rooftops along the route were crowded with spectators, and ten thousand people were believed to have swarmed out on to the streets to witness the event.

* * *

The fate of Eliza Fenning, a servant girl found guilty of attempting to poison her master and his family, aroused this extraordinary passion because she was far more than an icon of wronged womanhood and the heroine of a true-life melodrama. In the days following her conviction, she had become a political puppet, her cause taken up with alacrity by radicals and reformers, who used it, through the medium of the press and a flood of pamphlets, to attack the corruption of the law and rouse the public to indignation. The fact that none of the poison victims had died must have made it far easier for Eliza to gain widespread support. She had many notable champions amongst liberal politicians: Solicitor General Sir Samuel Romilly and advocate John Philpot Curran regarded Eliza as the guiltless victim of a prejudiced and badly conducted trial.

Throughout the nineteenth century, Eliza Fenning, her innocence an article of faith, was a symbol of the inadequacies of the law in cases of poisoning, which condemned to death a blameless young girl on the flimsiest of circumstantial evidence.[10] Indeed, one of the many pamphlets published on the case with the object of proving her innocence was entitled *Circumstantial Evidence*. Poisoning is a crime in which direct proof is rarely available, and, although legal writers had long argued that circumstantial evidence is not in itself inferior to direct – i.e. eyewitness or confession – evidence, in the mind of the public this is often not so.

Cited by the defence in later murder trials and appeals, romanticised by Charles Dickens, immortalised in song and on stage, resurrected in parliamentary debates on the death penalty, Eliza's memory was to weigh heavily on the minds of wavering jurors who were all too willing, when faced with an alleged poisoner in the dock, to give the accused the benefit of the smallest doubt.

In the numerous detailed contemporary analyses of the case, this controversial event provides us with a clear picture of the difficulties facing the law and medicine in dealing with the crime of poisoning at the dawn of the modern era of forensic toxicology.

CHAPTER ONE

The Devilish Dumplings

When surgeon John Marshall[1] arrived at 68 Chancery Lane, the home and office of law stationer Robert Gregson Turner, on the night of 21 March 1815, it was immediately obvious that he was dealing with a potentially fatal emergency.

At two o'clock that afternoon, Turner's two apprentices, Roger Gadsden and Tom King, aged seventeen and eighteen, and the servants, Eliza Fenning and Sarah Peer, had dined off a meat pie without suffering any ill effects. The family, consisting of Turner, his wife Charlotte, who was nearly seven months pregnant, and his father and business partner Orlibar Turner, visiting from his home in Lambeth, ate an hour later. Their hearty dinner consisted of beefsteaks served with potatoes, white sauce made with milk, and yeast dumplings. During the previous two weeks, Eliza had several times suggested to her mistress that she make yeast dumplings at which she said she was 'a capital hand'.[2] Charlotte said she preferred them made from bought dough, but Eliza, determined to make them herself, went ahead and ordered the yeast without her mistress's permission. The baker delivered the yeast on 20 March and Eliza used some to make a batch of light fluffy dumplings that were eaten by the servants and apprentices. The next day she made a second batch for the family dinner, but these did not turn out so well, the mixture failing to rise properly. Nevertheless, they were cooked and served.

Before the meal was over, Charlotte was afflicted by a severe burning pain and retired to her room, where she was violently sick. Soon afterwards, Robert and Orlibar were similarly affected. An apothecary, Henry Ogilvy of nearby Southampton Buildings, Chancery Lane, was sent for; he arrived at five o'clock, by which time Eliza, too, was unwell. Roger Gadsden was ordered to walk to Lambeth and fetch Orlibar's wife Margaret to help care for the family. By the time he arrived at Lambeth, he too had been taken ill, suffering so acutely that he thought he would die.

Margaret, accompanied by Gadsden, who vomited repeatedly all the way back home, arrived by coach at eight o'clock and took charge of the situation. It was probably she who sent Tom King to fetch the physician John Marshall, who had known the family for over nine years,[3] from his home in Half Moon Street just off Piccadilly. No one was in any doubt that all of those who were unwell had been poisoned, and King told Marshall that some of them were so ill that they might be dead before he got there.

A medical man called to a case of suspected poisoning in 1815 was obliged to perform three roles – physician, analyst and detective. Marshall had to decide very quickly whether the family had been poisoned or was suffering from disease, which would determine what treatment he would give.[4] Cholera can exhibit the same symptoms as poisoning: pain in the stomach and bowels, with vomiting and purging. It was usually treated by giving the patient fluids, with laudanum to stop the diarrhoea. In both cholera and irritant poisoning there is a burning acrid sensation in the throat, but in cases of poison this sensation is felt before the vomiting begins, whereas in cholera it appears afterwards as a reaction to the stomach acids in the vomit. Marshall would have been aware that cholera in Britain, when fatal, was rarely so in less than three days, whereas a victim of poisoning might die in hours.[5] In the Turners' case, several people who had eaten the same meal had been taken ill within minutes of each other and the pain in the throat had preceded vomiting. Marshall, with little time to spare, decided that the family had been poisoned, and the most probable culprit was arsenic, commonly used in homes as a vermin killer. His preferred treatment was to wash the stomach with copious fluid and give purgatives to eliminate the poison via the bowels as quickly as possible. Emetics, to induce vomiting, although used by other doctors, he regarded as unnecessary and tending to increase the painful irritation.

The first patient Marshall encountered was the cook, Eliza Fenning. She was lying on the stairs, flushed, retching and complaining of a burning pain in her stomach. He was told that she had already vomited a great deal. Margaret at once advised Marshall that the other members of the family were in a far worse state than Eliza and needed more urgent attention. Marshall ordered Eliza to drink water and milk, and she was taken up to bed.

Robert Turner was already in bed, prostrated by excruciating pain and appearing to Marshall to be on the point of death. The muscles of Turner's abdomen and viscera were contracting spasmodically, and so powerful were these movements that basins of his yellowish-green vomit included faecal matter driven up from the intestines. Turner was also suffering from violent diarrhoea, and his motions were a homogenous bright green like paint.[6] He complained of burning heat like a furnace or red-hot irons, which started at his tongue and went all the way down to his stomach, as well as insatiable thirst and a headache. His eyes were sensitive to light, and his extremities were cold. When he tried to walk to the commode he fell down in a fit and had to be helped back to bed.

The other patients showed similar symptoms, and it was feared that Charlotte would miscarry. Both she and her husband had lost the entire skin of their tongues, probably due to the corrosive effect of arsenic in the vomit. Over the next few days the cuticle of Charlotte's skin would peel from her body in flakes that resembled bran. Orlibar, while clearly very ill, was the least affected of the family. Marshall sent for Henry Ogilvy to find out what treatment had already been given and consult him on further action. Ogilvy confirmed that he had given no emetics, but had washed his patients' stomachs with sugar water or milk and water and given them doses of castor oil. Both men were convinced that they were dealing with a case of arsenical poisoning. Concerned that some of the poison had already left the stomach and was damaging the intestines, they decided to continue doses of purgatives until the faeces looked a more normal colour, while giving alkaline draughts to neutralise the effects of the arsenic. The patients were allowed to drink water, milk and mutton broth. Marshall and Ogilvy went to see Eliza in her garret, and, to their surprise, she refused to take any medicine, saying that her life meant nothing to her and she would sooner die. Only after very great persuasion did she consent to take something. They gave instructions to Margaret Turner about when the medicines were to be repeated and she agreed to stay up all night and ensure that this was done.

The next question to address was how the family had been poisoned, and the vital clue was the condition of apprentice Roger Gadsden, the only person taken ill who had not eaten with the Turners. At about twenty past three, at which time both Gadsden and Eliza were both well and the house had not yet been alerted to the condition of the Turners, Gadsden entered the kitchen and saw leftover dumplings that had been brought down from the dinner table. There was also some sauce at the bottom of the sauceboat. He was about to eat some dumpling, when Eliza exclaimed, 'Gadsden, do not eat that – it is cold and heavy, it will do you no good.'[7] He ate a piece the size of a walnut, but when she repeated the warning, he desisted, and sopped up the last of the sauce with a piece of bread. About an hour later he felt sick and vomited, but recovered and was well enough to start out on the journey to Lambeth to fetch Margaret Turner. It was only then that he became so ill that he realised he had been poisoned.

Margaret had tackled Eliza about the 'devilish dumplings',[8] to which Eliza replied, 'not the dumplings, but the milk, ma'am'.[9] The dumplings, which Eliza said she alone had mixed, had been made with flour taken from a storage bin (the same flour used to make the crust of the pie eaten earlier that day without ill effect), the remainder of the original batch of yeast, and milk left over from breakfast. Sarah Peer had gone out to fetch more milk after two o'clock, at which time the dumplings had already been mixed, and this new milk was used by Charlotte to make the sauce. Contrary to Gadsden's account of dipping a bit of bread in the dregs of the sauce, Eliza, still insisting that the problem was in the sauce and not the dumplings, claimed he had 'licked up'[10] three quarters of a boatful. Marshall's enquiries however revealed that Robert Turner, who was the most severely afflicted of all his patients, had eaten the largest quantity of dumpling but had had none of the sauce. The only dish eaten by all the sufferers was the dumplings. Eliza claimed to have eaten a whole one herself, although she was not seen doing so, and she made no mention of having had any sauce. The only two members of the household who had not been taken ill were Sarah Peer, who had gone out soon after serving dinner, and Tom King.

Suspicion was growing against Eliza, not only because she had made the dumplings and had offered no assistance to or shown concern for the Turners, but also because she was the only person in the household known to have a motive to harm the family. Eliza had been employed by the Turners from the end of January 1815.[11] About three weeks after Eliza's arrival, Charlotte saw her going into the apprentices' room in a state of partial undress, and spoke to her severely, giving her notice to quit. Eliza, claiming that she had only gone there to get a candle, was suitably contrite. Charlotte 'afterwards took compassion on her'[12] and allowed her to stay; nevertheless, the rebuke rankled. From then on, Eliza was sullen and less respectful towards Charlotte. Eliza told Sarah Peer that she no longer liked Mr and Mrs Turner, and was heard to say that '"she would have her spite out with her mistress"'.[13]

That night, John Marshall investigated the remains of the dumplings. He found them dense, tough and heavy, something he ascribed to a culinary failure and not poison.[14] He cut them into thin slices, and saw on the exposed interior 'white particles pretty thickly and uniformly distributed throughout the surface, which I conjectured to be white arsenic'.[15] This suggested to him that the powder had not merely been sprinkled on top of the dough but was very thoroughly and evenly mixed in.[16]

White arsenic in its powdered form is poorly soluble in cold water, rather better in hot, and was often to be seen in its undissolved state in the stomachs of murder victims, or lurking unnoticed as a gritty residue at the bottom of poisonous liquids. When no unconsumed poison was available for examination, a living victim of criminal poisoning whose vomit and faeces had not been preserved posed a conundrum even less soluble, and it was hard to obtain a conviction without a confession.

Chemical testing to show whether an unknown substance was arsenic was then in its infancy, and toxicology was only just emerging from darkness and superstition to become a science. Until the middle of the eighteenth century, medical men gave their opinions on whether or not someone had been poisoned based on symptoms, external appearance and, if the victim died, the condition of the organs. Grains of white arsenic found in the body or leftover food and drink were chiefly identified by appearance, but advances in chemistry were already suggesting that greater certainty might be found in a laboratory. Around 1710, Dutch chemist Herman Boerhaave experimented by subjecting poisons to the heat of glowing coals, and noticed that arsenic fumes smelled like garlic.[17] Over a century later, this was still a useful rule of thumb.

Marshall had no special apparatus for testing the dumplings, so he placed a slice on a polished halfpenny and held it over the flame of a candle on the blade of a knife. It gradually burned to a cinder, and gave out an unmistakable odour of garlic. When the coin cooled, the upper surface showed a silvery-white residue, which he believed to be a coating of volatilised white arsenic. This result, which became known as the reduction test, had first been noticed by a German professor of medicine, Johann Daniel Metzger, in 1787, who found that when heating compounds containing arsenic with charcoal, a deposit of white arsenic formed on a copper plate held over the resulting vapour. He further showed that when white arsenic was heated in a test tube with charcoal, shiny dark deposits of metallic arsenic, called 'mirrors', appeared on the upper cooler part of the tube.[18]

Although on his own admission the test carried out by Marshall was crude, it was an important step. Doctors and juries looking for evidence of poisoning wanted and expected to actually see the arsenic, either in powdered form or reduced to a metal. Failure to provide such evidence was a serious weakness in a criminal prosecution. In April 1806, when fifteen-year-old apprentice William Henry Wyatt was tried at the Old Bailey[19] for the attempted murder of his employer's family by putting arsenic in their coffee, the only proof that the white residue found in the coffee was arsenic was visual examination, and the evidence of an apothecary who placed it on a hot iron and testified that it smelled of garlic. Like Eliza Fenning, Wyatt had had free access to the poison kept by his employer and knew what it was. He had also drunk a small amount of the coffee, and said he had been ill. There seemed to be no motive for him to poison the family and he was acquitted.

Marshall also examined the knives used at dinner and found them so deeply tarnished that he asked if any vinegar had been used at dinner and was told it had not.[20]

The following morning, Marshall and Ogilvy returned to the Turners' house. Orlibar (who unfortunately never stated how much of the dumplings he had eaten) and Eliza were the most improved, even though Eliza had refused to take any further medication.

Orlibar was sufficiently recovered to make his own investigations. He questioned Eliza, who remained adamant that it was the milk brought by Sarah that was at fault, and he also looked for the only source of arsenic that had been in the house. A paper packet of white arsenic clearly marked 'Arsenic, deadly poison',[21] bought for the purpose of killing mice, had been kept in an unlocked drawer in the office, and he had last seen it on 7 March, about the same time that Eliza had started asking Mrs Turner if she might make some dumplings. The arsenic was now missing. Anyone might have had access to the drawer, and Eliza was known to go to it to get paper for kindling.

Eliza had not had the opportunity of cleaning the vessel in which the dumplings had been mixed,[22] and nothing suspicious was immediately visible; however, Orlibar added some water to the residue and stirred it up. Leaving the liquid to settle for half a minute, he held up the dish slantwise and saw a white powder settling at the bottom. He kept the dish in his own custody, allowing no one else access, and later handed it to Marshall and Ogilvy. Marshall diluted the scrapings with a kettle of warm water, stirred them, let them settle, and decanted most of the water. A white powder was distinctly visible. He washed the mixture again to remove the remaining flour, and collected and dried the powder that had fallen to the bottom of the pan, obtaining half a teaspoonful,[23] about thirty grains. If it was arsenic it was ten times the lethal adult dose. A few grains were placed between two polished copper plates, which were bound with wire and heated between the bars of a grate.[24] The garlic smell was distinctly noticeable and, when cold, 'each plate beautifully displayed the white fumes of the arsenic, with a resplendent silvery whiteness'.[25]

Marshall determined to eliminate all doubt by submitting the powder to further tests, which he believed to be the most sensitive and accurate ever devised, and for this he needed to consult an expert. Chemist Joseph Hume had been perfecting a new principle based on the pioneering work of Swedish chemist Torbern Bergman.[26] Hume first made a weak solution of the powder by boiling it in distilled water. Half an ounce of this fluid was put in a phial, and a glass rod dipped in a solution of silver nitrate and ammonia was applied to the surface. The result was an undulating yellow cloud staining the liquid, and a brown precipitate. Another portion of the liquid was similarly treated with a solution of copper sulphate and ammonia, producing a bright green precipitate.[27] Marshall and Hume were convinced. The powder was white arsenic. They also tested the flour in the bin and the yeast and found them both to be uncontaminated. Although they might have tested the vomit and faeces of their patients, the results from the dumplings were so conclusive that this was considered unnecessary.

Marshall was later to comment that there was enough arsenic in the dumplings to have killed a dozen families, and attributed the recovery of the victims to two factors: the rapidity with which the arsenic had acted as an emetic and purgative, carrying away a large part of the poison still with its coat of tough heavy dough protecting the stomach lining, and the plentiful fluids given early, as recommended by Mr Ogilvy, which had cleared the residue. All the victims eventually made a full recovery and Charlotte went on to give birth to a healthy daughter, Sophia.

On 23 March, Orlibar Turner again questioned Eliza, who was still in bed. Dissatisfied with her explanations, he told her he was going to have the matter investigated, and, together with John Marshall, went to the Public Office,[28] Hatton Garden, to make a statement to the magistrates. While they were gone, Eliza got up, dressed herself and tried to escape, but was prevented by Tom King, who fastened the outer and inner doors. An officer, William Thistleton, arrived to take her into custody. No poison was found on her, or amongst her possessions, but when her box was searched he found 'an infamous book, with a register on one of the pages, that explained various methods of procuring abortion',[29] then regarded as demonstrating 'the depravity of her morals'.[30] When questioned, Eliza now had two suggestions: the poison might have been in the yeast as it had had a sediment, or Sarah Peer, who was very 'sly and artful',[31] might have put it in the milk. Eliza denied ever going to the drawer where the arsenic had been kept, although the apprentices had seen her go to it often for scraps of paper to light the fire, and she claimed that she had never behaved improperly towards Gadsden but that it had been the other way about. On 30 March, her examination delayed until the witnesses were well enough to give evidence, Eliza was committed by the Hatton Garden magistrates to be tried for her life at the Central Criminal Court. The charge was that she 'feloniously and unlawfully did administer, and caused to be administered, to Haldebart [*sic*] Turner, Robert Gregson Turner, and Charlotte Turner, his wife, certain deadly poison, to wit, arsenic, with intent the said persons to kill and murder'.[32] Because of the violence of the victims' symptoms and the fact that there had been fears for Robert's life, murderous intent was assumed. Her only option was to plead innocence.

The offence was doubly despicable. A servant who killed his or her master or mistress or master's wife was held to be guilty of an aggravated form of murder known as petty treason, the act involving an element of betrayal and disloyalty in addition to violence.[33] Poisoners, as will be seen throughout this book, were often declared to be the worst kind of criminals. The secrecy of a poison attack, the idea that it had been planned in cold blood rather than happening in the heat of the moment, evoked a very particular horror. Poison threatened the established order of society. It placed a lethal weapon in the hands of those whom law and custom deemed should not be permitted to wield power. It enabled children to murder their parents, servants to murder their masters, and wives to murder their husbands. Magistrates, medical men and politicians, who were confident that they would never be stabbed to death in a drunken brawl, knew that they could still be poisoned by a daughter or a wife or a cook, and they felt threatened and vulnerable. This attitude was undoubtedly a powerful factor in Eliza Fenning's ultimate conviction, but the elevation of her cause to a matter of national concern enabled her to transcend such universal condemnation and win devoted champions, even amongst those who should most have feared her.

In most cases of suspected poisoning, there is no witness to the crime, so a trial must first demonstrate that poison has been administered, trace the possession of poison to the accused, establish that there was opportunity to poison the victim, and suggest a plausible motive. Eliza's trial was undoubtedly poorly conducted, hurried through before a brusque and unbending judge, with not all the relevant evidence or witnesses being produced. Tom King, whose location and actions at the time of the family's illness are largely unknown, and Mr Ogilvy,[34] the first medical man on the scene, were never called to give evidence, and Silvester refused to admit an affidavit brought by Eliza's father. Dr Marshall, while stating his certainty that arsenic was in the dumplings, based his opinion on the symptoms, and was not asked to give details of the tests that had been carried out. As one of Eliza's most passionate defenders said: 'There was not a particle of proof on [*sic*] the trial that any poisonous ingredient whatever was in the dumplings.'[35]

The jurymen, undoubtedly believing that society should be protected from cooks who put arsenic in the dinner, were convinced of Eliza's guilt by a strong chain of circumstantial evidence, and reached their verdict in a matter of minutes; but the public, moved by the prisoner's youth and piety and affecting declarations of innocence, distrusted the law and examined the chain for significant gaps, in the hope of winning a reprieve that could only be granted by a higher authority. The mandatory death penalty for attempted murder was part of the 'Bloody Code', as it came to be known, a system of harsh punishments intended to protect the lives and property of the landed classes. Attempted murder would remain punishable by death until 1861. It was not until 1823 that judges were given the authority to commute sentences, so even if Sir John Silvester had wanted to show mercy he was not empowered to do so.

Despite their sufferings, both Orlibar and Robert Turner were willing to sign a petition for Eliza's reprieve, but, after an interview with Silvester, not only did they find him steadfast in his opinion of her guilt, but he also advised them that if they signed the petition, then suspicion would fall on their family.[36] They did not sign.

As testimonials to Eliza's good character were collected, counter-balanced by allegations that she had tried to poison another family before the Turners,[37] public opinion was divided between firm belief in her spotless innocence and the call for her to be hanged as an example to all potentially murderous maidservants.

Eliza, languishing in Newgate and hoping for a reprieve, changed her story again. She asked to see the Turners on Monday 24 July, two days before the date set for her execution, and the family paid her a visit, in the hope of receiving important information. To their astonishment, Eliza turned on Charlotte. Alleging that there had been impropriety between her former mistress and Tom King, she said that Charlotte could account for how the arsenic had got into the dumplings better than she could.[38] On the following day, Eliza sent for Tom King 'to tell him that he had done it'.[39] In *Eliza Fenning's Own Narrative* (actual authorship unknown), published after her death, it was claimed that Tom had sneaked into the kitchen after the dumplings had been mixed, while Eliza was in an adjoining room washing knives.

On the day before the execution, information was received from a Mr Gibson of a firm of chemists and druggists, who stated that, in the previous September or October, Robert Turner had come to the premises in a 'wild and deranged state',[40] saying that he would destroy himself and his wife unless he was placed under restraint. Gibson told Silvester and Beckett that he had advised Orlibar not to proceed with the prosecution. Why he had left it until that late hour to make his statement was never explained.

None of these last-ditch accusations could save Eliza.

The furore that followed Eliza Fenning's execution focused much-needed attention on the inadequacies of the trial process in general, and her rushed and incomplete trial in particular, but did this necessarily mean, as was vociferously claimed then and is still claimed today, that she was innocent? A fact often put forward in her favour is that she ate some dumpling herself, claiming to have consumed almost a whole one, and was genuinely ill. Many of her defenders stated, without offering the source of their information, that she was in a worse condition than the Turners, but that claim is negated by the medical men on the scene, who reported that Eliza was less affected than the others and recovered quickly with almost no treatment. She can hardly have eaten as much as she claimed. Did she try a piece of dumpling to prove that it was innocuous? If guilty, did she do it to divert attention from herself, or even, since she refused medicines, to take her own life? It might seem unlikely, but it is not unknown: murderess Mary Wittenback went on to do just that.

Forty-one-year-old Mary had been on bad terms with her husband Frederick for some years. On 21 July 1827, he came home from work in good health, and ate two thirds of a suet pudding made by his wife. Soon afterwards, Frederick was taken very ill with burning pains in his stomach, and went into the yard to vomit. Mary showed the leftover pudding to Mrs Davis, in whose house the Wittenbacks lodged, asking if she thought there was anything in it. Some minutes later, Mary was also taken ill, having eaten the rest of the suspected pudding. The landlady's cat, finding Frederick's vomit in the yard, ate it, retched, crawled away and was later found dead. Mrs Davis called a surgeon, who examined Frederick and administered draughts of warm water and an emetic (John Marshall would have disapproved), which 'operated considerably'.[41] He gave no treatment to Mary and it is not known if he examined her. A second surgeon called round and applied the recently invented stomach pump to both patients. Early the following morning, Frederick died, but Mary, declaring that she would rather die than recover, was removed to an infirmary and survived. When Frederick's body was opened, the inflammation of the stomach lining suggested the action of a mineral poison, but the use of the stomach pump meant that no trace of poison could be found. Mrs Davis's husband had helpfully buried in a dust-heap any vomited matter not eaten by the cat, and presumably any faecal material had also been disposed of, since neither as available for testing. The presence of poison could, therefore, only be inferred from the symptoms of the deceased. Mary Wittenback might well have escaped justice but for one impediment – she confessed. Convinced that her husband was being unfaithful, she had bought threepence worth of white arsenic and mixed it into the flour from which she had made the suet pudding. After Frederick was taken ill, 'she ate the remainder of the pudding with the hope that it would prove fatal to herself, so horror-stricken was she at the excruciating pains the poison had brought upon her hapless husband'.[42] Mary was hanged for murder on 17 September 1827.

Eliza's motive to kill the Turners might, at first glance, appear inadequate. Frederick Hackwood, the headmaster and historian whose biography of her most ardent champion, radical publisher William Hone, was published in 1912, wrote: 'A crime of such enormity produced by so very slight a cause has probably never occurred in the history of human depravity.'[43] This only goes to show that Hackwood did not make a great study of murder trials. As barrister John Paget pointed out in 1876, 'all persons who have any practical experience of criminal courts know how slight and insignificant are the motives which sometimes impel to the commission of the most appalling crimes'.[44]

Paget's point is well illustrated by the case of twenty-year-old Isabella Hopes, which has some echoes with that of Eliza Fenning. On 5 November 1834, market gardener Elizabeth Cambridge noticed that there was money missing from her cash box, and, suspecting her servant Isabella, searched the girl's clothes and found a substantial sum in gold and silver. She tackled Isabella, who confessed that she had taken the money and repaid it. Remarkably, Elizabeth did not dismiss the thief, and agreed that the matter be regarded as closed, provided Isabella promised never to repeat the offence. Next morning, Elizabeth went out, leaving Isabella in the house; when she returned at one o'clock, the servant had gone. That evening, Elizabeth made tea for both herself and her lodger, Isabella's twelve-year-old half-brother John Walker. Soon afterwards, both were taken ill. The surgeon, Mr Moon, sent his assistant, who recommended they drink warm water. A Mrs Budd, who worked on Elizabeth's land, sat up with the patients and took care of them. When Moon called the following day, both were recovering and all the vomit had been thrown away. He took away a sample of the tea and, noticing that something was mixed with it, sent it to be tested. On 9 November, Isabella called to see Elizabeth, but was unable or unwilling to explain why she had left the house. Elizabeth questioned her about the tea, but Isabella denied putting anything in it.

Two months later, Isabella called again and told Elizabeth that Mrs Budd had confessed to her that she had put arsenic in the tea caddy. The only person with motive and opportunity was, however, Isabella, and she was arrested on 19 January 1835, by which time any medical evidence of poisoning was long gone and the original sample of tea had been thrown away.

At first, Isabella persisted in her accusations against Mrs Budd, but, the following day, she admitted that everything she had said about Mrs Budd had been false and that it was she herself who had poisoned the tea, explaining: 'In consequence of my having robbed my mistress, after she had gone to market on the 6th of November last, my mind was so uneasy, I could not stop there, and after she had forgiven me for the robbery I could never look her in the face...'[45] She added that she could not help herself, even though she knew her own brother would drink the tea: '[T]he reason I charged Mrs Budd was I thought it would save me.'[46] The source of the arsenic was not a mystery, as it was kept in the house to kill rats, and Isabella had been cautioned not to meddle with the little parcels because they were poisonous. Isabella's trial took place at the Old Bailey on 2 February. She was found guilty of attempted murder and sentenced to death, but was recommended to mercy by both the jury and Elizabeth Cambridge because of her previous good character and the fact that she had confessed. Isabella was reprieved and transported to New South Wales.

* * *

In the years that followed Eliza Fenning's execution, there were several rumours of deathbed confessions to the poisoning of the Turners. Although the yeast and flour used to make the dumplings had been tested and found to be uncontaminated, an unnamed baker was reputed to have confessed as he lay dying in a workhouse, claiming, in an unwarranted excess of zeal, to have murdered Charlotte.[47] Another confession was supposed to have come from an unnamed nephew of 'Mr Turner' (presumably Orlibar, since Robert was only twenty-four at the time of the events), who was said to have resided with his uncle and aunt (it should be recalled that, in 1815, Orlibar and his wife lived in Lambeth). Even if such a nephew existed, it has never been suggested that any member of the Turner family other than the ones who dined that day was in the Chancery Lane house when the poisonings occurred. In 1829, uncorroborated reports were published that Robert Turner had died in an Ipswich workhouse after confessing to the crime.[48]

The facts of the Eliza Fenning case and the comings and goings and statements of Eliza, the Turners and their visitors have been examined minutely many times, but, in the end, the simplest explanation is that Eliza, quite possibly not intending to kill anyone, had decided to revenge herself on her employers for being shamed and rebuked by taking the power of poison into her hands and giving them all a stomach ache. Terrified by the mayhem she had wreaked, she did everything she could to divert attention from her guilt by eating a small piece of dumpling and blaming everyone around her. The law did not then envisage that a poisoner had any intention other than murder, and, even if it had, the severity of the symptoms would not have allowed for a lesser charge. Had Eliza confessed straight away, however, and been suitably contrite, she might well have received a recommendation of mercy from the jury and been reprieved.

But why was there a gap of two weeks between laying her hands on the arsenic and using it? It is not known when Orlibar Turner had last dined at his son's house, but Charlotte Turner told John Marshall that Eliza had repeatedly asked her if her father-in-law was to dine there that day. Tuesday 21 March may have been the first opportunity that all the family were eating together.

CHAPTER TWO

Smothered in Onions

Arsenic was a particularly problematical poison to the nineteenth-century medical and legal profession. Cheap and easily available, its wide range of uses in medicine, trade, agriculture and animal husbandry, and its presence on the domestic scene in products such as flypapers and rodent poison, meant that it was responsible for a great many accidental deaths. During the first half of the century it was sold in its pure form without regulation. Almost tasteless and odourless, it was easy for a would-be murderer to mask the presence of arsenic from the unsuspecting victim, the cholera-like symptoms it produced affording a clever defence counsel ample opportunity to place reasonable doubt in the minds of a jury. Although many poisons were readily available, arsenic alone was responsible for nearly half of the homicidal poisonings brought to court. In the 504 criminal cases brought before English courts in the period 1750 to 1914,[1] arsenic features in 237. The second most frequently used poison was opium and its preparations, chiefly laudanum, which were responsible for fifty-two cases; then followed nux vomica and its powerful extract, strychnine,[2] forty-one; organic acids,[3] thirty-four; mercury, thirty-two; inorganic acids,[4] twenty-four; and other plant alkaloids,[5] nineteen.

Meanwhile, the efforts of chemists to isolate the active principles in poisonous organic compounds were unintentionally placing powerful new weapons in the hands of professional men – pure toxins that, for many years, defied the attempts of analysts to discover them in the remains of murder victims. When those with no access to such sophisticated poisons turned to substances other than arsenic, their choice was often limited to complex and distinctive materials with a higher risk of detection.

The poorer and lower middle classes chiefly obtained their medicines through apothecaries, who both prescribed and dispensed remedies. The profession was not legally regulated until the Apothecaries Act of 1815,[6] which laid down, amongst other things, that apothecaries must be over twenty-one years of age, must have been both examined and certified, and must have served a five-year apprenticeship. These requirements were to be managed by the Society of Apothecaries. A separation was beginning to take place between two professions: apothecaries, who were more nearly akin to today's general practitioners, and chemists and druggists, who were exempted from the 1815 act and became today's pharmacists.[7] Non-medicinal poisons such as cleaning products and rodenticides could still, however, be purchased from grocers. It was to be many more years before the sale of poisons was restricted to persons with a recognised qualification.

The first attempt to impose a legal control on the sale of poisons came in 1819, when a bill was brought before the House of Commons. The main concern at that time was not criminal poisoning, but fatal accidents caused by unlabelled poisons being consumed in error. It was proposed that every container in which certain stipulated poisons were sold – including arsenic, prussic acid and laudanum – should have a printed label affixed with the word 'POISON'. It was further recommended that arsenic should be mixed with carbon and oxalic acid (coloured pink), which was often mistaken for Epsom salts. The bill went to two readings and was referred to committee; however, when the Society of Chemists and Druggists met on 24 June to discuss the bill, it was felt that some of the provisions were 'likely to embarrass the dispensing of medicines and not calculated to effect the object intended'.[8] The society sent a petition to parliament opposing the bill, which was never debated in the House and was withdrawn. Perhaps parliament thought the chemists and druggists knew their business best and that the issue was better dealt with by a professional code of practice than legislation.

The society circulated a report to its members recommending the 'POISON' label idea. The report also proposed that any preparation 'likely to produce serious mischief'[9] should be labelled with its name and that no one should be allowed to sell poisons unless of sufficient age and experience to appreciate 'the great caution necessary in avoiding the sale of them to improper or ignorant persons'.[10] How many chemists and druggists complied with these suggestions is unknown, and, meanwhile, grocers continued to sell pennyworths of arsenic in unlabelled twists of paper to customers of all ages.

The unrestricted sale of poisons remained a matter of public concern. 'I observe you frequently publish letters addressed to you, proposing various precautions which ought to be taken by druggists in vending poisonous drugs,'[11] stated an anonymous correspondent of *The Times*, who went on to recommend that no one should sell poison to any individual unless that person was accompanied by another. This practice, which was already used voluntarily by some druggists, would, it was optimistically stated, defeat the purpose of the suicidally inclined, and either deter those intent on murder or lead to the detection of criminals. For the time being, parliament declined to act on all such advice.

Opium, the second most widely used poison for criminal purposes, is about 12 per cent morphine, a highly addictive analgesic. It had long been available in the form of laudanum, a preparation of powdered opium and alcohol in which the opium was about one twelfth by weight. Other more concentrated preparations were black drop and Battley's sedative liquor (in which the opium was dissolved in vegetable acids).[12] Toxicologist Sir Robert Christison described it as 'the poison most generally resorted to by the timid to accomplish self-destruction, for which purpose it is peculiarly well adapted on account of the gentleness of its operation'.[13] This very gentleness may also have recommended itself to a murderer who did not wish the victim to suffer. Its taste, however, was extremely bitter. Victims of opium poisoning tended to be frailer than the killer – those who could easily be induced or forced into taking a draught, usually children and sometimes the elderly. Robbers used it as a soporific, the plan being 'to deaden the sense of taste by strong spirits, and then to ply the person with porter or ale, drugged with laudanum, or the black drop, which possesses less odour'.[14] If the poison had not already been absorbed before death, which it often was as the victim might take many hours to succumb, the characteristic odour of laudanum would remain in the stomach.

In 1803, a crystalline extract was derived from opium by French chemist Charles Derosne, who believed that the extract was the source of the drug's properties, although German pharmacist Friedrich Wilhelm Adam Sertürner is credited with finally isolating the active principle in 1804. Sertürner experimented on himself and three of his pupils to determine the extract's effects on the human body and named it 'morphium' after the Greek god of dreams. Sertürner's work is believed to be the first ever isolation of an alkaloid, a class of poisons that would frustrate the best efforts of forensic toxicologists for very many years. Since the compound was insoluble, it was combined with acids to form soluble salts, the principal ones used being the acetate, sulphate and hydrochlorate.[15] Morphine, as it became known, did not attract a great deal of attention from the medical profession until Sertürner began to make it available in 1817. It was first manufactured as a commercial product in 1827, but was not widely used for another thirty years – the crucial event being the development of the hypodermic needle in the 1840s and 1850s, the use of which created a faster and more powerful effect.

With arsenic and opium so freely and cheaply available, it was only occasionally that other poisons took centre stage in a criminal case. It has been known since antiquity that the nux vomica bean, *Strychnos nux-vomica*, contains a powerful poison. In low doses it causes restlessness and muscle spasms, and it was, therefore, thought to be a stimulant. An extract of the beans was used to treat a wide variety of complaints, including fever, hysteria, paralysis and the plague. In 1818, French chemist Pierre Joseph Pelletier and pharmacist Joseph Bienaimé Caventou isolated the active ingredient, the formidable and intensely bitter toxin strychnine. This could be delivered in minute doses in pills, tinctures and mixtures, but, as late as 1829, it had not yet been introduced into British medical practice.[16]

Strychnine is not an obvious choice for murder because of its taste; however, it is toxic in very small quantities, approximately half a grain, and a victim given it in a liquid such as bitter beer might well consume a fatal dose before noticing anything amiss. It was occasionally used in suicides, or caused death after being taken in error.[17] The pure alkaloid was not to feature in a murder trial for another twenty-seven years.

However, the flat, hard, brown nux vomica beans, commonly called crow figs, are about 1.5 per cent strychnine, and had long been on sale to the public for the preparation of an infusion to kill vermin. Even in that form they could provide temptation to the criminal. On 13 August 1822, in Monmouth, Rachel Edwards was found guilty of the murder of her husband, William. While death was undoubtedly due to arsenic, Rachel had also prepared her husband's breakfast beverage by steeping sliced crow figs in the teapot, disguising the taste with sugar and cream, and serving it up with bread and butter.[18]

In August 1831, Eliza Harding, a fourteen-year-old servant who had lived with the family of Southampton tailor Edward Boyland for seven weeks, asked a chemist for something to kill mice, and was sold half an ounce of nux vomica beans. Shortly before breakfast on 5 August, Eliza came into the kitchen bringing with her a tin cup, which she put on the fire. Boyland asked her what it was and she said it was something she was going to take herself. He looked in the cup and saw five brown beans. As he left the kitchen, he heard Eliza say, 'I'll be curst if I don't.'[19] When he returned, the cup had been removed and, suspecting nothing, he made tea using water from the kettle. Mrs Boyland, a child and a lodger drank a small amount of the tea, which tasted bitter. Mrs Boyland checked the contents of the kettle, saw two beans and noticed a suspicious yellow tint to the water. Mr Boyland, however, had drunk a teacupful. He began to experience severe thirst, with heat in his throat and stomach, and started frothing at the mouth. A surgeon was called, who concluded that Boyland had been poisoned and ordered an emetic.

Eliza was closely questioned about the suspect beans. At first, she denied all knowledge of how they had got in the kettle, but then said that they had been dropped in by accident. Finally, she announced dramatically that she had drunk a pint of what was in the kettle and would be dead in an hour. Two more beans were found in her pockets and some pieces she had tried to dispose of in the fire were recovered.

Mrs Boyland took the terrified weeping girl and the beans to a magistrate, but nothing could induce her to say why she had poisoned the family. It was said that she had always been treated well, but was of 'a sullen and morose disposition'.[20] By the time Eliza was tried for attempted murder at the following year's Lent Assizes, she been in custody for seven months, and Mr Justice Gaselee must have thought she had learned her lesson. Citing some minor confusion in the evidence, he directed the jury to acquit, delivered a stern admonishment to the prisoner before dismissing her. He did, however, make a statement to which all juries should have paid careful attention: '[I]f a person administered poison, with intent to murder, it was not material that the party was ignorant of the exact strength of the poison, and that in consequence death had not ensued: the intent alone was sufficient.'[21]

Another notorious poison, hydrogen cyanide, commonly called prussic acid, is a compound derived from the dye Prussian blue. As little as one grain has proved to be a fatal dose. In its medicinal dilution, it was first popularised in 1819 as a cure for a variety of conditions, including nervous diseases and consumption. In overdose, it leads to collapse in seconds, the rapidity of its action making it especially appealing to suicides, who could purchase sufficient for their needs at any chemist's for sixpence. Available in a confusing variety of strengths, depending on where it was prepared, hydrogen cyanide's unforgiving nature meant that it sometimes featured in medical accidents and manslaughter trials of careless apothecaries. It was a less popular option for murder, since its distinctive bitter-almond smell and pungent taste made it hard to administer to victims without their noticing. At post-mortem, it was easily identifiable – the stomach contents, blood and body cavities exhaling the unmistakable odour.

In the rare prussic acid murders, the drug was usually administered either as a subterfuge, by persuading the victim it was medicine, or heavily disguised. In March 1829, Richard Freeman, a seventeen-year-old apothecary's apprentice of Leicester, was tried for the murder of his employer's servant Judith Buswell. Judith, who had been found lying on her bed, the bedclothes pulled tidily up to her chest, her arms under the covers, and a corked and paper-wrapped bottle of prussic acid by her side,[22] was seven months pregnant. Freeman, who was responsible for unpacking deliveries of drugs, was said to have boasted of his intimacy with the deceased and claimed that he knew how to obtain an abortion. When notifying the girl's family of her death, he had urged them to get the body moved quickly to prevent an inquest. Had Judith, either on her own account or abetted by Freeman, decided to take her own life, or had he murdered her, inducing her to swallow the poison by recommending it as a means of procuring an abortion? The discovery of six clean napkins and an apron in Judith's bed suggested that she was anticipating aborting the child. The verdict came down to whether the action of prussic acid was too rapid to enable Judith to cork and wrap the bottle, then arrange herself in a composed posture, drawing the bedclothes smoothly up, before she succumbed to its effects. Medical experts at Freeman's trial conducted experiments on cats and dogs to try to determine the speed of the poison's action; finding that the animals died in a matter of seconds, they declared the scenario to be so improbable as to be almost impossible, although the surgeon who had first been called to the scene was prepared to say it might just be possible. Freeman, perhaps because of his youth, was given the benefit of the doubt and acquitted. He later qualified as a pharmacist.

In January 1833, a tearful Monsieur Reges, who had tried to hang himself while awaiting trial, was brought to account in Paris for the robbery and murder of a Monsieur Ramus, who had been poisoned, beheaded and dismembered. The bloodstained hat, shoes and braces of his victim were found in the privy of the house where Reges was lodging, and bottles of laudanum and prussic acid were found in his possession, as was the stolen money. Reges had administered the prussic acid in brandy, but the effects were not as rapid as he had expected; indeed, it was believed that the victim, with care, might have survived. When Reges started dismembering the body in the hopes of concealing the crime, M Ramus was still alive. The prisoner was found guilty and sentenced to death.[23]

Murderers also made use of non-medicinal and industrial poisons when they were readily to hand. In April 1838, a seventeen-year-old servant,

Joseph Walters, was discovered by his employer, Elizabeth Kilsbey, in her stables 'in a certain situation'[24] with her mare. It was a hanging offence and, when she threatened to report him, he impulsively laced her tea with a stain remover containing nitric acid. She spat out the nasty-tasting liquid and was unharmed. Walters was found guilty of attempted murder. His death sentence was commuted and he was transported to Tasmania.

Richard Overfield was a thirty-five-year-old labourer in a carpet factory at Bridgnorth, where he earned eleven shillings a week. Between 11 and 12 on the morning of 21 September 1823, a neighbour, Louisa Davies, heard his wife, Anne, screaming and ran to see what was the matter. The hysterical mother was clasping her three-month-old son, Richard junior. The child was writhing in agony, his lips blistered and white with foam. Anne put her mouth to the child's lips and her tongue in his mouth and noted that something made her mouth smart, tasting hot, sour and bitter. Anne turned to her husband and demanded to know what he had done to the child. He protested that he had done nothing, but said that a black cat had got up to the child's mouth and brought up some foam and he had knocked it away. The anguished mother dashed out of the house, taking her baby to the nearest surgeon, Mr Spry; so violently was she crying that, for a time, she was unable to speak, while the child was throwing himself about in such paroxysms of agony that Spry could barely hold him steady. At length, Anne pointed to the child's mouth and asked the surgeon to taste his lips. Spry did so, and at once detected a pungent acid, which he thought was sulphuric, then known as oil of vitriol. The child's lips were white and shrivelled, and there were blisters on their inner surface, as well as in the mouth and on the tongue. His stomach was so damaged that the child was unable to vomit. All the surgeon could do was administer magnesia[25] suspended in water and copious fluid, but, suspecting foul play, he followed Anne back to her house. Spry asked Richard if he had given anything to his child by mistake, but Richard repeated his story about the black cat sucking the child's breath, a common local superstition. Spry had the cat brought to him, put his finger to its mouth and tasted it. There was no trace of acid. He also searched the house for acid and found none. The treatments continued as directed, but, by one o'clock that afternoon, the child could no longer swallow and, two hours later, he died.

Overfield, when questioned by Spry, had said he did not know what oil of vitriol was, yet it was used in the dye-house of the carpet factory where he worked. On the day of the child's death a neighbour, Mary Nicholls, had been looking out of her window and had seen Overfield dispose of something in his garden. She pointed out the spot to Louisa Davies, who recovered the object, a small phial. At the post-mortem examination of the child, almost a pint of blood-coloured fluid escaped from the abdominal cavity of the tiny body. The fluid was shown to contain sulphuric acid, as did the liquid in the phial. The stomach was so corroded that it had been reduced to a black substance the consistency of wet brown paper.

On the day after Richard junior's death, the superintendent at the carpet factory was astonished when Overfield calmly reported for work as usual. Overfield was arrested and, during the course of the inquest, he attempted to cut his throat with a piece of broken plate, which he had ground to a sharp edge.

He was tried for murder at the Shrewsbury Assizes on 19 March 1824, when the jury, after two or three minutes' deliberation, found him guilty. The only reason it had taken the jurors that long was their inability to imagine what his motive could have been. It was only after the verdict was announced that Mr Justice Park revealed something they had not known. Anne had been pregnant before the marriage and Overfield had often declared his hatred of both Anne and the child, saying he 'would not support her or her bastard'.[26] The parish had paid him to marry Anne before the child was born, in order to legitimise it, so that he, and not the parish poor rate, would be responsible for its upkeep.[27] Overfield was sentenced to hang three days later, and was taken away without displaying any emotion.

* * *

In 1815, John Marshall was satisfied that the chemical tests of Joseph Hume had provided the perfect method of identifying arsenic, and praised them for 'the accuracy, the elegance, and the certainty with which they serve to detect the minutest portion of the white metallic oxide',[28] asserting that they would find traces of arsenic even on the paper in which the poison had been wrapped, the pores retaining enough to prove guilt. Marshall's confidence must have stemmed from watching the tests being performed and assessed by someone experienced in their use, but any test that relied on judging colours by eye was open to doubt and therefore challenge in the courtroom. This weakness was to become apparent just two years later.

Mrs Elizabeth Downing was a sixty-four-year-old widow in sound health, with a fortune of £14,000. She lived in Falmouth, and her close-knit family consisted of four sons, two of whom, Edward and Samuel, lived with her, and two daughters, one of whom, Harriet, had married a young surgeon, Robert Sawle Donnall, in July 1816. Harriet had property of her own worth

£3,000, and Donnall had initially agreed that a settlement should be made before the marriage that would secure this fortune to his wife. But, when it came to be drawn up, Donnall changed his mind and said that he wanted to have more control over his wife's fortune during her lifetime.

Whether Donnall had ever intended to agree to the original settlement is unknown, for he was a creature of subterfuge. When paying court to Harriet, he had represented himself as a man of means, claiming that his medical practice was 'in a flourishing state',[29] when he was actually desperate for money. Since February 1816, Donnall had been borrowing from a moneylender, Gabriel Abrahams, saying that, although he was currently in financial distress, he would be able to repay the loans once he was married. Things must have been tight indeed when he borrowed £50 in May and told the moneylender it would be 'either his making or his ruin'.[30] After the wedding, Abrahams repeatedly asked for repayment of the loans, but was supplied only with promises. In mid-October, however, Donnall reassured his creditor that he would soon be able to settle his debts, which then totalled £125, as his mother-in-law, who was a wealthy woman, was very ill and would die soon, enabling him to make repayment in full before Christmas. At the time, Donnall also owed a substantial sum for goods to his brother-in-law Samuel, a mercer.

When Samuel dined with his mother on 19 October, he found her in good health and spirits. Later that day, she took tea with the Donnalls, who lived next door, and, soon afterwards, was seized with sickness, a burning pain in her stomach and violent cramps. The illness lasted for several days, but she made a full recovery. On Sunday 3 November, Mrs Downing was again invited to take tea with the Donnalls. Initially, she was reluctant to go, saying that the last time she had done so she had been taken ill. Nevertheless, she was persuaded, possibly because other members of the family would also be there, her sons Edward and Samuel and her other married daughter Mrs Jordan with her husband. That morning, Mrs Downing attended church, and she later dined heartily on a dish of rabbit smothered in onions,[31] accompanied by potatoes and washed down with beer.

That afternoon, the family gathered at the Donnalls', Harriet pouring the tea while seated at a table on which there was a plate of bread and butter. Mrs Downing was sitting to the left of the fireplace, and Robert sat facing her to the right. A servant brought in a large jug of hot cocoa, which Harriet and Mrs Downing preferred to tea.

No one could later recall who had poured the first cup of cocoa for Mrs Downing. Robert handed around the plate of bread and butter to the ladies, but, curiously, when he offered the plate to Mrs Downing, he did not go to her directly by the obvious and shortest route across the room. Instead, he took an unnecessarily circuitous path around the table, passing unseen behind the chairs of the other guests. As he did so, he knocked Edward's teacup, spilling part of its contents over Mrs Downing's gown. Edward exclaimed in annoyance, and wiped off the tea stains with his handkerchief. When Robert moved away, Edward saw that his mother had a cup of cocoa in her hand and had taken a piece of bread and butter.

Soon afterwards, the young surgeon was called away to see a patient and so was not present ten minutes later when Mrs Downing had finished her cocoa and was ready for another. Edward poured the second cup, but Mrs Downing had drunk only half of it when she complained of feeling violently sick. Mrs Jordan took her mother home, where she vomited into a basin. Mrs Downing went to bed, and Donnall, called to attend to her, informed the family that he had given her something to compose her stomach. It later emerged that he had given her tartarised antimony, an emetic normally used to treat parasitic infestations and create an aversion to alcohol. At three in the morning, Donnall, saying he was alarmed at a drop in the patient's pulse rate, called in another doctor, Richard Edwards, telling him that Mrs Downing had cholera.

Elizabeth Downing died at eight o'clock on Monday morning after an agonising illness lasting fourteen hours. Just two hours later, Donnall was doing everything he could to hasten the burial, which he wanted to take place on Wednesday, insisting that there was a danger that the body would swell up, with unpleasant consequences, and that there was already a discharge from the nostrils. Dr Edwards, on the other hand, the older, more experienced man,[32] saw no sign of swelling and no discharge. The body was still in the house on Wednesday when the family was shown an anonymous letter sent to the mayor of Falmouth. The writer stated that Mrs Downing had been taken ill twice after drinking tea [*sic*] at Mr Donnall's, adding: 'You will, I am sure, notice the similarity of her complaints to those of persons who have swallowed some corrosive substance. . .'[33] When Donnall read the letter, his hands trembled and he dropped it on the floor, exclaiming that it was a villainous thing meant to ruin him. He opposed the idea of having the body opened for an examination, saying he did not have time to attend to it, but, when the coroner asked Dr Edwards to do it, assisted by local surgeon, Mr Street, Donnall turned up and tried to take charge. He was tucking up his sleeves and preparing to open the body when Edwards sternly ordered him to have nothing to do with it. Donnall remained present as an onlooker, and, although he afterwards told Samuel that they had been satisfied that everything was all right, this was very far from the truth.

From the start, Edwards was looking for arsenic. He removed the stomach from the body and poured its contents into a basin; then he and Street explored the fluid with their fingers to see if there was any heavy or gritty substance that had sunk to the bottom. They found nothing. The stomach was badly inflamed and the internal coat so softened that, when Edwards examined it by scraping gently with his fingernail, it easily came away. Parts of the stomach wall had been nearly destroyed by the action of some corrosive substance. The contents of the basin, about half a pint in all, were poured into an earthenware jug, which Edwards placed on a chair, saying that particular care must be taken of it as he intended to examine it. The intestines were then removed and washed, and the water used for this purpose, about a pint and a half, was poured into a clean chamber pot. The intestines were found to be inflamed to a degree that Edwards was sure could not have been produced by natural causes in so short a time. When he next turned to look at the jug, it was empty. Donnall, for no reason he was willing to explain, had poured the contents into the chamber pot. Edwards angrily said that this would give him a great deal more trouble, and placed the now very diluted fluid into the custody of Mr Street. Despite being unable to find visible traces of arsenic, Edwards had no doubt that the condition of the stomach and intestines was proof that Mrs Downing had been poisoned, but, like John Marshall, he was to apply further chemical testing.

There was no vomit to test. Earlier samples had been thrown away before Donnall arrived to attend to the patient, and, later on, all voided material had been disposed of at his direction. Edwards did not try testing the stomach wall, which he could have done, since it was possible to separate a metallic poison from animal tissue using the method devised by German pharmacologist Valentin Rose in 1806, but he may not have been aware of this. The transmission of new techniques of analysis depended both on their publication and on their availability in translation. Texts such as Mathieu Orfila's groundbreaking two-volume work on toxicology, which described and improved on Rose's methods, had been published in English in 1816, but was unlikely to have been widely purchased and studied by provincial doctors.

Edwards concentrated his efforts on the stomach fluids, using Hume's copper sulphate and silver nitrate tests, and obtained the coloured precipitates that confirmed the presence of arsenic. He also compared the results of the same tests on samples of bile and tartar emetic. Mrs Downing's last substantial meal had been rabbit smothered in onions, so Edwards steeped some onions in water and tested this liquid. He was satisfied that he could distinguish the results from the precipitates created by the presence of arsenic.

Elizabeth Downing was finally buried on Saturday. The cocoa was already under suspicion, but it was obvious that there was no poison in the jug, as Mrs Donnall had also drunk some and the remainder had been warmed the next day for breakfast. Neither was there any poison in the second cup poured by Edward, as the servant had finished it with no ill effects. Attention, therefore, focused on the first cup, and, after the funeral, there was an animated family discussion as to whom had given this to Mrs Downing. No one would admit to it, and Donnall not only denied having been present at the start of the gathering, but even tried to persuade Edward to make a statement admitting that he had given his mother the first cup. By 14 November, Donnall, convinced that he would be arrested, was in a state of great agitation. After trying and failing to hide in the chimney, and making a number of attempts to escape, he finally gave himself up.

On 31 March 1817, Robert Sawle Donnall was tried for the murder of his mother-in-law, described by the prosecutor, Serjeant Lens, as 'a crime of the deepest magnitude, enhanced by the means which he used in order to effect it – means which can only be resorted to by the delusive but wicked hope that their secrecy may prevent the discovery of the act'.[34]

The defence was that Mrs Downing had died of natural causes, which meant, in all probability, cholera. Dr Edwards, while agreeing that cholera was a painful complaint that could produce the kind of violent cramps experienced by Mrs Downing, told the court that, in his experience, the disease was fatal in days rather than hours; he was sure that the cause of death was poison. Edwards described the tests he had carried out, but the defence had already seized on a serious problem in the prosecution case, the one that would undoubtedly appeal most to the jury: the fact that all Edwards could point to as proof of arsenic were coloured precipitates. There was no physical arsenic to see, either as a white powder or as a metal.

Asked if it was possible to produce arsenic in substance from a fluid, Edwards told the court: 'By evaporating the solution containing arsenic, and by exposing it to heat in a close vessel, you will produce it in a white solid state; and by mixing the residuum of a solution of arsenic with an inflammable substance, arsenic will be sublimed in its metallic state, by the same process.'[35] This was all very well, but Edwards had not performed this test because, after all his other experiments had been completed, there was only a small quantity of dilute liquid left, which he thought would be insufficient to produce a successful result. He had found, when conducting the previous tests, that arsenic was not present in a large quantity; this, he believed, was due to Mrs Downing's repeated vomiting, which would have thrown off a substantial amount.

It was remarkable for a doctor to assert with such confidence that arsenic was present in the stomach contents when he was unable to provide any physical evidence. Why had a suspicious-looking white powder not been found in the stomach fluids, or clinging to the inner coats of the stomach? John Marshall knew that, even after copious vomiting, caustic particles of arsenic would continue to stick to the villous walls, which was one reason he told the Turners to drink copious draughts of warm fluid to dislodge the damaging residue. Edwards thought he had the answer – the dose of arsenic had been given not as a powder, as was usual, but in the fluid state. If dissolved in hot water, one tablespoonful of the solution would be enough to cause death. Poisoners with no medical expertise, such as Mary Blandy (see Chapter 1, note 22), almost invariably overestimated the solubility of arsenic, but Donnall knew better.

The defence brought a number of witnesses to testify to Donnall's good character, and also some medical men, none of whom had seen Mrs Downing in her last illness or, it transpired, had attended a post-mortem examination of someone who had died of cholera. All were willing to testify, however, that the cause of death was cholera. More importantly, severe doubt was cast on whether the copper sulphate and silver nitrate tests were infallible. Based on the dinner eaten by Mrs Downing some five hours before she was taken ill, the defence witnesses had carried out the tests on an infusion of onions, and told the court that they had obtained the same result as that given by arsenic.

John Tucker, a surgeon of Exeter, said that he had once believed the silver nitrate test to be the most delicate test of arsenic, but had since discovered that this was a fallacy. Joseph Cookworthy, a Plymouth physician, said that the only conclusive test to show that arsenic was present was to reproduce it in its metallic state. He too had obtained a false positive result using the precipitate test on an infusion of meat and onions.

Mr Justice Abbott undoubtedly summed up for a conviction. He pointed out that the witnesses who had contradicted Dr Edwards had not seen the body, and advised that it was for the jury to decide whose opinion should be given greater weight. He thought that the suspicion of death by arsenic was strong and that there might not have been enough onion in the stomach fluids to affect the test results. If there was arsenic in the cocoa, then it could only have been put in the first cup drunk by the deceased. None of those in the room who were near enough to Mrs Downing said they had given it to her, and, Abbott hinted gently, it was up to the jury to consider whether the unusual route taken by Donnall had given him the opportunity to infuse her cocoa with arsenic.

Abbott also drew attention to Donnall's behaviour. Juries at murder trials are usually asked to consider three important aspects of a case: did the accused have the means with which to commit the crime, the opportunity to do so, and – though it is not necessary to prove this in order to obtain a conviction – a motive? Donnall undoubtedly had all three in abundance, but there are other behaviours exhibited by murderers, and by poisoners in particular, that are rarely analysed but that Mr Justice Abbott had observed in the defendant. Donnall, commented the judge, had urged haste in burying the body, claiming that there was a discharge when there was not. He had opposed the idea of a post-mortem, and then, when it was ordered, tried to interfere with its carrying out and with the preservation of evidence, the importance of which he well understood. The judge did not point out, but might have, Donnall's lie after the post-mortem, when he said that it had shown nothing wrong, as well as his attempt to divert suspicion from himself to Edward by trying to induce his brother-in-law to admit pouring Mrs Downing's first cup of cocoa. Abbott decided to give no weight to Donnall's comments to his creditor, as it was the kind of thing a debtor might have said to delay payment; however, suggesting that an intended victim is ill and expected to die soon, when he or she is actually perfectly well, is a ploy often used by someone plotting poison murder.

The jury took twenty minutes to return a verdict of not guilty. The lack of visible poison and the professional status of the accused may have been factors in the creation of reasonable doubt.

* * *

John Marshall's response was a vigorous defence of Hume's tests: he said it was impossible for the colours produced by onions and arsenic to be confused. He feared that where arsenic could not be displayed in its solid form, the result would be that murder could be committed with impunity.

During the next few years, evidence continued to accumulate that substances other than arsenic also created precipitates that could confuse the issue, but Hume's tests remained in use, although carried out in support of more traditional means of identifying arsenic. If solely relied upon, they were a snare for the inexperienced. In 1823, physician J. A. Paris claimed that the silver and copper tests 'are capable, under proper management and precaution, of furnishing striking and infallible indications . . . in most cases they will be equally conclusive, and in some even more satisfactory in their results, than the metallic reproduction upon which so much stress has been laid'.[36] In 1828, after the death of William Logan of Northumberland, Pennsylvania, following a fever, local gossip led to an exhumation. A board of physicians, relying on the confident assurances of J. A. Paris, conducted tests that suggested the presence of arsenic and that resulted in Logan's wife being arrested and charged with murder. It was left to the family physician, Samuel Jackson, to reveal the fallibility of the tests in inexpert hands and to secure Mrs Logan's acquittal.[37]

The Donnalls remained in Falmouth, where they raised a family and Robert continued to practise medicine. Harriet died in 1862 and Robert in 1874.

CHAPTER THREE

The Greatest Refinement of Villainy

In the 1820s, the increasing sophistication of substances available to the criminal poisoner posed fresh challenges to medicine, the greatest of which was identifying poisons of vegetable origin in the bodies of victims. Chemists knew how to separate metallic poisons, such as arsenic, from animal material by destroying the tissues, but such a procedure would also destroy the organic poisons they hoped to extract.

In 1823, a sensational murder trial took place in Paris. It was the first criminal case in which it was alleged that the instrument of destruction was morphine, and it pitted the machinations of a clever and resourceful poisoner against a new breed of scientist: the expert forensic toxicologist.

The modern science of forensic toxicology was born in the second decade of the nineteenth century, and its father was an influential, striking and controversial figure, Mathieu Joseph Bonaventure Orfila.[1] Born on the island of Minorca in 1787, he studied medicine first in Spain, then Paris, where he obtained his doctorate in 1811. He also acquired a thorough grounding in chemistry, anatomy and botany and, the following year, he started to lecture. In 1813, he told his students, with some authority, that the white precipitate obtained by adding limewater[2] to a solution of arsenic would be the same even if it were first mixed with organic material. He demonstrated this by introducing limewater to a solution of arsenic and coffee, and saw, to his astonishment, a violet-grey precipitate. This discovery stimulated many months of experiments with different reagents.

His first major work, a two-volume treatise on poisons, was published in French in 1814 and 1815, prefaced by the author's challenging declaration: 'Such a work is altogether wanting to science; I may affirm more; a very considerable number of the facts which should serve for its basis, are still unknown, or badly studied.'[3] The professor of medical jurisprudence called upon to advise a magistrate, would, Orfila asserted, find existing writings 'vague and unsatisfactory'.[4] Others recognised the value to more junior members of the medical profession: '[W]hat must be the confusion, embarrassment, and distress of a young practitioner often called on to exert a skill not to be acquired from any accessible source, to tender assistance not afforded by experience?'[5]

Although many works on the subject of poisons were available before Orfila's, there was, until his groundbreaking study, nothing incorporating the advances of the previous twenty years – no system of classification useful to the toxicologist, no thorough research on the action of poisons on animals and, importantly, no systematic testing of the currently approved antidotes. Orfila unified in one giant treatise the needs and findings of the naturalist, the physiologist, the medical practitioner, the chemist and the magistrate, not forgetting the concerns of the general public. 'The private individual,' he observed, '. . . is fond of discoursing on the fatal properties of various substances . . . Incensed at the most cowardly of crimes, he regards with horror the execrable assassin; and loudly demands the punishment of a monster, the more dangerous, as he always commits his ravages in silence, and often even upon his benefactors.'[6]

Since the first requirement in bringing a poisoner to justice is the recognition that a suspicious death has taken place that requires investigation, the textbooks of Orfila, and those who followed him as the century progressed, provided doctors with both the information and the confidence they needed to identify those deaths that should be reported to the authorities.

Orfila provided a classification of poisons by the nature of their action, showing how they were to be recognised, their effects on living organisms, how a practitioner should treat a patient who had been poisoned, and what analysis and examination should be carried out if the patient died. He believed that the only way 'to enrich the science of toxicology, and to raise it from the state of imperfection in which it at present lies',[7] was rigorous chemical research and, controversially, trials on living animals.

Orfila not only studied poisons in their natural state, but also examined how they were affected by bodily fluids such as bile, saliva and gastric juices, and he tested those substances that were already in use as antidotes. At the time of publication, he had been working for three years on further studies, carrying out over eight hundred experiments, many of which used dogs; he often sat up all night watching the animals. The procedures, carried out before the existence of anaesthetics on animals that could not be afforded chemical relief from suffering, as this would have confused the results, were not undertaken lightly. As one of his contemporaries observed, 'it has required no small degree of courage to overcome the disgust which accompanies so painful a task'.[8]

Orfila, a prominent figure at fashionable Paris salons, where he entertained gatherings with a singing talent considered good enough for the operatic stage, made many friends amongst the distinguished company, and his celebrity must have boosted the popular success of his publication. Other physicians, however, took great exception both to his criticism of their use of poisonous substances as medicine and to his contemptuous dismissal of the antidotes they recommended. Pierre-Toussaint Navier's[9] popular 1777 book on antidotes came in for a particular roasting as Orfila denounced his counter-poisons as both 'useless and dangerous',[10] pointing out that Navier had only conducted chemical tests, whereas he had tried the recommended substances on animals with either no or harmful effect. '*Ought we, in the present state of medical knowledge*, [italics in original] to persist obstinately in admitting a fact, not supported by precise experiments, merely because some illustrious men have advanced it, and it has been generally adopted?' asked Orfila. 'It is of importance not to embrace their opinions, when they are the declaration of incorrect facts, and tend rather to retard than accelerate the progress of medical science.'[11]

The youthful doctor, only three years qualified, was trying to upset established beliefs and practice, never an easy thing to do, and, inevitably, the 'illustrious men', not all of whom were acting out of jealousy or hurt pride, tried to administer caution as an antidote to Orfila's bombastic confidence. His two most vociferous critics were Charles-Hippolyte-Amable Bertrand[12] and Antoine Portal,[13] who were, respectively, ten and forty-five years older than the young pretender. Bertrand believed that the differences between animal and human physiology meant that the results of animal testing could not be applied to human subjects. Orfila disagreed.

'It is generally admitted', wrote Orfila, 'that, among the different means used to ascertain the existence of poisoning, that which consists of making dogs swallow the liquid found in the stomach of individuals who are thought to be poisoned deserves the preference over all others.'[14] It was a practice that 'had existed from time immemorial; it has been maintained by men who understood but little chemistry; and who have avoided, under frivolous pretences, endangering their reputation, in attempting the analysis of liquids'. It was a method especially favoured by physicians unable to detect vegetable poisons by experimental means, a skill that would elude researchers for many more years. Even this time-honoured test had its critics. Supposing, it was argued, the person believed to have been poisoned had not been poisoned at all, but had suffered from a disease that had altered the stomach fluids and made them poisonous – was it this that caused the death of the animal? How often was the experiment poorly done and the animal choked because liquid had run into its lungs? Could the struggles of animals being held down for the procedure be misinterpreted as convulsions?

Orfila carried out tests to confound the opposition. Previous experimenters had induced animals to swallow suspected substances, which were often mixed with food. Orfila wanted to avoid any suggestion of contamination and the risk of losing the sample either by it running into the dog's lungs and causing asphyxiation, which he believed happened in six cases out of ten, or by it being vomited. Using a fasting animal, Orfila separated out the oesophagus of his test dog, made a small opening, and inserted a glass funnel through which the liquid was introduced into the stomach. The oesophagus was then tied with a ligature to prevent vomiting. The entire operation, carried out skilfully, took two minutes, and the animal returned to its normal level of activity in an hour. Bertrand had suggested that the ligature could cause the animal's death independent of any poison; however, Orfila believed that death by poison would occur during the first forty-eight hours and be accompanied by symptoms that would never be seen after a simple ligature. To prove his point, Orfila carried out twelve further experiments in which he performed the ligature operation, but did not introduce any poison, and concluded that the operation produced no more than 'a slight fever and a little dejection'.[15] Six of his dejected dogs were left to grow feeble and die, which they did in a few days, but six others were killed by being hanged forty-eight hours after the operation, the post-mortem examination showing no lesions similar to those produced by poison. Orfila could, therefore, be confident that ligature of the oesophagus did not affect any conclusions drawn about poisoning, and asserted that it was 'impossible to write a complete work on poisons without often performing this operation'.[16]

Orfila had paid particular attention to the work of the English physiologist and surgeon Benjamin Collins Brodie (later Sir Benjamin, 1st Baronet), published in 1812. Doctors viewing the inflamed stomachs of animals poisoned with arsenic had, Brodie commented, assumed that the inflammation was the cause of death, the result of the 'actual contact of the arsenic with the internal coat of the stomach'.[17] Brodie's work, carried out on cats, dogs and rabbits, involved applying arsenic and other poisons to wounds, or injecting them into the stomach through a tube. He was able to show that poisons did not act upon the system until they had been absorbed into the bloodstream and circulated, and that they affected other parts of the body as well as the stomach. Brodie undoubtedly felt uneasy about the suffering of his animal subjects and he expressed the expectation that his work 'may hereafter lead to some improvements in the healing art. This consideration, I should hope, will be regarded as a sufficient apology for my pursuing a mode of inquiry by means of experiments on brute animals, of which we might well question the propriety, if no other purpose were to be answered by it than the gratification of curiosity.'[18] Orfila's detailed experiments confirmed Brodie's findings.

The question of absorption was tackled from a physiological rather than a toxicological point of view by Orfila's contemporary François Magendie, a skilled surgeon, who carried out experiments on dogs, working extensively with alkaloids. He was able to isolate different parts of a dog's system, leaving them connected only by blood vessels. By injecting poison into the isolated part, such as a thigh, he was able to observe the symptoms of poisoning appear, showing that the injected substance had been transported by the circulatory system.

It should be noted that most of the objections by doctors to these early animal experiments were made less on the grounds of cruelty, since the suffering of animal subjects was then considered regrettable but unavoidable, but more in the belief that the severe pain inflicted would obscure natural processes and that results obtained from animals could not, in any case, be applied to human physiology.[19] Many observers of demonstrations using animals did, however, feel an instinctive revulsion to the practice in principle, and this led to the founding and strengthening of an anti-vivisection movement, which was stronger in England than in France. There was to be no legislation controlling animal experimentation until 1876.

After the publication in 1818 of Orfila's second book, which dealt with the treatment of persons who had taken poison,[20] Antoine Portal went on the attack. Orfila believed in identifying counter-poisons that would react with the specific poison taken to produce a new and harmless product. Portal preferred a general treatment for all poisons. If recently swallowed, then the poison had to be expelled as quickly as possible with emetics and enemas. If its effects were already apparent, he advocated plenty of 'relaxant' and 'anodyne'[21] fluids to dilute its action, followed by a bath and those widely used and ineffective favourites of the previous century: bleeding and blisters.

In an address to physicians on the subject of antidotes to poison, Portal advocated general treatments with a long history of use by practising physicians on human patients in preference to the specific remedies recommended by Orfila, which had been tested only in the laboratory and on animals. He went so far as to doubt whether a specific antidote for each poison even existed. The subject remained highly contentious. While many of Orfila's translators worked under the guidance of the author and produced works faithful to the originals, J. G. Stevenson, who translated the third edition for an American publisher, actually deleted Orfila's criticism of Portal and his defence of animal experiments, and added material of his own.[22]

In 1819, Orfila, now considered one of the foremost authorities in his field, was appointed professor of legal medicine at the Faculty of Medicine in Paris, and soon became a most popular lecturer; in this, his elegant appearance and manners and melodious baritone voice gave him a natural advantage. To his young students, he must have seemed like the new hope of science, but to many of his elders, he appeared arrogant, even dangerous. Orfila actively promoted his books, using his contacts to obtain reviews and sending copies to London publishers, recommending that they should be made available in English translation. His work was also translated into German, Italian, Danish and Spanish, yet, despite this, it was some years before his approach was widely adopted. The young Scottish toxicologist Robert Christison,[23] who had heard Orfila lecture in Paris and had then returned to Edinburgh, where he became professor of medical jurisprudence in 1822, observed that Orfila's work 'had scarcely begun to be appreciated, or even known, in the British Isles'[24] and had 'found the entire subject of medical jurisprudence in a very unsatisfactory condition in every part of Britain'.[25] Changes were occurring, however. In 1823, Thomas Wakley, a surgeon and campaigning social reformer, launched a medical journal, *The Lancet*, which was to go on to international renown. In 1829, Christison published the first comprehensive nineteenth-century treatise on forensic toxicology originally written in English, and was to become a respected and influential figure in his own right.

Orfila, as he laboured and lectured and sang to the cream of Parisian society, had founded both a new integrated discipline and a new elite public role – that of the expert toxicologist. It is unsurprising, therefore, that many of the influential poisoning trials of the first half of the nineteenth century took place in France. These trials received extensive coverage in the French journals, notably the *Gazette des tribunaux,* which reported in depth on court proceedings; condensed accounts were published in English-language newspapers and journals.

Orfila, having developed his expertise using facilities unavailable to the doctors and chemists usually asked to give evidence at trials, was often consulted in cases of poisoning. His self-confidence rarely faltered, but his opponents believed that what happened in the laboratory was different from nature and that his results could not be relied upon. The stage was set for a public confrontation, but, when Orfila first gave evidence at a major trial, not only was his new discipline found wanting, but the defendant was one of his own former pupils.

Edme Samuel Castaing was born in Alençon in 1796. In 1821, after diligent study at the School of Medicine in Paris, he qualified as a doctor and established a medical practice. He also, however, formed a passionate liaison with a widow, who was struggling to support her children, and fathered two illegitimate children of his own, acquiring substantial debts. Some years previously, Castaing had become acquainted with the family of a wealthy lawyer called Ballet, and had developed a close friendship with his sons, Auguste and Hippolyte, who were of a similar age to him. On the death of their parents, the sons came into a substantial inheritance, but were happy to remain friends with Castaing despite the great discrepancy in their fortunes. Castaing, under pressure from his creditors and needing to support his mistress and children, might have used the ties of friendship to borrow from the Ballet brothers, but, instead, he saw the chance of realising his long-held ambition of becoming rich.

Auguste and Hippolyte were very different both in constitution and character. Twenty-five-year-old Auguste was the elder and in sound health, but lazy and dissolute. Hippolyte, with whom Castaing had established an almost brotherly relationship, was studious and respectable, but afflicted with consumption, although his doctor thought the condition was advancing slowly and he might live for many more years. Castaing had no expectation of inheriting from either brother, but was able to make use of Auguste's greed and Hippolyte's delicate health to form a scheme to acquire their fortunes.

Auguste would have expected to receive the bulk of Hippolyte's estate in the event of his brother's death, but, in 1821, concerned that Auguste was dissipating his share of their inheritance, Hippolyte spoke to a notary, M Lebret, saying that he wanted to make a will that would provide only a small annuity to Auguste, the rest of his money going to his older half-sister, Mme Martignon. It was a complication that Castaing, with a little ingenuity, was able to turn to his own account.

In the late summer of 1822, Hippolyte was taking the waters at a spa, where Castaing paid him frequent visits under the guise of a concerned friend and medical advisor. When Hippolyte returned to Paris in September, Castaing put his lethal plan into action, hoping to escape detection by using one of the new vegetable poisons, and making assurance doubly sure by giving his victim an emetic to remove every possible trace.

Castaing had attended Professor Orfila's lectures at the Faculty of Medicine, and had developed a particular interest in poisons. His study notes included the observation that, 'while mineral poisons disorganize the tissue with which they are brought into contact, vegetable poisons "act upon organs, without it being possible to find a single trace of disorganization – a final circumstance, which leaves us ignorant how they act"'.[26] He was, therefore, well acquainted with the new derivatives of opium, which included morphine acetate, a white, bitter-tasting salt made by combining acetic acid with the recently isolated wonder drug morphine and used medicinally since 1818. Its narcotic sedative action was very similar to that of laudanum, although without the side-effect of constipation, and it had the virtue of being more soluble than morphine and faster acting. It was given dissolved in water or syrup, or as a constituent of pills, for a variety of complaints, including stomach ache, neuralgia, period pains and cancer. It was yet to feature in a murder trial.

The previous May, Castaing had bought ten grains of morphine acetate from a Paris chemist, quite possibly to test its effects on animals, and, on 18 September, he went back to the same chemist for another ten grains. The normal medicinal dose was one eighth to half a grain, and three grains was a potentially fatal dose.

On the morning of 3 October, Hippolyte was very ill, having suffered a severe attack of vomiting during the night. His brother-in-law M Martignon visited him and found the invalid in bed with a swollen face and red eyes. Hippolyte's own doctor was not summoned; instead, Castaing took charge of his treatment and discouraged further visits. When Mme Martignon called the following day, the servants, on Castaing's orders, told her that Hippolyte was not to be disturbed. She offered to nurse her brother, but Castaing adamantly refused to allow it, saying that Hippolyte would be agitated by seeing her. She wept and pleaded, and, at length, she was permitted a brief glimpse of her brother's reflection in the looking glass of the dining room, on the condition that she first put on the bonnet of Hippolyte's servant Victoire, so that, if he were to see her, he would not realise it was his sister. Mme Martignon understandably did not want to go against the advice of a trusted doctor, but she was a persistent and resourceful lady. Still determined to nurse Hippolyte, she borrowed

Victoire's clothes, with the intention of disguising herself, hoping by that subterfuge to be allowed to go in and see her brother. Despite her most earnest pleas, Castaing again prevented her from entering the sickroom, saying that Hippolyte would guess it was she because she was taller than Victoire. The desperate woman was reduced to crawling on all fours into the anteroom of the bedchamber so that she might hear her brother's voice. That evening, at the insistence of M Martignon, a second physician, M Segelas, was called, but, at ten the following morning, Hippolyte's sufferings were at an end. The post-mortem was carried out by Castaing and M Segelas. The verdict was that Hippolyte had died of pleurisy, an inflammation of the membranes surrounding the lungs, made more serious by his consumption.

Hippolyte had left a fortune of about 260,000 francs.[27] He had been meticulous about keeping receipts and, on the day before he was taken ill, the sum of 6,000 francs had been in his desk to pay a bill; however, when M Martignon searched the desk, it was empty. No will was ever found and Hippolyte was assumed to have died intestate.

Auguste could not wait for his inheritance. On the day of his brother's death, he wrote a letter to an associate saying that he had an immediate need for 100,000 francs. Only later did the reason for this urgency become apparent. Castaing had boasted to Auguste that he had found Hippolyte's new will, in which Mme Martignon was the main beneficiary, and had destroyed it, but, on the same day that Hippolyte was taken ill, Castaing told Auguste that another copy of the will had been lodged with the notary Lebret, and that Mme Martignon had offered Lebret 80,000 francs to preserve it and ensure that it was valid. This was alarming news for Auguste, but Castaing reassured him that, for a counter-bribe of 100,000 francs, Lebret would destroy the will, and he offered to act as intermediary. On 7 October, two days after Hippolyte's death, 100,000 francs' worth of Auguste's effects were sold. The next day, he collected the proceeds from the bank and handed it all to Castaing.

On 10 October, the formerly impecunious Castaing invested 70,000 francs with a stockbroker, and, the following day, he lent his mother 30,000 francs. On the 14th, he gave his mistress 4,000 francs. The story of Mme Martignon's bribe to Lebret and the notary's willingness to destroy Hippolyte's will, which Auguste had been all too eager to swallow, was undoubtedly an invention of Castaing's.

Ultimately, in the absence of a will, three quarters of Hippolyte's fortune went to Auguste and the rest to their sister. Castaing then put the second part of his plan into action. Perhaps pleading poverty and the work he had done to enrich his partner in crime, he managed to persuade Auguste to make a will leaving him virtually all his property.

Castaing and Auguste were uneasy accomplices in fraud. Auguste, who probably did not suspect the young doctor of murdering his brother, but was irritated at paying out so much to obtain Hippolyte's inheritance, began to dislike his friend's company. By May 1823, Castaing, seeing that Auguste was spending lavishly on a new mistress, realised that he had to act quickly. He called on a cousin, notary's clerk Malassis, to check whether a will made in favour of a physician was valid. When Malassis confirmed that it was, Castaing said that he would be lodging such a will with him, adding for a little corroborative detail that the person who had made it was dangerously ill, having spat blood several times. On 29 May, Castaing left the will with Malassis, and he and Auguste went on a trip.

That evening, they took a room at a hotel in St Cloud, an attractive town some six miles from the centre of Paris. They spent the next day seeing the sights, dined at the hotel at seven, and, after another stroll, returned at nine, at which time Castaing ordered some warmed wine to be sent up to their room. Castaing had brought lemon and sugar with him, which he added to two glasses of the wine, and both men drank. Auguste thought his drink tasted sour, and complained that there was too much lemon in it. Unable to finish the drink, he offered it to the servant to taste, who agreed it was sour.

The friends retired to bed, but Auguste, who had been in good health when he left Paris, was taken ill with colicky pains, and spent a restless, sleepless night. Castaing also missed sleep. Rising at 4 a.m., he ordered a cabriolet and, two hours later, he was back in Paris, where he bought twelve grains of tartar emetic from a chemist. An hour later, he was visiting another chemist, Chevalier. This was an unwise choice as, not only had Castaing bought morphine acetate from Chevalier on a previous occasion, but he had also discussed with him the effects of vegetable poisons and which ones were most likely to escape detection. This time, he bought thirty-six grains. Castaing returned to St Cloud, where Auguste was ill in bed, his legs so swollen he was unable to put on his boots. Castaing, claiming that he had simply been for a walk, ordered a glass of milk for Auguste, then went out again. During his absence, Auguste was seized with violent vomiting and copious diarrhoea. Castaing, returning to find his friend in severe pain and distress, ordered that the vomited matter be thrown away and suggested calling a doctor from Paris, but Auguste insisted that he could not wait so long. A local physician, twenty-five-year-old M Pigache, was summoned, and he arrived at eleven that morning to be told by

Castaing that Auguste was suffering from cholera. Pigache prescribed lemonade and a soothing fomentation for the stomach. He returned at three to find the patient improved, and prescribed a draught. Pigache wanted to call again later that night, but Castaing advised him that it would not be necessary.

That afternoon, Castaing wrote to Auguste's manservant Jean in Paris, asking him to take the keys of his master's desk to Malassis, but Jean was distrustful and instead brought the keys to St Cloud. Jean was, therefore, present at eleven that night to hold a candle while Castaing gave his master a spoonful of the draught prescribed by Dr Pigache. A few moments later, Auguste started choking and gasped, 'I am a dead man, I cannot swallow my spittle.' Auguste's condition deteriorated in minutes. He appeared to be suffocating, gulping throatily like a man in his death agony. He then went into violent convulsions, and, soon afterwards, mercifully, lost consciousness. Pigache was sent for and found Auguste lying bathed in a cold sweat with clenched teeth and a fluttering pulse. The young doctor was given the alarming information that the change had occurred soon after Auguste had taken his draught. Still unaware that he was dealing with a case of poisoning, Pigache bled the patient and applied twenty leeches. Another physician was summoned, M Pelletan, who found Auguste unresponsive. He tried bleeding him again, but the blood ran slowly and he was obliged to rub Auguste's arm to stimulate the flow into little jets. When he stopped, the blood slowed to a trickle.

At about midday on 1 June, Auguste died. Many a poisoner sitting by the deathbed of his or her victim will demonstrate little, if any, concern, to the extent that this cold, unsympathetic demeanour is later commented upon. A few, however, show extravagant distress, and Castaing was of the latter kind. Perhaps in an effort to curry sympathy, he told Pigache about the unhappy coincidence of attending the deathbeds of both brothers, claiming that his position was the more distressing now, since he was Auguste's heir. Pigache revealed this startling conversation to Pelletan, who, suspecting poison, told Castaing that he would be obliged to order a post-mortem. Castaing burst into tears and said that he was overwhelmed with grief at the loss of his friend. He must have remained confident that no poison would be found, and – perhaps imagining that he would be able to participate in Auguste's examination as he had done with Hippolyte – he felt free to tell Pelletan that he would be doing him a great service by insisting on an investigation, saying, 'You will be acting as a second father to me in doing so.'[28]

M Martignon had not been advised of Auguste's illness until the morning of 1 June, and, by the time he arrived, his brother-in-law was dead. Castaing, at first, denied that Auguste had left a will, but admitted it the following day. By then, he had been confined to a room under guard, his state of agitation mounting as he waited for the result of the post-mortem being carried out in another part of the hotel.

M Pelletan separated the stomach from the rest of the intestines for a detailed examination. Five hours later, he gave Castaing the good news and the bad news. No evidence of suspicious death had been found in Auguste's body, but the Procureur du Roi (public prosecutor) had decided that Castaing could not yet be released.

Castaing was taken to Paris to await the results of further investigations. His behaviour was that of a desperate, guilty man. He tried to feign madness by refusing food and drinking his own urine, but gave this up after three days.[29] He claimed that he had received the 100,000 francs from an uncle, and tried, through a prison confederate, to write to the chemists who had supplied him with morphine acetate, asking them to say that it was harmless.

Meanwhile, doctors and chemists were doing everything they could to find traces of poison in the stomach contents of Auguste Ballet. They prepared an extract from the fluids and also made a separate solution of morphine, water and alcohol. The two liquids were compared by being tasted. The morphine solution tasted bitter, as expected; disappointingly, the extract from Auguste's stomach did not.[30]

On publication of the preliminary proceedings of the trial, the case became a sensation in Paris, and apothecaries' shops were overwhelmed with demands for poison. The Prefect of Police was, according to the Paris correspondent of *The Times*, obliged to issue an order for the return of the poisons, which had been sold in sixty times their normal amount. 'Let the *gourmands* beware.'[31]

Castaing's trial opened at the Paris Assize Court on 10 November 1823, when he was charged with the murder of both brothers and the destruction of Hippolyte Ballet's will. The crowds that attended the trial increased with each succeeding day, the action of the new poison arousing special interest. French journals, however, were coy about printing in full the results of the examination of medical men about the effects of vegetable poisons in general, and morphine acetate in particular; indeed, the *Journal des débats* made a formal apology to its readers for not doing so, saying that it abstained 'for more than one reason', adding: 'Besides, such details would be of no interest to the majority of our readers.'[32] As pointed out in John Buckingham's study of strychnine, *Bitter Nemesis*, not only was France more advanced than Britain in controlling the availability of poisons, but there was a deliberate policy of not revealing potentially dangerous scientific information to the general public.

Ultimately, the prosecution was forced to abandon the charge of murdering Hippolyte, since it was impossible to determine whether the inflammation of Hippolyte's membranes was due to poison or natural disease. There was a similar difficulty over Auguste. Castaing had undoubtedly acquired large amounts of morphine acetate, but, although the eminent medical witnesses testified that Auguste's symptoms led them to suspect that he had been poisoned, none of them could offer unequivocal proof that this was the cause of death. As medical man after medical man gave evidence, it became clear that the acknowledged experts in the field could not even agree on what quantity of morphine acetate constituted a fatal dose. On one thing, however, they could agree: certain vegetable poisons, even when given in fatal doses, might leave no detectable trace.

Orfila had taken no part in the examination of Auguste's remains, but was asked for his opinion on the post-mortem report. He told the court that it was impossible to say whether or not Auguste had been poisoned. A vital piece of evidence was wanting – Auguste's vomit. He felt sure that, had the vomit and stomach contents been submitted to him for examination, he would have been able to give a more positive reply. In the preceding two or three years, he announced, with a confidence that others could not share, progress in chemistry had meant that it was as easy to discover traces of vegetable poisons as it was to discover mineral.

While the accused in a British murder case would not be permitted to give evidence at his or her trial until 1898, this was an important feature in French trials, and did not always work to the benefit of the prisoner. When Castaing was first brought before the court, he created a favourable impression, his downcast eyes, gentle voice and decorous manners giving him the disarming outward air of a priest; but, when he was questioned at length by the presiding judge, that impression soon evaporated. Asked to account for his actions, the accused man was repeatedly trapped into obvious lies and unconvincing explanations, and was eventually forced to retreat into pathetic claims that he just couldn't remember.

The most compelling point advanced in his favour was the failure to find poison, which, it was suggested, meant that the prosecution had failed to establish that a crime had taken place. The prosecutor, Avocat-Général

Jacques Nicolas de Broë, pounced on this assertion and delivered a speech whose sentiments must have echoed throughout every courtroom in the land. Proofs that there have been a crime, he said, 'are infinitely variable according to the nature of things'.[33] The law laid down no rules by which the jury should decide whether there had been a crime, and it did not ask them to state the reasons for their conclusion.

'If', he went on, 'the actual traces of poison are a material proof of murder by poison, then a new paragraph must be added to the Criminal Code – "Since, however, vegetable poisons leave no trace, poisoning by such means may be committed with impunity."' In future, the message to poisoners would be: '"Bunglers that you are, don't use arsenic or any mineral poison; they leave traces; you will be found out. Use vegetable poisons; poison your fathers, poison your mothers, poison all your families, and their inheritance will be yours – fear nothing; you will go unpunished!"'[34]

De Broë, with no trace of apology, stated that the real evidence against Castaing was not medical but circumstantial, and that every fact in the case pointed to murder by poison. 'Where is the man who does not shudder at the idea of poisoning, a crime which unites the horror of homicide with the infamy of meanness?'[35]

On 17 November, after only two hours of deliberation, the jury found Castaing not guilty of the murder of Hippolyte, but guilty of destroying his will, and, by seven votes to five, guilty of the murder of Auguste. While he waited for his appeal to be heard, Castaing attempted suicide using poison smuggled to him in a watch. He failed. On 6 December, in the Place de Grève, Paris, he was guillotined. His guilt was never doubted, yet it was a matter of general regret, especially amongst the medical profession, that more satisfactory proofs had not been obtained.

The amounts and combinations of poisons given to the Ballet brothers by Castaing can never be known. Individuals vary in their tolerance, and Castaing's early morning shopping trip on 31 May may have been needed because he had unexpectedly run out of supplies when Auguste did not finish his wine, and dared not replenish them locally. The use of tartar emetic was a master stroke that nearly paid off; 'by the greatest refinement of villainy, the prisoner procured and administered the emetic for the very purpose of defeating that investigation which would have clearly established his guilt'.[36]

At first glance, the Donnall and Castaing cases have many similarities. In both, a young doctor employed his medical expertise to poison his victim in order to obtain an inheritance, and used his professional status to try to influence the outcome of the post-mortem. In neither case was the presence of poison successfully demonstrated in court. So, why were the verdicts so different?

In Castaing's case, he did not, like Donnall, expect to receive a legacy through a family connection, so he was obliged to carry out complex financial schemes that demonstrated a clear and calculated motive. Whereas no poison could be detected in the body of Auguste, the state of medical knowledge of vegetable poisons at the time suggested that this was not an unlikely result, whereas with arsenic, juries expected to see some physical evidence of poison. In England, Donnall was not permitted to enter the witness box and undergo questioning, which might well have exposed him as a liar. Castaing did, and showed the panic and evasion of a guilty man. And then there was M de Broë's terrifying speech, which raised the spectre of wholesale unpunished poisoning breaking out across the land.

Castaing's trial had given international publicity to the fact that there was a class of poisons that could defy the most eminent chemists, and other poisoners may have taken advantage of this information with more success than Castaing. The case stimulated new research into the properties of morphine and, in particular, the ways of detecting and quantifying it in the tissues, blood and secretions of murder victims. It soon became clear that Orfila's confidence had been premature. Twenty-seven years after Castaing's execution, vegetable poisons could be identified in their pure form, but, when they entered the organs of the victim, as they usually had by the time death ensued, the material could not be separated for testing.

CHAPTER FOUR

Arsenic and Old Graves

On 30 July 1823, M Demoutiers, an examining judge of Paris, concerned that a serious crime had taken place, sent for the celebrated Professor Orfila, and posed two questions. Firstly, was it possible to discover whether a person had been poisoned after the body had been buried for a month? If the answer was yes, the second question, an important one before the development of safe and effective antiseptics, was: would there be any danger to the investigator? Orfila replied with his accustomed confidence that it was easy to discover poison at the end of a month or an even longer period, especially if a mineral poison had been used, and, while the operation was potentially dangerous, it could be performed without risk if suitable precautions were taken.

Exhumation in a case of suspected murder was not something to be undertaken lightly, and was rarely contemplated unless only a few days had elapsed since burial. Until the beginning of the nineteenth century, it was generally believed that the bodies of poison victims decayed faster than usual, the hair and nails often falling off the day after death, the tissues rapidly liquefying into a pulp. In 1781, when Thomas Lonergan[1] stood trial in Dublin for the murder of Thomas O'Flaherty, who had died three years earlier from what was then assumed to be natural disease, there was never any suggestion of exhuming the body.[2] The openly intimate relationship between Lonergan and O'Flaherty's widow Susannah had led to rumours that her husband had been poisoned, and, when Lonergan tried to quash the accusations, he was arrested. The evidence presented in court consisted of the symptoms of O'Flaherty's fatal illness, the fact that other people had been similarly affected after sampling a custard pudding, which they had been warned was intended only for the deceased, and Lonergan's attempts to prevent any enquiry into O'Flaherty's death.

The Solicitor General for the prosecution, unfazed by a complete lack of forensic evidence and the usual absence of any witness to the actual poisoning, commented that, while 'this was a case in which positive direct proof could not from the secret nature of the transaction be adduced', nevertheless, 'a train of well-connected concomitant circumstances, established by a number of witnesses, must carry with them such an internal conviction as would by far exceed the most positive evidence'. Evoking the peculiar horror of poison murder, he told the court that, 'in all other modes of attack, the party assaulted has an opportunity to defend himself against open violence ... but in this dark and secret transaction, a man becomes prey to the machinations of those on whom he would have relied for assistance if attacked by any person'.[3] Lonergan was found guilty and hanged.[4]

The ability of arsenic to enable bodies both to resist decay and to decay in an unusual manner was first remarked upon by the medical inspector of Berlin, Dr Georg Adolph Welper, who made it his particular study. In 1803, Sophie Ursinus, the widow of a Prussian privy counsellor, made several attempts to poison her servant, with whom she had quarrelled.[5] Her scheme was exposed when the servant survived and took some plums she had given him to a chemist, who found they were contaminated with arsenic. Sophie was arrested and enquiries were launched concerning the fate of her husband, who had died in 1800, a lover, who had predeceased her husband by three years, and an aunt, from whom she had inherited a fortune in 1801. On investigation, it was decided that the lover had probably died of consumption, but Welper ordered that the bodies of the husband and aunt be exhumed. The remains, which were found to be not putrid, but dried up, were examined by chemist Martin Heinrich Klaproth and his assistant, Valentin Rose. Although they were unable to detect arsenic, the stomachs showed the visible evidence of contact with an irritant poison. The physicians who had attended Herr Ursinus were firm in the belief that his death was from natural causes. In the case of the aunt, however, the chemists were certain that she had been poisoned. Frau Ursinus, who had purchased arsenic under the pretext of needing it to kill rats, of which there were none in the house, confessed to the attempt on her servant and, at her trial, which took place in September 1803, she was found guilty of that offence and the murder of her aunt. She was sentenced to life imprisonment. This case may well have been the stimulus that prompted Valentin Rose to develop his method of boiling the stomachs of poison victims to extract arsenic from the tissues.[6] Working in Berlin in 1806, he cut up the stomach of a man who had been poisoned, and boiled the pieces in distilled water. The resulting soup was filtered, then all traces of organic matter were removed using nitric acid. Rose eventually produced a precipitate from the liquid, which he tested using Metzger's method, obtaining the telltale mirrors of arsenic.

Welper asked an acquaintance, Dr Klanck, to conduct experiments on animals. Klanck poisoned some dogs with arsenic and left the bodies in a cellar; some he buried, while others were left exposed to the air. Two months later, the flesh and intestines of the animals were red and fresh as if pickled. Eight months later, even after the cellar had been flooded, the intestines were entire and red, the muscles largely unaltered and the fat converted into adipocere, a waxy substance formed by the saponification of fatty tissue. Only those muscles directly affected by the water were soft and greasy. Klanck also carried out comparative experiments on dogs that had either been killed by blows or poisoned with corrosive sublimate[7] or opium, and buried them in the same place. After a similar passage of time to the previous experiment, the soft parts of the carcasses were converted to a greasy mass. In the following year, he repeated the experiment, leaving the bodies of poisoned dogs unburied in the cellar. Signs of putrefaction appeared in ten days, and the flies that settled on the carcasses died. The bodies remained unchanged for eight to ten weeks; then the soft parts became firmer and drier and the smell of decomposition was succeeded by a garlicky odour that became unbearably strong when the carcasses were removed to warm air. Three years later, the bodies were still dry and undecayed.[8]

The method devised by Valentin Rose proved crucial in the case of Anna Margaretha Zwanziger, a Bavarian housekeeper, who was arrested in 1809 on suspicion of having carried out a series of murders. Poisoning appears to have been a compulsion for her and, once she had got away with it a number of times, it became her preferred mode of removing people she found inconvenient. She disposed of women whose husbands she coveted, and it was thought that she also poisoned her employers non-fatally in order to nurse them back to health and thus gain their trust and affection. After being dismissed from one job, she took her revenge by administering poisoned drinks to the other servants, who recovered, and to a child, who died. When it was recalled that Anna had recently refilled a salt barrel, a task normally carried out by another servant, the salt was examined and found to contain arsenic. Three bodies were exhumed, respectively, five, six and fourteen months after death. 'In all of them, the external parts were not properly speaking putrid, but hard, cheesy, or adipocirous; in the last two the stomach and intestines were so entire as to allow of their being tied, taken out, cut up and handled.'[9] The Rose test revealed the presence of arsenic, and, confronted with this, Anna confessed. She was found guilty of murder, and beheaded with a sword on 17 September 1811.

In 1823, however, the work done on the effect of arsenic on putrefaction and the ability to find it in a long-dead body was little known outside Germany; when M Demoutiers sent for Professor Orfila, he was clearly unsure whether a month in the ground would render testing useless. Orfila had no difficulty in finding enough white arsenic to constitute a fatal dose in the stomach of the exhumed month-old corpse of M Boursier, a grocer whose wife had become enamoured of a handsome Greek adventurer. Both she and her lover were tried for murder, but, despite a strong case, during which the forensic evidence was not contested, both, to the astonishment of everyone, including the judge, were acquitted.[10]

In June 1829, Orfila was once again asked if it were possible to find poison in a long-buried corpse and, if so, how that might be achieved. In this case, the death had occurred seven years previously.

Orfila, unfamiliar with a poison case of that antiquity, said that the body had, by now, very probably, been reduced to ashes, but, if 'a sort of blackish coom' (dust)[11] could be found at the side of the spinal column, this would give material suitable for testing.

The deceased was Pierre Joseph Bouvier,[12] a wealthy and respected attorney of Bourg. A widower, he had one daughter, Josephine, who, in 1817, had married Alexandre Marie Henri d'Aubarède. Josephine should have been her father's sole heir; however, her father, for reasons that were never revealed, intensely disliked his son-in-law's family. In 1822, Bouvier was contemplating marrying a lady with whom he had a long-established liaison. Josephine feared being disinherited and must have made her feelings plain, as, some months before his death, her father had confided to an associate that he thought his daughter wished him harm. He does not appear to have taken seriously the idea that she might translate her desires into action. Josephine and her husband lived in Longchamps, about two miles from Bourg, and, that autumn, while her father was staying with them, Josephine travelled to Bourg to purchase arsenic, which she said she required to kill rats. The chemist refused to make the sale without a guarantor, so, on 7 September, she returned to Bourg accompanied by her husband, and, this time, was able to obtain the arsenic without difficulty.

Josephine arranged for a large dinner party to take place on 16 September. On the night before this event, she ordered the cook, Marie, to prepare a dish of bread and boiled milk for her father's breakfast the following morning. At 8 a.m., Josephine took some of the milk for herself, the rest being left for her father in the pantry. An hour later, Marie saw Josephine in the pantry with a paper in her hands. At ten, Bouvier called for his breakfast. A skin had formed on the milk as it stood, so, before serving it to her master, Marie skimmed it off and ate it. Soon afterwards, she was seized with painful colic and a desperate need to vomit. She rushed out into the garden where she vomited so noisily that Josephine heard, and, looking out of her boudoir window, asked the cook if she had been tasting her father's breakfast. This question struck Marie, at the time, as rather extraordinary, especially later on, when she learned that Bouvier was experiencing the same symptoms, only far worse.

After breakfasting on bread and milk, the unfortunate attorney began to suffer agonising stomach pains, which were soon followed by repeated and relentless vomiting. He demanded that a physician be summoned, and a local doctor was called, who recommended bleeding the patient; but Josephine refused to permit it. Another medical man, M Vermandois of Bourg, arrived accompanied by a nurse called Brun, who had previously ministered to the family. Vermandois prescribed a number of sedative and antispasmodic remedies and asked to be sent for early next morning if the patient did not improve. Nurse Brun had her own ideas as to how to treat the patients. She made Marie drink large quantities of milk, which had the effect of easing the vomiting and relieving the colic. Encouraged by this success, she hurried to give the same remedy to Bouvier, but Josephine was annoyed and spoke haughtily to the nurse, accusing her of giving useless remedies to the cook and forbidding her to give anything similar to her father. Meanwhile, the dinner party went ahead as planned and, before the large company, Josephine was able to spread the idea that her father was the victim of a natural indisposition.

Bouvier spent a night of sheer torture, with an inflamed face, pounding pulse and painfully tender stomach. His urine dried up and the irritated bladder contracted with such force that it ruptured. Josephine did not send for Vermandois again until midday, so he was unable to reach Longchamps until the afternoon, only to discover that the patient had not been given any of the remedies he had prescribed. Bouvier was, by then, beyond help and, shortly afterwards, became delirious. The violence of his convulsions was so extreme that four men were needed to hold him down on the bed. The torment continued until ten that night, when he died. Josephine did not remain to tend her father's body; having more important things on her mind, she hurried to his home in Bourg to ransack his coffers and gain possession of his papers. In her absence, the nurse who had been ordered to lay out the body found marks on Bouvier's chest and neck. These may well have been due to the attempts to hold him down during his convulsions; however, on seeing them, she became suspicious, and tried two or three times to suspend the preparations for the funeral. It was only with some difficulty that she was prevented from reporting her observations to the police. Rumours that Bouvier and the cook had been poisoned flew rapidly around the neighbourhood, but, ultimately, Josephine's fortune and position in society triumphed, and no enquiries were made.

A few days after the death of M Bouvier, Marie had a long conversation with her mistress, telling her that she was well aware of what she had done. She then uttered a terrible curse: Josephine would be deprived of health for the rest of her days and would find death preferable to life. At this, Josephine wept and promised to compensate Marie if she divulged nothing of what she knew. She gave Marie two bills, one for 4,000 francs and one for 2,000 francs, payable with interest nine or ten years after the date. This was to be in addition to the legacy left to Marie by M Bouvier of 4,000 francs, which was payable without interest five years after his death. This delayed recompense seemed to satisfy Marie, who agreed to remain silent.

Josephine took no notice of her public notoriety and continued to live in Longchamps. The years passed, but, when Marie was due to receive her legacy, Josephine, despite repeated applications, refused to pay. She had underestimated Marie, who revealed what she knew to the authorities. When Josephine, who had thought that the passage of time had made her safe from prosecution, found that she was in danger of arrest, she sold all her property and fled the jurisdiction of the French courts.

The body of M Bouvier was disinterred and found to be in such a remarkable state of preservation that it was recognised by the priest, the gravedigger and some of the National Guard, who had fired a salute at the burial seven years before. The chest had sunk in and the heart and lungs were blended together to look like a 'dark ointment',[13] but the stomach was intact, and there was no smell. Two expert chemists of Bourg, M Ozanam and M Ide, decided to remove only the trunk for examination, the head and remainder of the body not being considered necessary. A portion weighing nine pounds was reserved for experiments. It was assumed from the start that they were looking for arsenic. The material was boiled to obtain a solution, which was then subjected to a series of processes to remove the organic matter. Precipitate tests suggested the presence of arsenic and, when the solid residue was heated in a test tube with charcoal, 'small grey-coloured and brilliant points were seen'.[14] They had obtained a grain of metallic arsenic. This and other experiments more than proved that the deceased had ingested a considerable amount of arsenic, and, stated *The Times*, which reprinted the account of the case published in the *London Medical Gazette*, afforded 'a striking illustration of the importance of toxicology in forwarding the ends of justice'.[15]

On 20 and 21 November 1829, Josephine was tried in her absence for the murder of her father at the Ain Assizes held at Bourg.[16] Counsel for the defence asked for a six-month postponement in view of the prejudice against his client, adding that he had employed chemists at Lyon, Montpellier and Paris, who had said that the substance found in Bouvier's stomach was not arsenic. He also promised that, if the delay were granted, his client would come forward to stand trial. The public prosecutor observed that, since the accused had sold her property and fled justice, it was not credible that she would come forward. The court agreed, and the trial proceeded without the defendant. Josephine was found guilty. She was ordered 'to be conducted to the place of execution *en chemise*, with naked feet, and with her head covered with a black veil – to be exposed upon the scaffold while an officer shall read the sentence to the people – to have her right hand cut off, and then to be immediately executed'.[17]

Josephine d'Aubarède spent nearly three years of a more or less nomadic existence, travelling through England and Belgium, while her husband, to whom no suspicion attached, remained in France. In the summer of 1832, however, she was back in France, and, on 25 August, she appeared at the Assizes of Bourg to answer the charge.[18] It is unknown if any promises or inducements had been offered to ensure her return.

The new hearing lasted four days and, on the 27th, it transpired that, although Ozanam and Ide were still certain that the tests they had carried out had produced arsenic, they now disagreed as to its origins. M Ozanam stood by the conclusions of his original report, and believed that the arsenic came from the body of M Bouvier, but M Ide had retracted his original statement, and now thought that the arsenic had come from the charcoal used in the last experiment. A lawyer friend of M Bouvier testified that the dead man had led a life of licentiousness in which he had made use of strong liquor, stimulants and cantharides, a reputed aphrodisiac. When he had tackled Bouvier about this, his friend had replied that he would rather live an enjoyable, if short, life than a long, dull one. A doctor revealed that, some time before his death, Bouvier had suffered from colic, inflammation and retention of urine. Marie's version of events came under attack and it was alleged that she had been having an affair with M Bouvier. On 28 August, after listening to impassioned speeches on both sides, the jury retired. It took them half an hour to find Josephine not guilty. She walked from the court into a neighbouring room where her husband was waiting for her.

* * *

The first widely celebrated case where toxicology triumphed in court did not occur under the confident eye of Professor Orfila. Instead, it launched a new English courtroom star, William Herapath. Born in Bristol in 1796, Herapath was one of the founders of the Bristol Medical School, becoming professor of chemistry and toxicology when it opened in 1828. At the time, his chief claim to fame was the discovery of an improved process for the extraction of cadmium.

On the chilly 24 December 1834, Herapath, together with an undertaker, a sexton, the coroner, physicians Dr Henry Riley and Dr John Addington Symonds, and surgeon Joseph James Kelson, assembled at St Augustine's graveyard in Bristol to witness the exhumation of Clara Ann Smith, who had died aged sixty, on 26 October 1833. Mrs Smith, a widow, had once possessed the sum of £700 in gold coins and £100 in banknotes, as well as gold watches, jewellery and trinkets. Unhappy about placing her property in the hands of others for safekeeping, she had kept it all with her in a strongbox. Her nephew, Thomas Manley, had not seen her for four years, and did not hear of her death until 9 December 1834. He was able to trace his aunt's last-known address as Mrs Mary Ann Burdock's lodging house, 17 Trinity Street, and confronted the landlady, an attractive well-dressed woman of about forty, to be told that his aunt had died without leaving any property. Manley, who had expected to receive a substantial inheritance from his aunt, was not to be put off. Enquiries soon revealed that Mrs Burdock, who had been pleading poverty shortly before Mrs Smith's death, had soon afterwards deposited £500 in a bank and lent £400 to her common-law husband, Wade,[19] to set him up in trade. Wade, who was chronically unwell (the nature of this illness was never stated), died in April 1834, leaving his business to Mary Ann. The following month, describing herself as Mary Agar, widow, she married a tailor, Paul Burdock. She must have thought herself safe from discovery, but, seven months later, she was suspected of both robbery and murder, and, when Thomas Manley applied to the magistrates, they ordered an inquest.

Mrs Smith's coffin was lifted and placed on a gravestone. When opened, it was found to contain enough water to cover the lower body and legs of the corpse, although the head, neck and chest were above the surface. The physicians cut through the saponified outer layers of skin and fat from thorax to pelvis, exposing the interior. The stomach and intestines looked empty and were compressed and flattened, but were far better preserved than would normally have been expected after fourteen months. Riley thought the lower intestines were as firm as those of a person dead three or four days in cold weather. The viscera were of a dark blue colour, but here and there were spots of bright yellow. On separating the stomach from the duodenum, it was seen that the inner membranes were covered with a large quantity of a viscid yellow matter. The stomach was placed in a clean basin and the intestines in another. The organs were taken to the Bristol Medical School for closer examination, where it was found that 'the mucous membrane of the stomach, duodenum, jejunum, and some part of the ileum were thickly covered with an unctuous, yellow substance; it could be scraped off in quantities'.[20] Where there was no yellow matter, there was a thick dark fluid, which Dr Riley compared in consistency and colour to cocoa.

When the sticky yellow mass was removed, it was found that its corrosive action had, in places, penetrated the walls of the organs, leaving the heavily stained yellow spots that had been observed on the outer surface. The exposed inner mucous membranes were dark red, which suggested inflammation. All the medical men thought that the yellow substance was sulphuret of arsenic, or orpiment.[21] This compound was mainly used then as a colouring pigment, and was also believed to have antiseptic properties, which would preserve organic tissue.

Herapath carried out a series of tests, which were at once thorough and elegant. Washing out the stomach and duodenum, he collected seventeen grains of material from which he obtained four grains of pure yellow powder. This was dried and heated in a test tube with carbonate of soda and charcoal, producing metallic arsenic. He then oxidised the metal to white arsenic and made a solution, which he submitted to the precipitate tests. His master stroke was to complete the circle by bubbling a stream of hydrogen sulphide through the solution of white arsenic to reproduce the original yellow orpiment. These experiments were repeated five or six times.

The deceased's landlady, Mary Ann Burdock (for clarity she will be referred to by her surname) had been born Mary Ann Williams and was brought up in Kentchurch, Herefordshire, where her family worked on the land. At the age of nineteen, she came to Bristol to look for employment. The life that followed was one of pilfering and immorality. She later claimed that 'she had been beguiled at an early age'.[22] In 1819, she married a tailor by the name of Agar, who left her after a few months, and she then lived with a married man called Thomas. Not only was she reputedly dissolute, but she also compounded this enormity by allowing dissolute women into her house. Perhaps the arrival of children[23] led her to seek a more respectable trade, and she turned to running a lodging house.

Finances were at a low ebb, and Mrs Burdock was complaining about having to support Wade during his illness, when Mrs Smith and her trunk of gold arrived. It is doubtful that her landlady had a conscience to struggle with; she saw Mrs Smith as a useless individual, who owned something she needed and was best put out of the way.

She might have started by administering poison in small quantities, the better to evade suspicion, but the situation became urgent when Mrs Smith started to enquire after new lodgings. In October 1833, Mrs Burdock asked one of her lodgers, a seaman called Edward Evans, to buy some arsenic for her, saying that there were rats under her husband's bed. (No one in the house had seen any rats, but there were wire traps set for mice.) She gave him twopence, which he promptly spent on beer, and, when she asked about the arsenic, he was obliged to borrow the money from a friend. The chemist, Mr Hobbs, refused to sell it to him without witnesses, so Evans collected two friends from the nearby public house and all three returned to the shop. Hobbs was out of white arsenic at the time, but offered half an ounce of yellow instead. Evans asked if it was enough to poison a man, and Hobbs assured him that it was enough to poison half of Bristol. He sold Evans the arsenic for twopence.

Mrs Smith had been ill; either she had a cold or was suffering from the effects of her landlady's earlier attempts to poison her, or perhaps a little of both. She took to her bed and did not thereafter come down to the parlour for her meals, but had them carried up to her. Mrs Burdock did not feel inclined to look after the invalid herself, and, on Thursday 24 October, she went to the house of a Mrs Allen in Host Street and engaged her daughter, fourteen-year-old Mary Ann, as a nursemaid, promising her three shillings a week. It was stipulated that the girl would go home for her meals, and Mrs Burdock repeatedly cautioned her that she must not eat any of the food sent up to Mrs Smith, the pretext being that the patient was a dirty woman and spat in everything. Mary Ann accompanied Mrs Burdock to Trinity Street, where she found Mrs Smith not looking especially ill. The girl slept at the house that night then went home to have breakfast, returning to find Mrs Smith saying that she felt better and hoped to be up and about on the Sunday. The patient remained in bed, and, when Mary

Ann carried the slops downstairs, she was cautioned by Mrs Burdock that, if her charge were to ask her about other lodgings, she was to reply that she didn't know of any. There was no change in the condition of Mrs Smith during the day, but, that evening, Mrs Burdock came up to see how her lodger was doing, and said that she would make her some gruel. Mrs Smith protested that she didn't want any gruel, but her landlady insisted and went downstairs to make it. She came back with the gruel in a basin, but, instead of taking it directly to the patient, she went into her chamber, where Mary Ann, following her out of curiosity, saw her open a paper packet of yellow powder, put two good pinches into the gruel, and stir them in. The young nursemaid asked what the powder was and Mrs Burdock explained it was something to ease the patient, then she washed her hands well and scrubbed her nails with a nail brush. She told Mary Ann not to tell Mrs Smith there was anything in the gruel or she wouldn't take it. Mary Ann saw that the gruel was now a strange reddish colour, but did as she was told. Mrs Smith drank half the gruel and the basin was removed, but, five minutes later, she was moaning in agony saying she felt very ill.

Mary Ann suggested fetching a doctor but Mrs Burdock told her that Mrs Smith wouldn't want one. Half an hour later, the patient was in great pain, rolling about the bed so violently that she struck her head on the headboard. After that, she was quiet. Mrs Burdock and Mary Ann continued to sit by her, and, in the early hours of the morning, the girl went to see how her charge was, and found that she was dead. Mrs Burdock cautioned the girl that she was to tell no one that anything was put in the gruel and, if asked about the lady, to say that she was a foreigner. The room was then locked.

Next morning, Mrs Burdock told Edward Evans that her lodger had died poor, but there was some old plate she would dispose of to pay the funeral expenses. When Evans was at breakfast, however, he saw Wade with gold rings, which he said had belonged to Mrs Smith. A low-cost funeral was arranged, and took place two days later. No one from the lodging house attended.

When the inquest opened, both Mary Ann Burdock and her maid were under suspicion, but, on 30 December 1834, the coroner's jury found that Mrs Smith had died from arsenic administered by Mrs Burdock, who was arrested and charged with murder.

Feelings against Mrs Burdock ran so high in Bristol that an application was made for permission to transfer the trial elsewhere, on the grounds that she could not obtain a fair trial in the city. The request was refused. When she arrived at the Broad Street courthouse on 10 April 1835, she was greeted by crowds that almost blocked the street, uttering 'the most discordant yells and groans'.[24] Despite this, the prisoner 'appeared to be perfectly composed and confident'.[25]

In court, William Herapath's quiet assurance, good reputation and the fact that he brought exhibits to show the jury must have been very convincing. A report by an expert is all very well, but jurymen undoubtedly like something they can actually see, and these jurors got a treat. Herapath showed them the metallic arsenic he had obtained from the yellow powder. He displayed the bright stain of Scheele's Green precipitate he had collected on blotting paper, and declared that no substance other than arsenic would produce the results he had obtained and had so thoroughly and repeatedly verified. The defence cross-examined, suggesting that other materials might have caused the same colour, citing cadmium, but this idea crumbled under Herapath's calm expertise. 'There was no cadmium there. I was the English discoverer of cadmium.'[26] Dr Symonds brought the preparations made from the stomach, duodenum and jejunum of the deceased, preserved in spirits of wine in three glass vases. The yellow spots on the interior of the intestines were clearly visible. There was some excitement in court as Dr Symonds perched Mrs Smith's skull in a conspicuous position on the front of the witness box. The skull, which was 'beautifully white',[27] clearly showed that there was no injury from the blow to the head.

When the Recorder summed up the evidence, there was special praise for William Herapath, who 'had manifested great nicety in making his experiments, which were not only satisfactory to himself, but also to every competent judge who had witnessed them'.[28] While Mrs Burdock had undoubtedly appropriated Mrs Smith's property, this was not, in itself, proof of murder, since she might have done so even if her lodger had died of natural causes. The thoroughness of Herapath's work, however, complemented by the examples shown in court, gave the jury the confidence, even in a case where the body had been exhumed fourteen months after death, to find the prisoner guilty. Their deliberations took just eighteen minutes. The case was also remarkable in that, unusually in a poisoning case, someone had actually seen the poison being given. Mary Ann, who became so exhausted at the trial that she had to be carried out and given time to recover, had not mentioned this detail at the inquest, but the Recorder very sensibly pointed out her youth and the fact that she was obviously nervous, which he thought was more than adequate explanation for the omission.

Less than two days after the conclusion of the trial, Mary Ann Burdock was executed in front of a crowd of some 50,000 eager onlookers.

There remained one mystery. The nursemaid had said that the gruel was red. William Herapath later addressed this issue. He purchased some arsenic from Mr Hobbs, asking for the same kind sold to Evans, and found it was composed of three kinds mixed together – white, yellow and red, the two former in small lumps, the latter in powder. He thought it was the red powder, realgar,[29] that had been administered and that the process of putrefaction, which involved the production of ammonia and hydrogen sulphide, would convert the realgar to orpiment. He tested his theory by poisoning an animal with the same kind of realgar, and found that, after putrefaction, its internal appearance was the same as in the body of Mrs Smith. Realgar, like white arsenic, also has a tendency to control putrefaction and convert bodies into adipocere. Some doubts had been cast on Mary Ann's evidence, partly because no one could believe that anyone would be so foolhardy as to put poison in a dish in front of a witness,[30] but Herapath's tests confirmed the nursemaid's observation as to the colour of the gruel, something he thought she could not have known unless she had seen it.

Over the next few years, William Herapath, now 'the celebrated analytical chymist of Bristol',[31] was frequently called upon to perform his tests in cases of suspected poisoning in the West Country, especially when a body had been long buried, and his methods were being widely reported. His fame, said *The Times*, 'is not confined to England, but is well known to the scientific world'.[32] Over the years, he was to examine many sets of intestines coated with pasty yellow matter, and, eventually, he concluded that, in any body that had been buried for more than twelve weeks after arsenic, including the white oxide, had been taken, the hydrogen sulphide produced by putrefaction always resulted in the formation of orpiment. His testimony is notable for its clarity, for, even though he addressed principles of chemistry with which the jury would not have been familiar, he had the knack of engaging their interest and leading them carefully through the procedures he adopted, illustrated by samples he showed in court, bringing them irresistibly to the same conclusions as he had reached. If the accused was acquitted, as often occurred, it was not the fault of any failure in the medical evidence, but because the lapse of time had made it especially difficult to prove administration. He was particularly concerned with what juries would find convincing, as is shown by his comments to the British Association for the Promotion of Science in August 1836. A recent development in tests for arsenic 'was the most elegant that could be conceived;

at the same time that it was the most sensitive: but it would require a few modifications to make it the best for exhibition to a jury'.[33] In 1839, he was performing the same tests as in 1834, declaring, 'I have never found any thing equal to these tests in efficacy,'[34] and, as late as 1843, he still preferred his own methods. The years 1835 to 1840 were, however, to be a significant period both in testing for arsenic and in the emergence of new courtroom experts.

CHAPTER FIVE

The Suspicions of Mr Marsh

On 5 February 1844, thirty-three-year-old James Smyth was tried at the Central Criminal Court, London, for demanding money with menaces. Smyth had struck up a casual acquaintance with butler Thomas Robinson, and, after paying him a visit, sent him two letters demanding money, alleging that Robinson had made improper advances. Robinson immediately took legal advice and, as a result, Smyth was arrested. His handwriting was identified by his brother-in-law, from whom he had stolen £50 only a few months earlier. Smyth was found guilty and sentenced to be transported to Tasmania for twenty years, but, before he departed, he had a confession to make. His real name was John Bodle[1] and, more than ten years previously, he had poisoned his grandfather. Bodle felt safe in making this confession because he had already been tried for the murder and acquitted. His trial highlighted the inadequacies of arsenic testing and led directly to one of the most significant advances in toxicology.

If the confession was genuine, and there is no compelling reason to doubt it, then John Bodle was a member of a rare and select class of poisoners: those who, while targeting a specific person, do so careless of any harm they may cause to others.

On Saturday 2 November 1833, eighty-one-year-old retired farmer George Bodle, of Plumstead in Kent, sat down to a breakfast of coffee and toast. His seventy-four-year-old wife, Ann,[2] was still in bed and was brought breakfast by Mrs Elizabeth Evans, a daughter by a previous marriage, who called regularly to look after her. Once breakfast had been served to the Bodles, Mrs Evans and the other members of the household, Mrs Bodle's granddaughter, Elizabeth Smith, and servant, Sophia Taylor, had theirs. Soon afterwards, everyone in the house was seized with identical symptoms – burning pains in the stomach and repeated vomiting.

Judith Lear, a poor widow who worked as a charlady, came to the Bodles' farmhouse every morning to get milk, and was allowed to take away the leftover breakfast coffee. She would tip the liquid and grounds into a mug she had brought for the purpose, and give it to her daughter, Mary Bing. Mary and her labourer husband had seven children and were struggling to manage on a low income, so the second-hand coffee would have been very welcome.

Mrs Lear had not been back home long when she was sent for to care for the suffering Bodle family; she returned to the farmhouse to find George sitting in the kitchen looking very ill. He protested that he had had nothing but a roasted potato for supper the night before, and coffee and toast for breakfast. Mrs Lear suggested that a toad or an insect could have got into the kettle, so, when George wanted tea, he ordered that the kettle should be cleaned very thoroughly before it was used again. The kettle was taken outside and its contents emptied on to the ground; then its interior was repeatedly washed and scrubbed with a brush. Mrs Lear was understandably worried that the coffee she had given her daughter was contaminated, and sent Mary a warning not to touch any of it. Fortunately, although Mary's fourteen-year-old daughter Eliza had already poured the coffee into a cup, she had thought it rather thick and had not drunk any, pouring it back into the mug instead . There it remained until Mrs Lear came and took it away.

It wasn't until the evening, when the occupants of the Bodle house were still very ill, that they sent for Mr John Butler, a surgeon of Woolwich, saying that they thought they were suffering from cholera. The symptoms and the fact that all had been similarly and rapidly affected suggested to Butler that they had swallowed a mineral poison, and he administered one of the standard treatments: whites of egg (albumen having been shown by Orfila to impair the action of corrosive poison). George was the worst affected and, several times, placed his hands on his stomach, saying, 'I know poison has been given me.'[3] He asked who had been to the house and was told that his grandson, John junior, had been there that morning.

John junior's father, John senior, occupied a property at the end of George Bodle's garden, together with his wife Catherine, two sons and servant Mary Higgins. John senior was George's under-bailiff, tending the garden and a small farm, but his younger son, John junior, had had no gainful occupation for some time. His grandfather disliked the youth and had refused to allow John senior to employ him. John senior admitted not only that his son was 'fond of his bed',[4] but that he 'lived like a gentleman'[5] and was content to be supported by his parents. There may have been more to old George's dislike of his grandson than the young man's laziness. In view of his later history, it is possible that John junior may already have shown a tendency to petty crime.

On the morning when George Bodle's household was taken ill, John junior had behaved in a manner that was noticeably out of character. Mary Higgins had come downstairs early, only to be greeted with the unusual sight of young John awake and sitting before the fire, which he had made up himself. She had never before known him to get up so early or light the fire. At daylight, he took the milk can and went to his grandfather's house, arriving to find Sophia lighting the fire. She too was surprised to see him. Suggesting to her that she was up late and behind with her work, he took the tea kettle, a substantial container that was later estimated to hold 266 fluid ounces (over thirteen pints or about six litres), and went outside to fill it at the pump; then he hung the kettle over the fire. Sophia gave him the milk he had called for, and he returned home.

At half past six that evening, John senior went to his father's house for his wages, and it was then that he learned that his family was ill and poison was suspected. He informed his own household of the situation, and, that same evening, young John abruptly left home and went to London, where he stayed with his married sister, Mary Andrews, who kept a coffee house in Clerkenwell.

Old George was an obstinate man and could only be persuaded to take a small amount of medication. The following day, all the patients were improved except George, who, in defiance of his doctor's advice, had drunk a pint of ale. Butler's early concern was to identify the source of the poison, and he soon realised that the probable vehicle was the coffee, as it was the only item all the victims had consumed.

Coffee was an expensive commodity, and it was kept in a glass jar in a locked cupboard to which only George and his wife had the key. On the Saturday, old Mr Bodle had unlocked the cupboard and taken out the coffee needed for that day in a cup, which he then gave to Sophia. This left the jar almost empty, and it was later refilled with fresh supplies. Butler poured out the fresh coffee and, finding a coating of old coffee at the bottom of the jar, scraped some out, getting about a teaspoonful. He also took a sample of the fresh coffee. The coffee given to Mrs Bing was recovered and given to Mr Butler. Mrs Lear was asked to keep samples of the patients' vomit, but she was looking after four invalids and was later unable to attribute the samples she collected to specific individuals.

On Monday, with Old George still suffering from stomach pain and a racing pulse, another medical man, Dr Sutton of Greenwich, was summoned, and he too was convinced that the patient had taken poison. Next morning, George, suspecting that young John had put something in the water, asked if his grandson had been to the house again, and gave orders that, if he did come, he should be turned out of the kitchen. When John senior visited that afternoon, the old man's condition was deteriorating and he probably knew that he was dying. He whispered to his son that he was sure that John junior was the cause of his illness.

On Tuesday night, George Bodle died, and Butler notified the coroner, who opened an inquiry.

No one had to look far for motives to murder George Bodle. Although frugal with his coffee, he had been a wealthy man. His estate, which consisted of forty acres of freehold land, leasehold properties and bank stocks, was valued at £20,000. George had made a will, and John senior, as the only living son, believed that he would inherit all his father's property; however, only days before George was taken ill, he had drawn up and signed a new will. How much the family knew about its contents remains in doubt.

In the new will, John senior was not even named as executor; the family members trusted with this office were Samuel Baxter, the husband of George's daughter Mary Ann, and William Baxter, Samuel's eldest son. The distrust that existed between George and his son became even more apparent in the bequests. The will left all George's personal and household effects and the rents of his properties and interest on his investments to his widow. On her death (or in the unlikely event of her remarriage), John senior was to inherit the farmhouse, the cottage he currently occupied, a market garden and an orchard. He was not permitted to sell them and on his death the property was to be divided equally between his children. The valuable freehold properties, remaining land, interests in the bank stocks, and any goods and chattels not otherwise mentioned were to go unencumbered to the Baxters.[6]

The post-mortem examination of George Bodle was carried out by Butler, Mr Samuel Solly, lecturer in anatomy at St Thomas' Hospital, and Dr Bossey, a surgeon of Woolwich, who discovered clear evidence that some irritating substance such as arsenic had been taken into the stomach. The amount of poison consumed had probably not been great, and might not have killed a younger person, but, for elderly obstinate George, it had been fatal. John junior, it was soon learned, had purchased arsenic from a Woolwich chemist only a day or two before his grandfather was poisoned, saying he wanted it to kill rats. A warrant was issued for his arrest, and Constable James Morris went to Clerkenwell to apprehend him.

On being told the reason for the constable's arrival, John junior protested that it was his older brother who was wanted; but, on being assured that there was no mistake and he was under arrest, he fainted. Once he had recovered, Morris accompanied him back to Plumstead. On the way, Morris searched the prisoner and found a key, which John said would open his box at his father's cottage.

Next morning, with young John duly delivered to the coroner's court, the box was unlocked and was found to contain two packets of arsenic, both, marked 'POISON'. One of the packets appeared to be intact, but the other had been opened, and some of the poison was missing. Morris was very excited by this discovery; so much so that he took the packets with him when he later visited two public houses, and showed them to the other drinkers. The evidence was passed from hand to hand, and one man dipped his fingers in the exposed arsenic and rubbed the powder on his chin.

All the evidence collected was passed for analysis to the distinguished man of science,[7] Michael Faraday, professor of chemistry at the Royal Military Academy, Woolwich. Faraday, who was occupied with other pressing work, took no part in the analysis and passed the task to his assistant, thirty-nine-year-old chemist James Marsh. The problem facing Marsh was similar to that in the Donnall case. Arsenic had made its presence known by the symptoms of the victims and the damage to the dead man's stomach, but it was invisible. If, as was suspected, it had been put in the kettle, it had dissolved in the hot water, and no telltale crystals remained. The amount in the kettle might not have been very great, as only eighty-one-year-old George had consumed a fatal dose. No trace of arsenic could be found in the stomach contents, the vomit, the new coffee, the crust of the old coffee or scrapings from the scrubbed kettle. There remained only the leftover grounds and coffee liquor given to Mrs Lear, a piece of evidence so compromised by its handling that it would probably not be admitted in a modern court. There was little enough material for Marsh to work on, though the details of the tests he carried out were not recorded by the reporter at the inquest. He might have heated the sample to produce a deposit, dissolved it and applied the precipitate tests. Or he might have used another test, described by Christison in 1829,[8] passing a stream of hydrogen sulphide gas through the fluid sample, which should be either neutral or slightly acidulated, and producing the characteristic yellow precipitate of orpiment. Marsh's comment, however, was that he had detected the arsenic 'from its peculiar smell, not from its presence'.[9] This suggests that he simply applied heat and produced the telltale odour of garlic.

The inquest opened at the Plume of Feathers, Plumstead, on Wednesday 6 November. The jury viewed the body, which still lay in the deceased's bedchamber and showed no unusual signs other than being very swollen about the abdomen. There were to be six days of hearings, in which the only serious suspect was John Bodle junior. Mary Higgins testified that she had overheard a disturbing conversation between John junior and his mother, a conversation that Catherine later denied had ever taken place. Mary said that her mistress was arguing with her son over something he had had simmering in a cup over the fire. John junior, she claimed, had said that he wished his father might die first, and then he wished his grandfather dead, after which he would expect to have between £100 and £1,000 a year. Catherine, far from being shocked by this statement, replied with calm practicality that one had better die one week and the other the next, to allow time to settle business. John had added that he would not mind poisoning anyone who affronted him.

At the inquest, John junior explained his purchase of arsenic by saying that he used it to treat pimples, either shaking up the powder in water or mixing it with lard to make an ointment, a practice that Mr Butler thought dangerous. The coroner observed that, if the prisoner had indeed used arsenic, as he claimed, he might himself have shown symptoms of poisoning, and he referred the jury to cases quoted by Robert Christison in which this had occurred. Unfortunately, Constable Morris's drunken spree had made it impossible to know how much arsenic had been in the opened packet when collected from John junior's box, and the coroner believed that this misbehaviour alone had effectively destroyed any case against the suspect. Despite this, the inquest found John junior guilty of the murder of his grandfather and he was committed for trial. This took place at the Maidstone Winter Assizes in December.

It was never seriously claimed at the trial that George Bodle had not died of arsenical poisoning; the main difficulty was proving who had administered the arsenic. John senior, who benefited under the will and might, at the time of his father's death, have believed he was the main heir, also had motive, although there was no evidence that he had been to his father's house on 2 November until the evening, and he had had no opportunity to poison the water used for the breakfast coffee. Marsh told the court that he estimated that the coffee liquor, of which there was about five ounces, contained four to six grains of arsenic. He had experimented by infusing seven quarts of water with half a grain of arsenic per ounce of water and comparing the appearance in the test tube with the coffee liquor, which he thought was very similar. He was obliged to admit, however, that he had very seldom made such experiments and had never before tested for arsenic. Ultimately, he said, he would rely on no test except that of 'reproduction'[10] – the extraction of visible arsenic from the sample.

John junior, in an eloquently argued defence statement, had given up trying to transfer his guilt to his brother, and now accused his father of the murder, suggesting that he could have been to the house the day before and put arsenic in the coffee jar. Commenting on the practice of execution following soon after conviction for murder, he raised the innocent virginal spirit of Eliza Fenning, warning the jury that wrong judgments had followed from reliance on circumstantial evidence, and, in that case, 'a short time would have brought the error to light'.[11] There were character witnesses for young John, including a Mrs Brett, who said she had seen him mixing arsenic with lard and knew he wanted it to use as an ointment. The jury, told they must acquit if they had any reasonable doubt that the prisoner had administered poison, found him not guilty, to popular acclaim.

Despite the doubts, no further action was taken over the death of George Bodle. John junior's 1844 confession may have been due to an attack of conscience, assuming that he had such a thing, but it might also have been prompted by his family urging him, before he departed on the convict ship, probably never to return, to finally remove any residual suspicion from his father, who had died the previous year. John arrived safely in Tasmania, where he had obtained work as a butler. In 1852, he packed a bag, used his master's ticket to board a ship to Australia, disembarked at Melbourne and disappeared.

It is sometimes suggested that John Bodle's acquittal was partly due to the failure of Marsh to pinpoint the means by which the arsenic was administered, and the chemist is variously said to have been frustrated, angry or stung by the result. Whether or not Marsh was angry at Bodle's acquittal is unknown, but, undoubtedly, his interest had been aroused. He set about devising an improved test, one that, when he revealed it to the world of science, was so simple and elegant that more eminent men were astonished they had not seen the answer themselves, especially as it was based on existing knowledge. Marsh's starting point was the work of Carl Wilhelm Scheele, who, in 1775, had shown that, when zinc and arsenic were mixed in a solution of nitric acid, the reaction produced zinc nitrate, water and a gas, arsine. Although in itself odourless, arsine oxidises in air to produce a garlicky smell.

Marsh advised that, if the suspect items to be tested were solid, such as pudding, they should first be boiled to produce a solution. Thick liquids, like soup or stomach contents, were diluted with water and filtered, but fluids like tea and coffee could be used without any process being applied. Marsh added zinc and acid to the liquid to be tested, which, if arsenic was present, would produce arsine. His apparatus directed the gas up a curved glass tube, and the gas was ignited as it emerged, the hydrogen burning off first to produce water vapour. A piece of cold window glass was held in contact with the flame, on which there formed a thin film of metallic arsenic. If the flame was directed into a glass tube open at both ends, the exposure to air oxidised the arsenic, which formed 'a white pulverulent sublimate of arsenious acid'.[12] After the apparatus was used, Marsh cleaned it thoroughly and repeated the test with zinc and acid as many times as were necessary to reassure himself that no residue remained. The test was remarkably sensitive, and he obtained distinct metallic deposits from one drop of Fowler's solution,[13] which contained just one twentieth of a grain. Marsh had used a specially constructed apparatus, but, when the results were published in October 1836, he assured his readers that this was not essential and, in cases of necessity, a phial with a cork and piece of tobacco pipe or a bladder might be pressed into service. His only warning was that they must be certain of the purity of the zinc by first testing it in the presence of acid.

The Marsh test, as it came to be known, was quickly hailed as a major breakthrough in forensic chemistry, and Marsh was awarded the Large Gold Medal of the Society of Arts.

The scientific community must have known about the test well before actual publication. In May 1836, a woman died in Guy's Hospital after swallowing arsenic, and it was decided, even though she had confessed what she had done, to use the new test to verify the nature of the poison. One can sense from his subsequent report that twenty-nine-year-old Mr Alfred Swaine Taylor was eager to seize on the first opportunity offered to try out 'the ingenious apparatus which had been then but recently proposed'.[14]

Taylor was already showing the brilliant promise that would later result in his being dubbed 'the father of British forensic medicine'.[15] After studying medicine and chemistry, he was appointed lecturer in medical jurisprudence at Guy's Hospital in 1831, at the age of twenty-five, teaching the first ever course on the subject in England. The following year, he became joint lecturer in chemistry. Taylor performed many experiments with the Marsh apparatus, both on artificial mixtures and on the contents of stomachs of persons poisoned with arsenic, and was impressed by the delicacy of the tests.

The Marsh test was not without fault. The cold glass could shatter in the heat of the flame, and other chemists preferred mica or porcelain or a heated glass tube.[16] It could fail to produce results in the hands of the inexperienced, and, even when it showed the presence of arsenic, it did not quantify it. The latter issue was rectified in 1837 by Jöns Jacob Berzelius, a Swedish chemist, who modified the apparatus so that the deposit could be weighed. In the same year, Lewis Thompson found that antimony could also produce results that resembled those of metallic arsenic, a serious problem if a poison victim had been treated with tartar emetic.[17] Because of these concerns, even analysts adept at the Marsh test did not use it exclusively, but subjected samples to a series of tests, starting with the more familiar ones.

In April 1839, William Herapath demonstrated the detection of arsenic to the Bristol Philosophical Society, starting with the traditional tests, ending with 'the elegant method proposed by Mr Marsh', though with some additional reservations. If examining a stomach, which was 'the case of most common occurrence, a tenacious froth arises, which impedes the flow of the gas, and causes jets of fluid to be thrown upon the glass, thereby interfering with the result',[18] adding that 'the whole process becomes inapplicable'[19] if tartar emetic was administered. The overall reaction of chemists to the Marsh test was, however, warm approval, and many started exploring its possibilities and refining and improving the original process. Marsh devised a means of distinguishing arsenic and antimony with Hume's solutions, publishing his methods in 1839.[20] In 1840, German chemist Karl Gustav Bischof showed that an alkaline hypochlorite solution dissolved arsenic but not antimony.

Due to the lack of detail in trial reports, it is hard to know when the Marsh test was first used in a case of murder; however, the importance of the process as a deterrent to criminals was soon recognised by courts. In April 1838, at an inquest into the death of a butcher, John Bruce, surgeon John Hewson, who had examined the remains and tested the suspect contents of a tea kettle, was asked by the coroner to demonstrate his methods for the jury, 'that it might be generally known that detection was inevitable'.[21]

Hewson first performed the precipitate tests, one of which at the jury's request he carried out a second time on a sample of pure water. He then produced orpiment, and ended with 'the best and final'[22] process of all, that of Mr Marsh. The culprit was Bruce's fifteen-year-old apprentice, Samuel Kirkby, who was taking his revenge for a beating. Kirkby was tried in July and sentenced to death, but this was later commuted to transportation.

In May 1838, two French chemists, Thinus and Mollier, were asked to examine the remains of Mme Défourneaux, whose husband, an ill-tempered drunk who had subjected her to more than twenty years of violent abuse, had been accused of poisoning her.[23] Having discovered arsenic in the stomach fluids by the established methods, they decided to try the Marsh test and declared it the most suitable test for arsenic, due to its ease, simplicity and the certainty of the results. Not knowing about the Kirkby trial, they claimed to be the first ever to employ it in a serious forensic case. The accused man was tried, found guilty and condemned to death the following August.[24]

The more sensitive and certain a test, however, the greater the expectations of a jury. In August 1839, twenty-two-year-old John Clifton was tried at the Old Bailey for the murder of his wife Caroline, who was six months pregnant. A daughter had been born to the couple in December 1837 but died aged eleven months. When surgeon Mark Brown Garrett was called to the desperately ill woman, who was suffering intense pain with repeated vomiting and diarrhoea, she told him that her husband had given her a powder because 'he did not want any more b—— children',[25] but she begged Garrett not to tell her husband she had said this or he would kill her. She died soon afterwards.

The post-mortem was carried out by Garrett and another surgeon, John Adams. This showed that the stomach, intestines and rectum were highly inflamed, which, they both believed, could only be accounted for by a powerful poison. Caroline's agonies had ruptured the membranes surrounding her foetus, and the umbilical cord was torn.

Clifton had recently purchased arsenic and cream of tartar (potassium bitartrate), the latter being what he said he had given to his wife.[26] When a shoemaker who lived in the same house told Clifton that the death was 'a bad job',[27] Clifton declared that he was innocent but, if the worst came to the worst, 'they could only bring in manslaughter'[28] and he would be transported.

The dead woman's stomach contents were sent to an experienced analytical chemist, Thomas Lunn, who was acquainted with the Marsh process. He applied all possible tests to the samples but found no trace of any mineral poison. Despite this, both Garrett and Adams were sure of their opinions and suggested that the poison might have been vomited or absorbed. It is possible, judging by Clifton's comment to the shoemaker, that he had coerced his wife into taking arsenic, hoping to procure a miscarriage without intending to kill her. The judge advised the jury that a verdict of manslaughter was not an option; they had to either find him guilty of murder or acquit. They acquitted him.

Meanwhile, in France, the Marsh test had led to a dramatic shift in the focus of forensic examination, and, inevitably, at the head of the movement was Professor Orfila, his confident authority and taste for long hours of research undiminished. No longer the young pretender daring to criticise his elders, he was in his fifties, Dean of the Paris Medical Faculty and acknowledged as the pre-eminent toxicologist in France. He had also acquired a great many rivals unafraid to challenge him.

Orfila had seen possibilities in the Marsh test that its inventor had never envisaged. Previously when chemists had looked for arsenic in cases of suspected poisoning, they had mainly examined the intestines, stomach contents and vomit of victims, and any uneaten suspect foodstuffs. The sensitivity of the Marsh test gave Orfila a more refined method of studying how arsenic was absorbed by and transported throughout the body, and he detected it in blood, organs or tissue with which it had not initially come into contact. Orfila was to spend many months and sacrifice over two hundred dogs in this new work. An issue he was unable to resolve, however, one of vital importance in criminal trials, was that of defining the time when poison had been administered.

He killed his dogs by inserting arsenic into either the stomach or a wound in the thigh and, after some early failures, was able to detect arsenic in the brain, blood, viscera, skin, bones and muscles. Marsh had not removed organic material from his test samples, but Orfila did, treating them with nitric acid. He confirmed that when arsenic is absorbed by the body it passes into the blood and is carried to all the organs. To prove that this occurs during life and not by post-mortem processes, he hanged a dog that was already dying from arsenic and produced the same test results. The crucial forensic advance was that the Marsh test enabled Orfila to demonstrate the presence of the arsenic that had actually killed the victim, as opposed to earlier tests, which had acted only on the unabsorbed residue.[29]

The new science also prompted studies to determine whether the flesh of poisoned animals was safe to eat and the studies supplied useful information concerning the speed with which arsenic leaves the body. Taylor reported an experiment in which a non-lethal dose of arsenic was introduced into the stomach of a sheep, and daily records were kept of its elimination.[30] Arsenic was detected in the animal's faeces for the next fifteen days, but it was not until the thirty-fifth day that it no longer appeared in the urine. On day thirty-eight, the sheep was killed and no arsenic was found in the animal's flesh. The meat was then cooked and eaten by six people, who suffered no ill effects.

The sensitivity of the Marsh test was, however, creating its own problems, and, once again, Professor Orfila was in the eye of the storm. Early in 1838, a young physician and chemist Jean-Pierre Couerbe[31] brought a very worrying question to the great doyen of French toxicology. Couerbe had been carrying out research on the decomposing bodies of people who had not been poisoned and had found arsenic both in the bones and in the putrefying flesh. Was arsenic produced during the process of putrefaction, he wondered, or could it be a normal component of a healthy body? If the latter, how could this 'normal arsenic' be distinguished from traces resulting from criminal poisoning?

Couerbe and Orfila worked on the problem together. They used the bones of humans, some of whom had been dead for only a few days and others who had been buried for some months, and their test results demonstrated the presence of arsenic. It was only detectable, however, when the bones had been calcined, not when they were boiled in distilled water. No arsenic could be found in the viscera, and, while at that stage in their endeavours it could not be positively shown that it was in the muscles, they could not rule out the possibility that it might be discovered in future research.

In October that year, Orfila thought it necessary to send a sealed letter to the Academy of Medicine to inform them of his results. Work continued, and, by January 1839, when he presented a paper to the Royal Academy of Medicine in Paris, he was able to announce that 'normal arsenic', as it became known, could easily be distinguished from ingested arsenic as it was not soluble in boiling water. Samples of grave soil had been tested, with only slight traces of arsenic found in some and none in others, and Orfila thought it highly improbable that soil could yield its arsenic to a body so as to give misleading indications of poisoning. He had also considered the possible presence of arsenic in the sulphuric acid and zinc used in the Marsh test, and suggested ways of ensuring that they were free of impurities.[32] He was happy, therefore, to herald a brave new era of medical detection in which the toxicologist would wield the power to stamp out poisoning. 'From now on, crime will be successfully hunted down to its last refuge. . .'[33]

But having revealed the problem of 'normal arsenic', albeit while showing how it could be addressed, Orfila had introduced a potentially damaging and controversial concept into medico-legal science. He could not persuade his critics that innocent contamination of samples was unlikely, or that what occurred in his laboratory was identical to the processes of nature.

The widespread use of arsenic in agriculture and manufacturing made it almost omnipresent in the nineteenth century. It was in soil, paint, fabric, paper and metal, and could well be in the very material from which the chemist's apparatus was made. The sensitive and delicate Marsh test, which provided evidence of small quantities of arsenic, was open to the accusation that it had simply detected something that had crept in from materials used for the transportation and storage of a body, or even during the tests themselves, or that the arsenic had developed in the grave by some process as yet not understood. The wonderful new tool to trap poisoners could, therefore, also play into the hands of a clever defence counsel, who could see a world of fresh opportunity for inserting reasonable doubt into a case.

Throughout 1839, battle lines were being drawn that were less to do with science than with personalities and politics. One of Orfila's most outspoken critics was Francesco Rognetta,[34] a noted Italian physician then exiled in Paris, where he specialised in toxicology with particular reference to arsenic. Rognetta was a Republican activist while Orfila was a Royalist with strong connections to the political and academic establishment, and there was no love lost between the two men. Rognetta took every opportunity to attack Orfila's work, claiming that the discoveries were not original, although it was pointed out by others that, despite the fact the absorption of arsenic had been suspected by other researchers, it was Orfila who had actually demonstrated it.[35] This enmity finally exploded into an incident in the summer of 1839, when Rognetta was arrested and Orfila arrived at the police station as he was being questioned. According to Rognetta, Orfila demanded that he write a letter retracting his criticisms. When Rognetta refused, Orfila responded by threatening not only to prevent his rival lecturing at the School of Medicine in Paris, but also to use his political influence to have Rognetta's authorisation to practise medicine in France revoked, and then see him expelled from the realm.[36]

M Couerbe, meanwhile, had read Orfila's series of memoirs, published in 1839, on the subject of 'normal arsenic' and the work they had carried out together, and had found, to his astonishment and deep mortification, that he received no credit for his important discovery, which he had imagined would be published in a joint paper. In December, Couerbe wrote an open letter to Orfila, bespattered with multiple furious exclamation marks, in which he accused him of plagiarism. The public journals cautiously reproduced only part of the letter. It was left to Rognetta to happily publish the whole letter in a volume of his own,[37] which he mischievously dedicated to Orfila as a token of his unchanging beliefs. Orfila's response was to admit that, while Couerbe had made the initial suggestion, he had not provided the ultimate experimental proof.[38] It was a petty and rather unedifying squabble that might have left many with the impression that the powerful Orfila had simply squashed the lesser man, but Couerbe had attracted to his side two eminent supporters, Rognetta and François-Vincent Raspail,[39] the latter a talented and forward-thinking chemist, biologist and political activist, who had once been imprisoned for his Republican beliefs.

The stage was being set for a bitter and acrimonious confrontation that would be the first major public test of the Marsh process.

CHAPTER SIX

French Porcelain

Louis Mercier was a widower and the father of four children, one of whom, twenty-year-old Nicolas, was of restricted intelligence and needed constant attention. Nicolas was unable to keep himself clean and, according to his father, was uncontrollably greedy, overeating until he made himself sick. In 1837, Louis remarried without fully advising his second wife Marie of Nicolas's condition. Nicolas had been staying with a brother, but, on 7 December 1838, he returned to live with his father at Villey-sur-Tille, just north of Dijon. Marie did not hide her disgust at her stepson, and objected strongly to being obliged to live under the same roof as 'the idiot boy'.[1] She threatened to leave Louis or to stab him. Louis, always ready with an easy excuse, later claimed that he hadn't paid much attention to his wife's complaints because she had been ill; also, he explained, women were often capricious.

Louis had, in fact, taken his wife's objections very seriously and was prepared to do something desperate to placate her. He was overheard reassuring Marie that she would not have to put up with the situation much longer. On 13 December 1838, Louis Mercier visited an apothecary and bought an ounce of arsenic, which he said was for the purpose of killing rats. He put it in a drawer. Two days later, Nicolas was taken ill with vomiting, raging thirst and inflammation of the throat. No medical aid was summoned, no samples of vomit were retained for testing, and the youth's only attendants were his father and stepmother. On 21 December, when Louis went to meet his wife as she was returning from a fair, she at once demanded to hear his news and he replied, 'It has been taken.'[2] That evening, the sufferings of Nicolas were redoubled, and he died at 5 a.m. the next day.

The agonising end of a young man who had aroused such hatred in his stepmother naturally led to suspicion that he had been put out of the way with poison. Fourteen days after his death, the body of Nicolas Mercier was exhumed. An examination showed acute inflammation and ulceration of the intestinal canal and ulceration of the mucous membranes of the stomach. Three chemists of Dijon tested the liquid found in the intestines with hydrogen sulphide, and later with Marsh's apparatus, but they were unable to discover any arsenic. They next cut the stomach and intestines in pieces, boiled them in nitric acid, and passed a current of hydrogen sulphide through the resulting liquor. Once again Marsh's test was employed, and once again it failed to reveal any arsenic.

The body was reburied, but the circumstantial evidence and what were described as 'moral proofs'[3] of poisoning were becoming more compelling by the day, and it was still believed that Nicolas had died as a result of ingesting an irritant poison. Professor Orfila was consulted, and said that the experiments of the three chemists of Dijon 'had not been pushed far enough'.[4] He asked for the body to be disinterred and sent to him in Paris for further examination. It was now four months since it had first been buried. Nicolas's body, packed in a barrel, made the eighty-mile journey to Paris, where Orfila performed a new examination of the rotting corpse together with associates Devergie, Le Sueur and Ollivier. It was no easy task. The remains of the intestinal canal no longer showed any organic structure, and the other parts were disintegrating, forming a 'mass of flesh, half decomposed'.[5] In an important departure from earlier procedures, Orfila tested Nicolas's liver and limbs and the putrid fluid that had surrounded the body in the barrel. He found arsenic.

Mercier and his wife were arrested and Louis ungallantly wrote a letter to the Procureur du Roi blaming his wife for the death of Nicolas, an accusation he later regretted, claiming (clearly nothing was ever his fault) that he had only done so after being given bad advice. Meanwhile, Mercier's representatives consulted Rognetta who, ever keen to oppose Orfila, advised them that the evidence of the three chemists of Dijon was conclusive proof that arsenic was not present, and that the appearance of the intestines on dissection indicated nothing more than disease. Rognetta proposed consulting Raspail, whose arguments in Mercier's favour were so convincing that he was summoned as the principal witness for the defence in direct opposition to Orfila. The trial of Louis and Marie Mercier took place in November 1839 and was an exercise not in science and law, but in personal, professional and political enmity on both sides – something that should have had no place in court.

Louis Mercier made a very poor impression under questioning, giving evasive and inconsistent replies. He said that when Nicolas vomited he thought that his son, being greedy, had taken the arsenic himself, thinking it was sugar, which did not explain why no medical help had been called when the youth was so desperately ill. Mercier even tried to convince the court that he had not realised his son's condition was serious when, two days before his death, Nicolas had made efforts to vomit that were so violent he had collapsed. Tasked with the incriminating statements he had made to his wife, Mercier said he couldn't remember them.

Orfila's evidence was crucial in helping the court understand why his analysis of the remains differed from that of the chemists of Dijon. He pointed out that, until the publication of his memoir in January 1839, chemists suspecting poison had simply been in the habit of examining vomit, matter found in the intestines, and the condition of the intestinal canal itself. If the morbid appearances corresponded with poisoning, and poison was found, then they declared it to be a case of poisoning. If none was found, then they gave the contrary opinion. He had shown, however, that arsenic could be discovered in organic tissue even when no trace could be found in the alimentary canal. His first experiments had been confined to dogs, but he had recently been able to confirm his findings using human subjects. He had boiled the stomach of M Soufflard, a condemned murderer who had poisoned himself with arsenic while in prison, and then tested the resulting liquid for arsenic. He had subjected the liver of another suicide, M Lorrin, to a similar process, and tested blood taken from a living woman who had poisoned herself. By those means, he was able to confirm that if no arsenic was found in the tissues, then none had been taken. With respect to the negative findings of the chemists of Dijon, he said with unusual generosity that he did not blame them, as the methods he adopted had not been published at the time of their report.

He showed the jury porcelain plates with deposits of metallic arsenic obtained from his tests on Nicolas's remains and, for comparison, arsenic extracted from the remains of a dog that had been poisoned with white arsenic. He assured the court that the materials used for his tests were pure.

Raspail had a difficult case to argue and he did his best, making a three-hour address to the jury in which he maintained that powerful subterranean and largely unknown forces of nature could not be reproduced in the theoretical experiments conducted in Orfila's laboratory. Where arsenic was found in a solid state in a body examined soon after death, there was a strong presumption of poisoning; otherwise, he argued, it was not possible to be certain. Arsenic extracted from the tissues of an exhumed corpse might, he claimed, have come there accidentally from the environment, from the many manufactured products that contained it, from grave soil – or even from the containers in which the remains had been transported. He alleged that Orfila's examination of samples of grave soil was imperfect as it should all have been analysed. He suggested that there was no proof that the material deposited on Orfila's porcelain plates was arsenic, as other substances, such as coffee and onion juice, could give the same appearance.

Orfila calmly refuted Raspail's objections one by one, and, in doing so, performed an elegant and expert dissection of the defence's arguments. He had already established that arsenic would not be found in the tissues if none had been taken. By exhuming and analysing two bodies, he had disposed of the idea that arsenic might be generated by decomposition. He had analysed four samples of the earth surrounding Nicolas Mercier, each several pounds in weight, and on only one occasion had he found a very slight trace of arsenic. The spots on the porcelain plates had, he said, four distinct characters showing that they were arsenic. Nicolas Mercier had therefore been poisoned with arsenic. He concluded with a firm offer: '[I]f M Raspail could name a single body, other than arsenic, in which were assembled the four properties mentioned in the report, he would instantly tear his report in pieces and abandon his opinion.'[6]

Raspail refused to retreat. He replied that there was a vast number of substances that gave the same test results; for example, certain volatile oils mixed with a phosphate gave a yellow precipitate with the nitrate of silver tests similar to that of arsenic. Orfila disagreed, pointing out that arsenic treated with nitric acid gave a brick-red precipitate and not a yellow one. The debate between the two witnesses was starting to get heated when the court decided to call a halt to the discussion and directed that Orfila and his associates go to the laboratory of the Paris Academy and, in the presence of M Raspail, examine the nature of the spots supposed to be arsenical on the porcelain plate. Raspail agreed but stubbornly protested that even if the spots were shown to be arsenical this didn't establish a charge of poisoning.

The following day, the court reconvened. The spots found by Orfila had been compared with the results of tests carried out on a known quantity of arsenic and they were identical. The court held that the spots were arsenical. Raspail objected and made another long speech, but to no avail. It was decided not to question Mme Mercier, although the president invited the jury to consider the question of her complicity. The jury deliberated for half an hour and Mme Mercier was acquitted. Louis Mercier was found guilty of the murder of his son, with extenuating circumstances, and condemned to hard labour in the prison hulks for life.

Raspail never forgave Orfila, but the Mercier case was just a skirmish. What followed was a full-scale battle, and the scene was central France, far from Paris, where the intricacies of the Marsh test had not yet penetrated. It was a case with all the heady sensationalism that appealed to the public, and divided society as to the guilt or innocence of the central figure, Marie Lafarge, who was accused of that most dreadful of crimes: the murder of her husband by poison.

* * *

Born Marie-Fortunée Capelle in January 1816, she was descended from a natural daughter of the father of King Louis Philippe I of France. She had wealthy relatives but only a modest fortune of her own, and was not considered beautiful. By 1835, she was an orphan in the care of her maternal uncle and two aunts, her ambition stimulated by a taste for romantic novels and the society of girls who had all the advantages she lacked.

Marie's relatives, anxious to find her a suitable husband, approached a matrimonial agency and came up with twenty-eight-year-old Charles Lafarge from the Limousin. He owned an estate, Le Glandier, near the town of Uzerche and managed an extensive ironworks, but he was in debt and looking for a wife with a dowry. Marie was told of the young landowner, and, charmed by descriptions of his beautiful country estate and prosperous business, agreed to an introduction. Lafarge thought Marie attractive and cultured, she thought him ugly and boorish, nevertheless life as the mistress of an estate appealed to her. On 10 August 1839, within days of their meeting, despite the shock discovery that her future husband was a widower, the civil contract was signed. Two days later, they were married. That night, Marie, claiming exhaustion, slept apart from her husband. At dawn next day, the couple set out for Le Glandier. The estate was a long carriage ride away, and though there were stops at hotels, Marie, rushed into marriage with a man she hardly knew, continued to resist all Charles's demands for intimacy.

When they arrived on the evening of 15 August, Marie looked in vain for the lovely chateau she had been expecting and found instead a rambling farmhouse badly in need of repair, its aged and faded furnishings having nothing in common with the refined and fashionable tastes of Paris. Her new family, while cheerful and welcoming, she found loud and vulgar. Desperate to avoid the fast-approaching horrors of the nuptial bed, she decided to escape. To do so alone was impossible, so she decided to make herself so repugnant to her husband that he would voluntarily send her away. Marie wrote a long letter to Charles saying that she did not love him but another, a handsome nobleman who was even then waiting for her in Uzerche, and she would become an adulteress if she did not leave at once. She asked for horses to be saddled for her departure. She said that she was prepared to take arsenic, which she had with her, since she would rather die than receive her husband's embraces.

Marie's maid handed this melodramatic letter to Charles, and there was a hysterical scene when the anguished husband tried to break into the bedroom where Marie had locked herself. Eventually Charles was permitted to enter, and the couple were able to discuss their ill-starred marriage, finally agreeing that if Marie remained she could do so on her own terms. Marie stayed, and Charles ordered repairs and decorations of the dilapidated property. As the weeks passed, they became friendlier and even intimate. In October, they made wills in each other's favour, but the damaging letter Marie had written on her arrival was not forgotten.

At the end of November, Charles went to Paris to try to obtain a patent and raise business loans. His letters home, while affectionate, revealed his financial anxieties and Marie became increasingly disillusioned with her situation. On 12 December, she wrote to a chemist in Uzerche asking for arsenic to kill rats and, soon afterwards, Charles received an unexpected gift from his wife: a packing case containing a miniature portrait of Marie and a cake.

Charles ate a piece of cake, and suffered a violent attack of vomiting and diarrhoea with pains in the head, which lasted for a day. When he returned to Le Glandier on 3 January, he was still very unwell and went straight to bed. Marie sat beside him and tempted him to share her dinner, after which he succumbed to a sudden attack of colic and vomiting he was unable to shake off. The family physician, Dr Bardou, was called and, at first, he did not believe the patient's condition to be serious, but, a few days later, he was told that Charles was worse and had been vomiting faeces. Bardou suspected an obstruction of the bowel and prescribed accordingly. Marie had another request to make. Although she had spread arsenic paste around the house, they were still troubled with rats, and the local chemist had refused to sell her any more. Could Dr Bardou oblige her? He could, and wrote a note for four grammes.[7] Marie obtained this the following day, together with a few ounces of gum arabic, which was then used medicinally to make a gentle mucilaginous mixture.

Charles's condition deteriorated daily and Marie moved into the room next door to his so she could care for him. The whole family rallied round with nourishing and soothing drinks. On 10 January, Marie obtained another packet of arsenic, and this was placed in the charge of a manservant. Another doctor arrived and, believing that Charles was suffering from a nervous complaint, recommended a strengthening eggnog. The drink was prepared and Marie was seen stirring some white powder into it. Charles could take only a little, and it was set aside. Dr Bardou, noticing white flakes on the surface, wondered if ash or plaster had fallen on to the drink, and touched a speck to his tongue. It tasted unpleasant and later raised a blister. Bardou was himself unwell at the time, and when he got home a bad chill forced him to take to his bed, where he remained for several days. When the rest of the eggnog was thrown on to the fire, a gritty white residue was noticed at the bottom of the glass. Marie was later seen standing by a cupboard stirring something into a mixture of wine and water made for Charles. When he tasted it he said it burned him.

The family, who had never quite warmed to Marie after the terrible scenes that had taken place on her arrival, by now suspected that Charles's illness was being caused by something she was giving him. Marie was prevented from giving any further medicines or drinks to Charles and, on 12 January, the dregs of the eggnog were sent to the chemist in Uzerche for analysis. He was able to isolate metallic arsenic from the material and sent a note warning that the patient should only take food and drink prepared by people he could trust.

At midnight on 13 January, another doctor, M Quentin-Lespinasse, was summoned from his bed and told that it was thought that Charles Lafarge was being poisoned with arsenic. He found the exhausted patient suffering such agonies that he kept losing consciousness, and administered iron peroxide, then used as an antidote, as it combined with arsenic to produce a less poisonous compound.

Lespinasse looked about him carefully and found some white powder scattered on a chest, and more in a pot in a drawer. When he threw some of the powder on the fire, he smelt garlic. He warned Charles that he was being given something harmful, but all the remedies had come too late and, at 6 a.m. on 14 January, Charles Lafarge's sufferings came to an end.

The post-mortem was carried out in nearby Brive[8] two days later. Red raw areas and haemorrhages were found in the stomach, and a patch of gangrene in the duodenum. Orfila's observations on the importance of examining all bodily tissues for arsenic had not percolated to Brive, and only the stomach, the duodenum and their contents were preserved for analysis.

Nothing could be discovered in the vomit, but the stomach contained a yellow flaky matter believed to be sulphate of arsenic. Unfortunately, when this was subjected to further tests, the glass tube broke and the contents were lost. Sugar water and toast and water prepared for Charles yielded arsenic. On the other hand, the packet of arsenic Marie had given to a servant for safekeeping and the rat-killing paste she had made previously were found not to contain arsenic at all but bicarbonate of soda and flour. Marie had presumably made a substitution. The conclusion was that Charles Lafarge had been poisoned by arsenic administered by his wife.

There was another shock in store for Marie, when she learned that, shortly before his departure for Paris, Charles had made a new will in favour of his mother and sister.

On 25 January, Marie was arrested and placed in a cell at Brive Prison. The family's private agony became a public scandal. It was rumoured and widely believed that M Orfila himself had been called in on the case and had given a contrary opinion to the doctors of Brive, something he was obliged to deny in a letter.

While Marie waited for her trial, an unsavoury incident from her past created a new sensation. In 1838, she had spent the summer with a friend, a beautiful heiress who had recently married and was now Vicomtesse de Léautaud. Her friend was especially proud of a fine necklace of diamonds and pearls, which, trustingly, she kept in an unlocked drawer. A few days later, it went missing. Only the family, Marie and the servants had been in the house at the time, and once the servants' quarters had been searched without result, suspicion turned to Marie, who had by then left. The vicomte had discussed his concern with M Allard, chief of the Paris police, who was especially interested to learn that, when in Paris, Marie lived with an aunt who had an apartment above the Banque de France. The police had once been called there to investigate a robbery in which silver coins had been taken from a drawer, and gold pieces replaced with copper discs. The thief had not been found. Despite his suspicions, the vicomte decided not to take action against his wife's friend.

In the wake of Marie's arrest for murder, however, the vicomte suggested to M Allard that Le Glandier should be searched. On 10 February, a number of diamonds and pearls were found in a drawer in Marie's bedroom and were recognised as belonging to the vicomtesse's missing necklace. Marie, an imaginative liar unafraid to change her story as it suited her, claimed that her friend had given her the necklace to pay for the recovery of some indiscreet letters. She was tried for theft at the Palais de Justice in Brive, and, on 15 July 1840, her hastily manufactured and unconvincing stories caught up with her and she was sentenced to two years in prison. Shortly afterwards, however, the verdict was overturned on a technicality and a retrial ordered.

The murder trial opened on 3 September at the law court in Tulle, about fourteen miles from Brive. When Marie's unfortunate letter was produced as proof that she disliked her husband, she claimed that she had panicked and allowed her imagination to get the better of her. M Paillet, who acted for the defence, had consulted Orfila, and, after the report from the doctors of Brive was read, Paillet responded triumphantly with the professor's words, which he said should be given great weight, coming as they did from an illustrious expert, the 'prince of science'.[9] Orfila was adamant that the presence of a yellow precipitate was not in itself evidence that arsenic was present; it was necessary that it should be reduced to the metal. There was a profound sensation in court,[10] since the only test that had produced metallic arsenic was one conducted on the eggnog, a drink that Lafarge had barely sipped.

Three new experts from Limoges were appointed to make fresh experiments using the remains of Charles's stomach and four bottles of stomach fluids and vomit. On the following day, the senior chemist, M Dubois, reported the results, emphasising that the most modern methods, including the Marsh apparatus, had been employed. To the astonishment of the court it was revealed that no trace of arsenic had been found either in the stomach or in the vomit. The confusion and uncertainty was such that the court ordered the exhumation of Charles Lafarge's body. The chemists and doctors of Brive and Limoges who had been involved so far, together with two more experts from Tulle – nine men in all – were directed to form a commission to examine the remains.

The trial was by now attracting national and international attention, and interested parties from all over France were arriving in Tulle, while an immense crowd assembled at Beyssac for the exhumation. When the coffin was opened, the remains were found to be so decomposed that the usual instruments could not be used to remove samples and it was necessary to send to the village for a spoon to scoop what was effectively a paste into some earthen pots. At Tulle, an open-air laboratory was created outside the Palais de Justice, with a circle of furnaces fed with charcoal from a brazier kept red-hot. The area was crowded with spectators but the dense and fetid vapour prevented them from seeing most of what was happening. The stench was so powerful it permeated the court and initially it was thought that it would be impossible to hold the afternoon session, but determined crowds still assembled, armed with handkerchiefs and smelling bottles of which it was said that five hundred had been sold that day.

On 9 September, M Dupuytren, brother of the celebrated surgeon, rose to present the report of the commission. The chemists had tested the liver, stomach, heart, intestines, brain, bladder and lungs, and they had found no arsenic. The crowds, many of whom were supporters of Marie, had been listening with breathless attention, and, on hearing this, they broke into loud cries of 'Bravo!' The ladies, and not a few men, shed tears of joy and relief.

The prosecution was not, however, about to abandon the case. The new Marsh test might have failed to find poison in the body, but the old-fashioned methods, judging by symptoms and the internal appearances, still strongly suggested that Charles had taken arsenic, and enough samples remained for further tests to be carried out. There was only one thing to do. A telegram was sent to Professor Orfila. While his arrival was anticipated, the trial continued, and the Marsh test, which had recently demonstrated the presence of arsenic in a box of gum arabic in Marie's possession, was the subject of newspaper articles, probably its first major exposure to the attention of the general public. The mood in Tulle was confidence that the experiments performed by the commission were conclusive and that Orfila would do no more than confirm its findings.

The professor and his associates arrived on 13 September, and retired to a separate room with a number of sealed jars. The hearing continued, and, with unpleasant odours already filling the courtroom, a new and almost intolerably foul stench announced the fact that the experts were hard at work.

At the end of the following day's session, after a series of defence witnesses had attested to the good character of the accused, M Orfila finally addressed the court. In a room that must have vibrated with emotion, Orfila and his associates were perfectly calm, and he delivered his evidence to the silent and attentive audience in a precise, orderly manner. The tests had been conducted in the presence of members of the commission and had employed the same methods, but with the additional use of potassium nitrate, which he had brought from Paris. His analysis showed that there was arsenic in Charles Lafarge's body. There was a general sensation in court. The arsenic, he went on, did not come from any of the reactives used in the experiments or from the earth surrounding the coffin or from that portion of the arsenic existing naturally in the human body.

He had found arsenic in the vomit, the stomach contents, and part of the stomach. He had taken organs from the thorax and abdomen, as well as the liver, a portion of the heart, intestines and brain, boiled them for four hours, and cleaned them of organic matter with nitric acid. This also yielded arsenic. The residue, burnt for seven hours with potassium nitrate, gave the largest amount of arsenic. He pointed out that the original tests, done in Brive, had been compromised by the bursting of a test tube. In Tulle, the chemists had worked on individual organs, whereas they should have worked on them together, and they had used quantities too small to give results. He also spoke of the difficulty of efficiently manipulating the delicate Marsh apparatus, which might not work at all if the flame was too high, or allowed to waver in a draught, or if the dish was held too far or too near. Nor had the chemists of Tulle carried out the potassium nitrate process that had yielded the largest amount.

Orfila was asked how much arsenic he had found. His estimate was a total of half a milligramme.[11] Lafarge, he added, had vomited continuously and had also been made to drink large quantities of water, which explained the small amount found. Marie's supporters were dismayed, and a barrister was sent in haste to Paris with a letter from Marie, pleading for the help of Orfila's great rival, M Raspail. When Raspail received the message, he at once set out for Tulle, but there were mountain roads on the way and the weather was against him. While Marie's supporters waited in desperate hope for the arrival of M Raspail, M Paillet tried to persuade the court that Marie was incapable of murder, and that the tiny amount of arsenic found by Orfila was as good as nothing at all. Raspail, whose carriage broke down three times en route, had still not arrived when, on 19 September, the jury found Marie guilty of murder with extenuating circumstances. She was sentenced to hard labour for life, and public exposure in Tulle. When Raspail arrived, it was too late.

Next morning he visited Marie and, like so many men who had known her, found her, despite her lack of beauty, fascinating. He examined the residue on the plates left by Orfila, and said dismissively that he thought there was no arsenic at all on the first two, and any on the third probably resulted from contamination by the potassium nitrate brought by Orfila from Paris. He declared that he could obtain more arsenic from a piece of old furniture than Orfila had produced from the remains of Charles Lafarge. Raspail returned to Paris, published a memorandum for use in the appeal that was already being planned, and aired his opinion in the press. Public opinion remained divided, fed by the new dispute between Orfila and Raspail, and M Paillet's distrust of the half-milligramme.

Orfila's response was to invite the members of the Academies of Science and Medicine and the elite of practitioners resident in Paris to a series of meetings held at the Faculty of Medicine in October and November to watch a demonstration of his methods. Saying that he had tested two hundred specimens of potassium nitrate and found not a particle of arsenic in any of them, he showed how both arsenic and tartarised antimony were absorbed and circulated in the blood to the organs of the body, and expelled in the urine. A number of dogs had been poisoned for this procedure, and in some cases their penises had been firmly tied to prevent them urinating so that the urine could be collected.

He described the recovery of arsenic from the blood of the assassin Soufflard, and also demonstrated to everyone's satisfaction the existence of 'normal arsenic'. At the special request of a critic who had suggested that arsenic might be formed by putrefaction, he processed a liver that had been in the dissecting room for twelve days, and was in an advanced state of decay, and was not able to detect any arsenic.

The Lancet, reporting on these demonstrations, was in no doubt that the old days of judging by appearances were over. 'But we must never rely upon aspect in medico-legal inquiries. Chemical tests are our sheet-anchor.'[12]

Marie Lafarge's appeal was dismissed the following December, but her sentence was commuted to life imprisonment without hard labour. In the following year, she was found guilty of the theft of the necklace.[13]

On the other side of the Channel, however, some experts were expressing doubts about the existence of Orfila and Couerbe's 'normal arsenic'. Dr Charles Schafhaeutl writing in *The Lancet*[14] said that he had analysed some ancient bones and found no trace of arsenic. Others, however, accepted Orfila's findings and were very concerned about the implications. If 'normal arsenic' did exist, 'it is difficult to imagine any circumstances under which a counsel, acting in a criminal case, would fail to obtain the acquittal of the prisoner'.[15] *The Year-book of Facts in Science and Art,* published in London in 1840, included a report on Orfila's game-changing findings quoting the opinion of the *Literary Gazette*: 'It is considered that he has thus rendered great service to the study of forensic medicine.'[16] The Lafarge case had, however, thrown a searching new light on the Marsh test and the black stains on Orfila's porcelain plates. The first serious crack in the 'normal arsenic' theory was a report by pharmacist Charles Flandin and glassblower Ferdinand Danger, who had experimented with the Marsh apparatus and concluded that the carbonisation of animal matter created traces on the plates that resembled those left by arsenic. The ensuing controversy led to official reviews by the Academy of Science and the Academy of Medicine. When Orfila was asked by a special commission of the Academy of Science to demonstrate the presence of 'normal arsenic', he carried out the same experiments as before, but, to his surprise, found nothing. The Academy of Medicine, where Orfila had many supporters, had only praise for his work, but the Academy of Science concluded that there was no such thing as 'normal arsenic'. Orfila had overreached himself and been deceived by his own refined tests. By the time the 1842 edition of *The Year-book of Facts in Science and Art* had been compiled, which reversed its previous conclusions, Orfila, perhaps regretting his earlier eagerness to wrest priority from M Couerbe, had admitted that he had made a mistake.

There was an inevitable impact on Orfila's own professional reputation, and he was criticised for a dogmatic belief in his infallible ability to interpret the nature of deposits left by the Marsh test by their appearance. It was also a warning to the emergent science of forensic toxicology, since the refinements of chemistry were providing ever more delicate traces, which were hard to identify. The 'normal arsenic' error had brought a strong note of caution against over-enthusiasm, and more sober commentators pointed out that it was the duty of the medical witness to satisfy not only himself, but also the court. 'Until chemistry becomes a fixed science,' observed the *London Medical Gazette*, 'and the action of every possible combination of substances has been tried, how can we be sure of our facts . . . every test is valid, till a fallacy is discovered in it.'[17]

CHAPTER SEVEN

This Troublesome World

In August 1841, a trial took place at the Cheshire Assizes in which a family was charged with what *The Times* called 'this shocking and hitherto unheard of crime in England',[1] revealing a scheme of poison murder that would have been thought unimaginable by many who heard about it. The accused were Ann Sandys and her husband Robert, both aged twenty-five, Robert's brother, twenty-eight-year-old George, and his wife, Honor, twenty-seven. Ann was already married to a Thomas Devaney in Ireland, but had left him six years previously and, not troubling herself to discover whether or not Devaney was still living, she had married Robert Sandys two years later. All the accused followed the occupation of doormat making, and the two households lived in adjacent cellar rooms in Stockport.

Bridget Riley, who lived nearby with her parents,[2] was a friend of Ann's, and godmother to one of her daughters, Mary Ann, but, after quarrelling over a shilling, the two women did not see each other for eight months. The breach was healed in September 1840 when Ann's six-month-old daughter Elizabeth died. Bridget attended the wake and heard Ann make the ominously prophetic comment that 'a corpse being so limber was a sure sign that another would follow'.[3]

In October 1840, Robert and Ann had four surviving children: Robert, six;[4] four-year-old Mary Ann;[5] Jane;[6] and Edward. On the morning of 12 October, all four were well when Bridget Riley called to see Ann and, finding that the couple were quarrelling, took Mary Ann and Jane to her home. She gave the girls tea, cold cabbage and bread and milk. An hour later, the two girls returned home and, soon afterwards, they were both taken very ill with vomiting and diarrhoea. Ann and Honor took the children to a surgeon, Daniel Reardon, who thought that they had been given poison, and offered to supply an emetic. He does not appear to have suspected criminal intent; poisons were a commonplace thing in the home, and accidents all too frequent. He supplied a powder with directions for its use, but the women looked doubtful, saying that a child had died three weeks before after being 'physicked'.[7] His impression was that they were indifferent to the sufferings of the children and disinclined to give them any medication; nevertheless, they took the powder away with them.

Ann and Honor, now accompanied by Robert, took the girls to another surgeon, John Ryan, telling him that Mr Reardon thought they had been poisoned. Ryan asked if the Sandys family kept any poison in the house to destroy vermin, but they said they had never done so. Ryan told his assistant to give the children an emetic, and, once this was done, left the surgery to fetch a basin for them to vomit in. When he returned, the visitors had gone, taking the children with them.

By the next day, Jane's condition had improved, but Mary Ann grew weaker and died that afternoon. A neighbour, Elizabeth Orrell, called on Ann and found her sitting by the body of her child, lamenting 'Oh! Dear me, why did Biddy Riley poison my children?!'[8] adding that the girls had told her that Bridget 'had given them tea which had thickened their bellies'.[9] When Bridget came to visit the stricken family soon afterwards, George Sandys turned her out of the house.

Henry Bowers, constable of Stockport, arrived to look into the death, and Ann told him about the children's visit to Bridget Riley's house, alleging that Bridget had poisoned another child, called Mary O'Neil, eight months before. He asked to see Mary Ann's pinafore and any vomit that might still be about, but the pinafore had been washed and a dog had licked up the vomit. Bridget Riley was questioned and admitted that there had been arsenic in her house, but it had been thrown away. She and her parents were arrested on suspicion of murder.

Doubts were beginning to emerge, however. There were witnesses who had seen the two little girls some time after they had left Bridget's house and both had appeared well. Then there was the death of little Elizabeth, which had taken place when Bridget had not been friends with Ann, and there was also a chilling motive for the parents to commit murder: almost inconceivably, it was thought possible that the Sandys parents had murdered their own children in order to obtain their burial money.

Life assurance, originally a means of protecting business interests, was a rapidly growing resource in the nineteenth century, available to all ranks of society. It was brought directly to the home by agents, who could insure individuals of both sexes and any age. For those able to afford large premiums, the rewards were substantial. Thomas Griffiths Wainewright, who was transported for forgery in 1837, was thought to have poisoned several members of his family, including his twenty-year old sister in law Helen Abercrombie. He had taken out two policies on her life in the sum of £5,000 and £14,000 and, soon afterwards, Miss Abercrombie was seized with a sudden attack of illness with vomiting and died. The insurance companies refused to pay, but it was impossible to prove that she had been murdered.[10] Alfred Swaine Taylor was later to comment that he believed this to be the first ever strychnine murder.[11]

Friendly societies had been under legal protection since 1793, but further legislation in 1829 led to the creation of many new mutual assurance societies. The poor joined burial clubs, which, for a penny a week, would pay out a sum on the death of the insured to enable the family to provide a decent funeral. The high rate of infant mortality meant that many children were entered in burial clubs. The cost of burying a child was twenty or thirty shillings, and the clubs, instead of simply meeting the funeral costs, paid out a fixed sum, which might be three to five pounds. Some babies were entered in more than one club, and a case was cited in which a father had made payments into nineteen different clubs.[12] At best, this was a cynical gamble on the death of a child; at worst, it was temptation for those living in straitened circumstances to commit murder.

All of the Sandys children had been entered as members of the Stockport Philanthropic Burial Society. A payment of a penny a week was required for each, and, at the end of seventeen weeks, the child became a full member. If a child died after that date, the parents would be entitled to a sum that, after deducting expenses, amounted to £3 8s. 6d. Robert and Ann had claimed funeral money for both Elizabeth and Mary Ann. In July, George and Honor had received a payment following the death of their daughter Catherine.

The body of Mary Ann was taken to the police office and, on 15 October, baby Elizabeth was exhumed. Both the girls had been healthy, and neither had external signs of injury, but in both cases the intestines were inflamed, and the examining surgeon, Mr Rayner, using both the Marsh and precipitate tests, was able to recover visible arsenic in lethal quantities.

Bridget and her parents were released and Robert, Ann, George and Honor apprehended. Their cellar dwellings were searched but no arsenic was found. Two pieces of paper were recovered and submitted to Mr Rayner but, despite the sensitive nature of the tests, he could not detect arsenic. The emetic supplied by Reardon was discovered unused. Rayner also examined the body of Mary O'Neil who, Ann had alleged, had been murdered by Bridget, and found nothing to suggest that the child had been poisoned. George and Honor Sandys's daughter Catherine was also exhumed, and Mr Rayner examined the remains, finding white specks on the interior of the stomach, which was very inflamed. He was in little doubt that the child had taken a mineral poison, but was unable to state positively that this was the cause of her death.

While the Sandys relatives were detained in the gaol of Chester Castle,[13] the cellars they had occupied were cleaned and whitewashed and let to new tenants. In March 1841, Michael Coughlan, who had moved into the cellar previously occupied by Ann and Robert, was standing on a stool to clean the windows when he found a piece of brown paper concealed in a small niche between the top of the window and the woodwork. The paper contained a powder that Mr Rayner was able to show was common white arsenic.

Ann and Robert Sandys were charged with 'the horrible and unnatural crime of poisoning their offspring to obtain a paltry pittance from a funeral society',[14] and George and Honor were charged with being accessories both before and after the fact. The trial opened on 2 August 1841, and all the accused pleaded not guilty.

Six-year-old Robert Sandys junior was a vital but terrified witness. He had been carefully coached about being truthful in court and testified that he and his sisters had been given slices of bread and butter from the same loaf, but his sisters had had salt added to theirs. His testimony came to an abrupt end when he became confused, appeared to faint and was carried from the court.

Counsel for the defence, Mr Welsby, delivered an eloquent and forceful address that occupied two hours, in which he suggested that Mary Ann and Jane had taken poison by accident at Bridget Riley's house. The idea that his clients were guilty was 'too horrible', the motive 'insignificant',[15] and the arsenic found in the cellar could either have been put there during the six months after the Sandys family had vacated the property or have already been there for many years. The jury, after a short deliberation, acquitted all the prisoners. Robert Sandys prayed audibly, but his ordeal was not over. The following day, he and Ann stood trial for the murder of Elizabeth. The baby had been entered into the burial club on 7 June and, twelve weeks later, Robert had asked the society collector if she was a full member so that her parents would be entitled to the burial money if she died. He was told that she was not. A week later, on 6 September, he made the same enquiry and was told that his daughter was now a full member. A neighbour saw the child on 20 September. She was very ill and, so the neighbour was told, had been ill for a fortnight. Elizabeth died three days later, and Robert provided conflicting reports about the cause of death to the sexton and the registrar. A statement made by Ann at the inquest on the children was put in evidence and read. In it, she claimed that she had not given her baby any food apart from the breast, but that, on one occasion, Robert had put powder into some milk and given it to the child, who had been taken ill. This time, Mr Welsby's eloquence failed. Ann was acquitted, but Robert Sandys was found guilty of murder and sentenced to hang. His sentence was later commuted and he was sent to a prison hulk. In 1842, he was transported to Tasmania.

Ann was fortunate to escape conviction for murder. Her lies, evasions and attempts to blame the death of one child on her friend and another on Robert strongly suggest a guilty and callous disregard for the truth. George and Honor Sandys, though suspected of murdering their daughter Catherine, were never charged with the crime.

Two of Ann's surviving children (she and Honor had both given birth while awaiting trial in Chester Castle) had been taken to the workhouse while she was in custody, and, on her release, she refused to remove them or contribute to their support. In September 1841, charged with neglecting her family, she was sentenced to a month in prison with hard labour.

In 1843, social reformer Edwin Chadwick's influential report on burial practices in towns,[16] while concentrating on sanitary issues, denounced burial clubs as a stimulus to the morally depraved to commit murder. These grim cases were a long way from tales of ancient Rome and the Borgias, poisonings carried out in the highest echelons of society, which provided a suitably distanced frisson to the Victorian reader. 'In our day, it is the lowest of the low who do it,' commented the prolific sociological and historical writer Harriet Martineau. 'They want something – money or a lover, or a house, or to be free of the trouble of an infant; and they put out the life which stands in the way of what they want . . . The guilt and the shame lie with the whole of society, which has permitted its members – hundreds and thousands of them – to grow up as if they were not human beings at all, but a cross between the brute and the devil.'[17]

The Friendly Societies Act of 1846 attempted to address the problem of infanticide for insurance by prohibiting the entering into a burial club of any child under the age of six, but, in 1849, *The Times* revealed that the practice of insuring children as young as sixteen weeks of age continued, since the 1846 act did not apply to societies formed before that date.[18] This situation was not corrected until the 1870s, when limits were also set on the amounts that could be received. A doctor or coroner had first to provide a report, and, if this were satisfactory, the registrar would issue a certificate. Only then could the insurance money be paid. Stricter laws and the work of the National Society for the Prevention of Cruelty to Children had all but eliminated child murder for insurance by the 1890s.

The public outcry against child murder for insurance highlighted a long-standing problem concerning the investigation of suspicious deaths in England and Wales. Coroners principally relied on the medical profession and the public for information, and also received reports from local police and parish officers who dealt with the dead, but they were not always able to hold inquests or gather the evidence they needed. Control of the sums paid to coroners for the expenses of inquests was vested in Justices of the Peace, whose attempts at economy restricted the availability of medical examination and even imposed limits on both the number and the complexity of cases that could be reported to a coroner. The expense of the new chemical tests for poison and fees demanded by the emergent expert analysts formed a considerable obstacle to enquiry. During the 1850s, there was, for this reason, a substantial drop in the number of inquests held, and the decade was to end without reform. This was a serious matter since, in the first half of the nineteenth century, the coroner was the main mover in cases of alleged poison death, and the justices in accusations of attempted poisoning. Only later in the century did the police take over the central role of investigating suspicious deaths. This lack of medical evidence as to cause of death in a substantial proportion of cases had not gone unnoticed by the Registrar General, and the public and press inevitably assumed that large numbers of poison murders went undetected.

Although this book will not be undertaking a detailed analysis of the motives for poison murder,[19] it is worth observing at this point that the cases discussed so far involving arsenic, the most widely used of all poisons, reveal an interesting pattern. Eliza Fenning, Isabella Hopes (Chapter 1), Anna Zwanziger (Chapter 4) and Samuel Kirkby (Chapter 5) were servants with a grudge against their employers. Mary Wittenback (Chapter 1), Susannah O'Flaherty (Chapter 4), Madame Boursier (Chapter 4), M Défourneaux (Chapter 5), Louis Mercier and Marie Lafarge (Chapter 6) wanted to remove hated or inconvenient spouses or relatives, while Robert Donnall (Chapter 2), Sophie Ursinus, Josephine d'Aubarède,

Mary Ann Burdock (Chapter 4), John Bodle (Chapter 5) and the Sandys family (above) were motivated by financial gain.

All of these cases have an important factor in common: the active dislike of and/or indifference to the sufferings of the victim, and the essential selfishness of the perpetrator. Although not involving arsenic, Richard Overfield's (Chapter 2) callous removal of his son falls into this category. Even if the arsenic poisoners were not initially fully aware of the effects of the poison on their victims, several of them, having administered it once and seen the results, continued to administer it to the same victim or went on to poison others. But not all poison murders were like this; there was another area of murder – one using different methods and being differently motivated – that society, science and the law seemed powerless to address.

Infanticide, the murder of a child less than a year of age, was the most common homicide charge brought against women in the nineteenth century. In the years 1801–1900, 199 cases of infanticide were tried at the Old Bailey and almost all of the accused were the mothers of the dead children. The victims were usually newborn or only a few days or weeks old. It was often hard to determine the exact cause of death, or even whether the child had been born alive, and, for these reasons, 112 of the prosecutions ended in acquittal. When murder was proved, the accused had usually drowned or smothered their babies. In only one instance was it certain that poison was used. It is often said that infanticide was the only method of birth control then readily available to the poor and, while there is some truth in this, such a claim does not evoke the tragic reality of women who were often forced to choose between destitution and committing murder.

The year 1834 saw the passing of 'An Act for the Amendment and better Administration of the Laws relating to the Poor in England and Wales'.[20] It swept away old legislation and provided a unified national system for dealing with poverty, which it was believed was caused not by economic or social conditions, but by the indigence of the poor. The new act called for the establishment of Poor Law Unions with parish workhouses, where conditions were kept deliberately harsh to discourage all but the most wretched to enter. It became unlawful to give poor relief to able-bodied persons other than in workhouses (this provision was later modified) or to women with illegitimate children. Mothers were no longer allowed to apply to the courts for maintenance from the child's father and were solely responsible for the upkeep of their bastard children until the child was sixteen. If the woman was unable to support the child, the parish could make an application to the father for maintenance at the Quarter

Sessions, but it was necessary to prove that the man really was the father of the child, and any money received could not be paid to the mother.

The 'bastardy clause', as it came to be known, must have looked good on paper: no longer would the parish have to support the children of unmarried mothers; women would be discouraged from entering immoral relationships; and there would be fewer illegitimate births. That must have been the theory. The actual outcome was a rise in the incidence of infanticide and female suicide.

Women with no family to help them did fight against the odds to support their babies, but desperate parents of older children with whom they had already bonded sometimes turned to poison and, when they did, the poison of choice was usually laudanum. The soporific analgesic effect of the drug would have recommended itself as a kinder way of death. Child murder was often the last resort for those with no future and nowhere to turn for help.

In 1838, twenty-three-year-old Catherine Michael, employed as a servant in London, was dismissed when found to be pregnant. In the following year, she gave birth to a son, George, but the father of the child abandoned her, leaving her solely responsible for George's maintenance. Dr James Reid, a physician who had known Catherine since her confinement, found her work as a wet nurse with one of his patients, and, when George was a fortnight old, Catherine gave him to be minded by Sarah Stephens of Praed Street. Catherine paid Sarah seven shillings a week for the first five months, but struggled to meet the sum out of her wages. Even though the weekly fee was later reduced to five shillings, Catherine was finding it impossible to keep up the payments. George was a healthy ten-month-old child when, on 27 March 1840, Catherine sent for Mrs Stephens and gave her a frock for the baby and a bottle of medicine, saying it would do the baby's bowels good. Sarah said that the baby was well and didn't need any medicine, but Catherine persisted, saying it had done her mistress's baby good and it would do hers good too. She told Sarah to give George a teaspoonful every night, adding that if anything happened to the baby she was to be informed at once.

Sarah took the bottle home and looked to see if it had any directions on it, but there were none. She put it on the parlour mantelpiece, intending to consult a doctor about how it should be given.

Four days later, George, who had been happily playing during the afternoon, was found lying on the sofa unconscious and unresponsive, and Sarah sent her eleven-year-old daughter Jane to fetch a doctor. A surgeon, Mr

Philpot, and his assistant, Mr Sellick, arrived, and, although the contents of the suspect bottle had been thrown into the fire, they were able to determine from the odour on the child's lips and the dregs in the bottle that George had been given laudanum. Everything that could be done was done, and Catherine was sent for, but George died the following morning. Catherine told the two medical men that she had obtained the bottle from Dr Reid but that he had since admitted to her that his young man had given her medicine intended for another patient by mistake. This was a blatant and damaging lie, as Catherine had bought the laudanum from a chemist in mid-March. It had then been labelled 'Laudanum – POISON'.

But who had given little George the laudanum? Sarah Stephens said that she had never even uncorked the bottle. At the time when George had swallowed the fatal dose, she had been working in the kitchen, leaving George, Jane and her two sons, five-year-old William and three-year-old Richard, in the parlour. Jane said she had not dosed the child either – she had gone out for ten minutes to get a loaf and, on her return, William told her he had given George some medicine. The bottle was then only half full. Sarah said that she was so upset by what had happened that she impulsively threw the bottle on to the fire. A cynical person might be inclined to wonder if it was either Sarah or Jane who, in all innocence, had given the laudanum as directed by Catherine, and, terrified at the repercussions, placed the blame on little William.

That evening, Catherine went to see Dr Reid, telling him that her child had died after being given laudanum by mistake. 'If I am hanged for it, I could not support the child on my wages,' she said,[21] admitting that she had told Mrs Stephens that the child should be given a teaspoonful at night. (A teaspoonful is sixty drops and Sellick later said he thought it dangerous even to give one drop to a child.) Alarmed by this revelation, Reid felt it was his duty to inform Catherine's employers, who immediately dismissed her.

On 6 April 1840, Catherine was tried for the murder of her child. Mr Baron Alderson, presiding, told the jury they must be satisfied that there was an intention to kill the child, and that if they were, even though the poison was administered by another, it was the prisoner who must bear the guilt. The jury deliberated for an hour and found Catherine guilty of murder but with a strong recommendation to mercy: '[T]he fact of the prisoner having been seduced, and thrown upon the world is an extenuating circumstance in this case.'[22] On 6 October, she was transported to Tasmania.

Ann Rothwell was twenty-four, married but separated from her husband. In December 1843, she had been cohabiting for three years with a man called Charles Turner, a hat-finisher at Oldham, and the couple had a two-year-old son, Edwin. Turner treated Ann harshly and, just before Christmas, when she was away for a short while, he stripped their cottage of almost every article of furniture and moved to Manchester. When Ann returned to find her home bare and Turner gone, she was devastated. With no means of income, she began selling her few remaining possessions in order to buy food. On Christmas Day, she wrote a letter to her sister, Mrs M'Lain, in Manchester. 'Dear Sister, I write to let you know that Charles has done his utmost. He has belied me, and now left me destitute. I cannot beg, nor yet see my child starve. If you will come today, you will find me and Edwin a corpse. You must get a wire to lift the latch. See you have the tickets. Farewell.'[23]

On receipt of this letter, Mrs M'Lain's husband traced Turner and they both set off for Oldham at once. Turner managed to open the cottage door but found everything in darkness. He called out to Ann but there was no reply. He went to get a light and, as he returned with it, Ann opened an upstairs window and asked feebly who was there. Turner rushed up the stairs and, throwing open the door, he gave a loud shriek. On seeing him, Ann collapsed. M'Lain followed and saw Ann lying on the floor looking haggard and wild, and Turner holding his son in his arms. The boy, who was wrapped in his mother's petticoat, was dead and cold. Turner went to get a doctor and, returning to Ann's side, begged her to forgive him, which she did.

Ann had given her child laudanum and taken it herself. It was only by the strenuous efforts of the medical man who attended her and used an emetic and a stomach pump that her life was saved. The conditions in the house revealed Ann's fatal desperation. There was no furniture, and very little fire. The only food was some bread sopped in water and a little sugar wrapped in paper. She had spent her last few pennies on laudanum.

Ann was tried for the murder of her son the following March and there was only one possible defence. Turner and M'Lain both testified that Ann's mind had been affected by the death of her father, who had hanged himself, and she had been subject to mental aberration and depression ever since. The idea that she had been deserted was, claimed Turner, a delusion. Counsel for the defence told the jury that, even if the child had died from laudanum administered by the prisoner, she had, at the time, been incapable of telling right from wrong. While Edwin had died from congestion of the brain, one of the effects of poisoning by laudanum, the medical witnesses professed themselves undecided as to whether or not the laudanum had actually been the cause of death. Ann was acquitted.

Some of the mothers who committed infanticide, and it is impossible to know how many, must have been suffering from post-natal depression, then known as 'puerperal mania' and recognised as an extenuating factor beyond the woman's control. Twenty-six-year-old Jane Harrington was the wife of a carpenter's labourer of 5 Harrow Street, Southwark, 'a place closely inhabited by very poor people',[24] where they occupied one room. Already the loving mother of two small children, Jane gave birth to a son on 20 August 1854 and welcomed the new arrival. Five days after the birth, she was out of bed and doing the washing, but the baby would not suckle at first, and there was a noticeable change in Jane's mood. Mary White, who lived upstairs in the same house as the Harringtons, said, 'When the infant does not take its milk properly, it very often flies to the head of the mother; it did in this case.'[25] Mary thought that, after the birth, Jane was 'not in her right senses'.[26] Sometimes, the new mother was distracted and bewildered, at other times, wild and strange. She neglected her home, her children and herself. On 4 September, Ann Evans, a girl who had looked after Jane during her confinement, found the baby lying outdoors on the cold stones and begged Jane to pick it up or it would catch cold and die. When Ann went to pick it up, Jane prevented her, saying, 'Oh, let it lie there; I shall be hanged for it.'[27] Ann went to tell Jane's husband what she had found, and Jane brought the child indoors.

On 5 September, after a sleepless night, Jane told Mary White that she had given her child half a teaspoon of laudanum. The baby looked very ill, and Mary rushed it to a nearby surgeon, William Rendle, who was sure from its appearance that it had been given a narcotic poison. He administered medicine, but the baby had become unresponsive, and he thought it was dying. Mary took the child back home, but Rendle followed, and demanded to know what Jane had given it. She seemed dazed and indifferent, as if she had taken a narcotic herself. A policeman, Sergeant Golding, was called to the house, and Jane weakly confessed that she had mistakenly given the child some medicine sent for her during her confinement – laudanum. Golding removed the child to the infirmary at St George's Workhouse, and took Jane into custody.

At the infirmary, it was discovered that the baby was still alive, and it showed promising signs of recovery, but, two days later, it suffered a fit and died. The post-mortem was carried out by William Rendle. He was unable to find any laudanum in the body, but concluded that, in the time that had elapsed between the dose and death, it had been absorbed. He had no doubt that the baby had died from the effects of a narcotic poison.

Jane finally admitted that she had purchased laudanum and given a substantial dose to her baby. At her trial for murder, which took place on 18 September, Rendle's report seemed to make a conviction inevitable, despite the efforts of the defence to claim that, because Jane had returned to household duties within five days of the birth, 'there appeared to be very little doubt that this premature exertion had excited her and affected her brain'.[28] The evidence from the infirmary, however, changed everything. A nurse told the court that when the baby was first admitted she had suckled it, and surgeon Edward Evans said that he had seen no sign of the effects of laudanum. In his opinion, even if the baby had been given a narcotic, by the time it arrived in the infirmary it had entirely recovered from the effects, and the actual cause of death was convulsive fits. It had died as he had seen hundreds die before. Jane was acquitted.

It can be seen from the Michael, Rothwell and Harrington trials (above) that Victorian courts were not, as is so often supposed, heartless. Jurymen were moved by genuine distress and often entreated mercy for prisoners, even acquitting in cases that excited their pity. Some cases, however, evoked only horror.

Eliza Joyce was not in financial want, or suffering from depression, neither did she use poison as a primitive form of birth control. She had no feelings for her victims, whom she dispatched without pity or reason, using either arsenic or laudanum as convenient, and would probably have gone on killing if she had not been caught when she was. She may well be an early example of the psychopathic serial poisoner.

William Joyce, born in 1800, was a market gardener of Boston in Lincolnshire whose first wife had died in 1840 shortly after giving birth to their fifth child, Emma. This was not an impoverished household. William employed a gardener and a servant[29] and also kept a small beer-house. In 1841, William married twenty-seven-year-old Eliza Chapman. Eliza was expecting a child when, in October that year, her stepdaughter Emma died, aged eighteen months. Eliza gave birth to a daughter, Ann, on 1 January 1842, but the baby died three weeks later.

Eliza's eldest stepson, William junior, fifteen, became very ill in the autumn of 1842 and was left in Eliza's care. On 16 September, she asked Mr Simons, the local druggist, to sell her some 'white mercury', as arsenic was sometimes popularly called, to poison mice. Simons recommended another compound, made with nux vomica, but Eliza was insistent that she preferred the white powder and so was sold two ounces. The next day, Simons mentioned to William that he had sold arsenic to his wife. William was clearly very disturbed by this news, because a quarter of an hour later he returned it, but, when Simons weighed the packet, half an ounce was missing.

On the same day, William junior became very ill with sickness and diarrhoea and, when this continued for several days, he complained to his father that his stepmother had given him medicine that had caused his illness. The concerned father called a surgeon, and a pitcher of William's vomit was sent for analysis. Mr Green, a chemist of Boston, saw white sediment in the vomit and used the precipitate tests on some of it; the rest he heated with charcoal to produce a metallic crust, which he heated again and obtained the telltale smell of garlic. He estimated that the sample he had been sent contained between one and two grains of arsenic. Young William made a statement before Mr Adams, the mayor of Boston, to the effect that his stepmother had given him something that had made him ill, and Eliza was arrested. She could hardly deny buying arsenic, so tried to avoid a criminal charge by offering an unlikely scenario, claiming that she had spilled some on the floor and used a spoon to scoop it up, then used the same spoon to stir her stepson's medicine. William junior recovered, but died of liver disease the following December.

As a result of these events, suspicions had been aroused concerning the death of little Emma, and her body was exhumed. A surgeon, Edward Coupland, carried out a post-mortem examination but found no trace of arsenic in the stomach contents.

Eliza was tried for the attempted murder of William junior at the Spring Assizes of 1843, and pleaded not guilty. Due to an administrative error, she was discharged and ordered to be retried at the Summer Assizes. In a lengthy speech, the defence pointed out that there was no evidence of any ill will between Eliza and her stepson, and no motive for her to wish his death, and suggested that the arsenic might have been taken by accident. Perhaps the factor that swayed the jury was the apparent lack of motive. Eliza was acquitted.

William senior, described as 'a remarkably inoffensive man',[30] nevertheless remained convinced that Eliza had harmed his son and he refused to have her back in the house. She was obliged to go and live in the Boston Union Workhouse, where her husband paid for her support.

In the summer of 1844, Eliza was taken very ill and, perhaps fearing that she was in danger of death and facing a higher judgement, she decided to unburden herself, confessing to the Union surgeon Mr Coupland that she had given arsenic to William junior. He asked her about the deaths of the two younger children, Emma and her own baby Ann, and she told him that she had given them laudanum. Eliza then made a confession to James Wilson, master of the workhouse. He asked her what could have induced her to poison the children and she said, 'I don't know, except I thought it was such a thing to bring a family of children into this troublesome world.'[31]

On 8 July 1844, she made a full voluntary statement to John Sturdy, the new mayor of Boston, saying she made the confession because her mind was 'so burdened that she could not live, and hoped, as she had confessed, she should be better'.[32]

Eliza, described as 'a mild and not uninteresting-looking woman',[33] was tried at Lincoln Crown Court on 22 July 1844, pleading guilty to two counts of murder. She was quickly found guilty and sentenced to death. Society recoiled in horror from Eliza's crimes, and tried, in vain, to understand them. However, in doing so, society gave no constructive thought as to how to prevent them. 'Few probably of those present left that court with hearts untouched, or spirits capable of enjoyment for some time,' commented *The Times*.

> Is it too presumptuous to hope that ... some course of enquiry will be instituted, and diligently followed up, into the whole conduct, over the long space of time in question, of the husband of this unhappy convict? Men feel that women do not commit horrible crimes such as this from sheer depravity of their nature, and expect ... that the fullest inquiry should have been made into all the circumstances, and her real impulses become ascertained beyond a doubt.[34]

Eliza was publicly hanged on 2 August, the first woman to be hanged at Lincoln in twenty-seven years.

There was no renewed campaign for a restriction on the sale of poisons to the public, following these seemingly senseless murders, and no questions were asked in parliament. That movement would follow the actions of another woman, crueller and more cold-hearted even than Eliza Joyce.

CHAPTER EIGHT

Getting Away with Murder

The first challenge in a trial for murder by poison is to demonstrate to the satisfaction of a jury that the deceased died from the effects of poison. However, even supposing this difficult task is accomplished, the secretive nature of such a crime means it is often impossible to prove that the poison was administered by the accused. Defence counsel will suggest to jurors that the deceased might have committed suicide or taken poison by accident. When the gallows beckoned, as the following two cases show, a prosecution often foundered on a jury's reluctance to convict, even where there was opportunity and strong evidence of motive. Well into the nineteenth century, the pale shade of Eliza Fenning still hovered reproachfully over the Old Bailey.

On the evening of Saturday 17 September 1842, twenty-six-year-old Jane Bowler, wife of a Southwark chair-maker, was wandering the street near her home in a highly distressed state. Encountering a friend, James Thomas, she took him by the hand and begged him to come at once as her husband Joseph was dead. At the Bowlers' home in Union Street, James found twenty-nine-year-old Joseph lying on the floor, his face bruised, dark phlegm oozing from his nose and mouth, the floor stained with faeces and mucus. He helped lift the body on to the bed. John Dunster, a young clerk who lodged in the same house, fetched a surgeon, James Coulthred, who confirmed on arrival that Joseph Bowler had been dead for about an hour.

Coulthred told Jane it would be necessary to conduct a post-mortem, and she raised no objections. She said that her husband had been taken ill on the previous Wednesday. He had eaten an apple pudding and immediately gone out into the yard and brought it up again. Some of the pudding crust remained and she gave it to Coulthred. It was later revealed that

Joseph had been continuously ill for three days before his death, was dosed with rhubarb and castor oil, and vomited everything he was given.

Powdered arsenic is more persistent than most murderers realise, and at the post-mortem Coulthred found twenty grains of it in Joseph Bowler's stomach. He thought that it had been taken six to twelve hours before death.

The Sunday before Joseph was taken ill, the Bowlers' fellow lodger Sarah Morley had told Jane that she had read in a newspaper about a lady and gentleman who, together with their servant, had been poisoned with arsenic by their daughter. Jane's response was to ask Sarah if she knew whether the poison was sold as a liquid or as powder. The following Saturday, Sarah, seeing that Joseph was very ill, asked Jane why she did not call a doctor. Jane protested that they could not afford it as they owed money, but Sarah pointed out that there were parish and dispensary doctors. The dying man demanded to know why he could not have a doctor, but Jane dismissed the idea, saying, 'You don't want to be physicked and tortured about.'[1]

On the morning after Joseph's death, Sarah helped lay out the body and cleaned the room. Finding a half-pint mug of tea, she emptied the contents into the drain and saw white sediment in the bottom. She showed it to her husband, saying, 'See what a dirty mug the poor fellow had to drink his last tea out of.'[2] He told her to put it in the kitchen. She never saw it again. When Sarah mentioned to Jane's mother that Joseph had asked for a doctor Jane denied it, adding, 'If you say that, you will hang me, for you will be called on the jury.'[3]

Joseph and Jane Bowler had not been on good terms, although their marriage was not of the quarrelling kind. Mary Musgrave, who had attended Jane at her last lying-in, commented, 'There was never hardly any words between them – more of silent contempt, sulky like, to one another – they had no words before anybody.'[4] Not only that, but Jane was casting her eyes elsewhere. Jane had told Sarah that she and John Dunster were romantically involved, and he would 'stop five years for her',[5] although Sarah had never seen them going out together. The clerk was already betrothed to another, and denied that there had been any relationship between him and Jane. The romance may have been no more than a flirtation, or even wishful thinking on Jane's part. Between inquest hearings, Jane revealed what was on her mind by saying, in a jocular fashion, 'I am in it, and must get out of it in the best manner I can; there is one thing, they cannot hang me, on account of my children.'[6]

The defence at Jane's trial for murder was that Joseph had committed suicide. The chemist from whom it was thought Jane had bought arsenic was unable to supply a positive identification and witnesses testified that Joseph was a gloomy, sullen fellow who had more than once tried to throw himself in the river. Jane was acquitted.

In January 1845, twenty-year-old Thomas Dickman of Uxbridge, and his wife Mary, who was twenty-two, had been married for a year. A baby had been born in July 1844, and Mary also had a two-year-old child who was illegitimate. The pair cohabited in a state of ongoing domestic misery. Dickman was keen to get out of the marriage and had often threatened to 'do' for Mary and her child and then 'go for a soldier',[7] while Mary had been obliged to use a poker to defend herself from his violence. On the morning of Wednesday 15 January 1845, Mary drank some coffee with milk and sugar. The sugar curdled the milk, and the drink made her ill with stomach pains. Mary tasted the sugar in the basin, which had previously been used without ill effect, and found it acidic. She recovered after taking an emetic. Recalling that she had heard the lid of the sugar basin rattle before Thomas went out that morning, Mary had the contents analysed. The sugar was found to contain oxalic acid.

A friend of Dickman's, Frederick Showring, was a shoemaker who used oxalic acid in his trade, although he was adamant that he only kept it in its dissolved state and had never missed any. At the Old Bailey, Dickman was charged with the attempted murder of his wife and child, but the judge stopped the hearing on the grounds that there was no evidence of administration. He was discharged.

* * *

Jane Bowler and Thomas Dickman were fortunate to escape conviction. If guilty, and they probably were, their methods were crude and unimaginative. A medical man, however, could, and had to be, more subtle. He could not plead ignorance of the effects of poison and had to appear to be doing all a physician could to help the victim. As he was able to obtain poisons other than arsenic without arousing suspicion, it was often to these powerful, rarer and more refined toxins that he turned, the trust of the victims ensuring that there would be no difficulty in persuading them to swallow something of unusual appearance, taste and smell. The poison did not necessarily have to be concealed in food or drink, but could be given openly and taken willingly in the guise of medicine. The legitimate presence of medicinal poisons in the home of a doctor also provided the opportunity for him to claim that the deceased had committed suicide or taken the poison in error.

James Cockburn Belaney was a different kind of murderer. He had not tired of or fallen out of love with his comely young wife or sought to replace her with a mistress. He had married her with the sole intention of acquiring her fortune and, having done so, he disposed of her.

Rachel Skelly was born in Northumberland in 1823, and, as she grew to womanhood, her grace and beauty attracted many suitors and the soubriquet the 'Rose of the North'. Just as rosy was the value of the Skelly family property in Seahouses. Her widowed mother held a valuable copyhold property,[8] which produced a rental income, and an interest in a lime-works and colliery. On 23 February 1843, Rachel, still only nineteen, married Belaney, a thirty-two-year-old surgeon.[9] The groom was rather stout and plain but had an intelligent face and a lively, talkative manner. Soon after the marriage, he gave up his medical practice and took over the management of the Skelly properties.

On 1 July, Rachel's sixty-two-year-old mother, who had previously enjoyed good health, died suddenly.[10] Her death was reported by her son-in-law, who gave the cause as bilious fever. The circumstances were not thought to be suspicious. This may well have been Belaney's first murder.

On Mrs Skelly's death, her properties passed to Rachel, who soon afterwards surrendered the copyhold to the joint use of herself and her husband. Belaney was entitled to control his wife's personal and copyhold properties and enjoy the income, but his right over them was not absolute; he could not dispose of the personal properties without his wife's consent, and the feudal copyhold would not pass to him until his wife's death. Nevertheless, Belaney, now considering himself to be something in the community, set about constructing a house for his enjoyment, a romantic dwelling at the end of the village of North Sunderland, overlooking a cliff.

In August 1843, Rachel suffered a miscarriage but, early in 1844, she would have known that she was pregnant again. The Belaneys planned to visit London and, on 31 May, before setting out, two mutual wills were drawn up, each half of the couple leaving their property to the other.

On 4 June, they arrived in Stepney and took lodgings in Tracey Street, near the home of a friend and former patient of Belaney's, master mariner Captain William Clark. Belaney was intending to proceed on a tour of the Rhine while Rachel was to remain in England to await her husband's return, with Clark's daughter Mrs M'Eachern, as her companion. On the same day, Belaney went to a surgeon in Stepney and bought, amongst other items, a one ounce bottle of Scheele's prussic acid, a preparation supplying about 5 per cent prussic acid and usually taken in doses of a few drops for indigestion. That night, he and Rachel went to the theatre.

Early the next morning, Belaney sent for a black draught (a common purgative mixture). Soon afterwards, Rachel started to vomit and continued to do so throughout the day, spending much of her time lying on the sofa resting. By the following day, however, she was well again. Early on the morning of 8 June, Belaney asked the landlady Mrs Matilda Heppingstall to bring him a jug of hot water, a tumbler and a spoon. Less than an hour later, he called for her urgently, saying that his wife was very ill. Mrs Heppingstall went up to the bedroom and was horrified to find Rachel lying on the bed insensible, and gasping for breath, the upper part of her body convulsed. She was foaming at the mouth and bringing up some chewed biscuit. On the table was a tumbler containing some liquid and a paper of salts (these were later identified as Epsom salts) and another tumbler, which was empty.

Mrs Heppingstall exclaimed that she had seen her daughters in fits but had never seen fits like this. Belaney said his wife had had fits before but never previously one that she would not come out of. The landlady begged him to send for a doctor. He replied that he was a doctor, but she insisted, saying that two heads were better than one. Finally, with Belaney's consent, her servant was sent out, not for a doctor, however, but to the home of Captain Clark. Rachel was struggling for breath and had almost no pulse, but Belaney did not call for medicine; he simply rubbed his wife's hands, feet and face with warm water, while, at Mrs Heppingstall's direction, the young woman's temples were bathed with vinegar. Belaney said that he thought his wife had a disease of the heart and that her mother had died some months ago in the same way.

Mrs M'Eachern arrived and Mrs Heppingstall implored her to go and get a doctor. She at once went to fetch her family medical attendant Mr Garratt, who lived a mile away, while another servant was sent for Captain Clark. The captain did his best to assist, applying a mustard poultice made by Mrs Heppingstall and suggesting they rub the patient's legs with rough towels and a clothes brush. Understandably, neither of these measures had any effect on the patient, who, after a few more gasps, expired. The new widower shed tears, wailing that he had caused his wife's death through neglect, which his listeners took to mean that he regretted taking her on a long journey while only two months from her confinement. Her mother, he said, had died from a similar attack. Mr Garratt arrived and asked Belaney if Rachel had taken anything and he said she had only taken some salts.

The inquest opened on Monday 10 June and was adjourned for a post-mortem examination to be carried out by Mr Garratt and Thomas Vizard Curling, house surgeon at the London Hospital. Belaney must have been hoping that his wife's death would be passed off as due to natural causes, and he knew that the examination would reveal a fatal secret. He had to rethink his story.

The procedure was carried out at the lodging house. Belaney and Clark hovered nearby, and the two surgeons were obliged to express their concerns to each other quietly to avoid being overheard. As Belaney must have feared, Garratt and Curling established that there was no disease of the heart or lungs that might have caused Rachel Belaney's death. As soon as the stomach was opened, however, the strong smell of bitter almonds announced the presence of prussic acid. The specimens were sealed for later examination. Both Belaney and Clark asked the surgeons to state the cause of death, but they would not make any pronouncement. That evening, Belaney called on Garratt at his home, and again asked him the cause of death, but the surgeon still refused to comment.

Later that evening, Clark and Belaney called on Garratt together, but the surgeon was not in. Belaney suggested they retire to a more secluded street, where he wept again and repeated his previous regrets at his negligence. When Clark queried him further, Belaney explained that his wife had taken the wrong medicine. 'What was it she had?' asked Clark and was shocked at the reply. 'Poison.' Belaney said that he had obtained prussic acid for his own medicinal use, although he denied giving it to his wife. Clark naturally demanded to know why Belaney had not said this from the very start, and suggested they go at once to a mutual friend, Mr Hobson, to make a statement before a witness. Belaney protested that he was afraid of the disgrace it would bring to his family, but, eventually, he went with Clark to Mr Hobson, made a statement and then appeared to have a fit. At Clark's suggestion, they all went to see Mr Garratt, who was now at home, and Belaney repeated his story.

Belaney said that he was in the habit of taking three drops of prussic acid to correct his stomach. It had been in a phial but the stopper was stuck, and, in his efforts to remove it, he had spilt some on the floor and the neck of the bottle had broken off. Not wanting to waste it he had poured it into a tumbler in his wife's bedroom, but placed it on the chest of drawers on the side of the room furthest from the bed. He then left the room and on hearing a scream, returned and found his wife in convulsions. She told him that she had swallowed the 'hot drink'[11] from the tumbler and asked him for some cold water. His next actions were inexplicable except as a sign of guilt. He had seized the fatal tumbler from his wife's hands and emptied the contents into the chamber vessel, and thrown away the phial.

Belaney was arrested on suspicion of murder and, when questioned under caution by a police sergeant, was obliged to admit that, though he had practised as a surgeon in Edinburgh, he had not obtained a diploma – a lack of qualifications of which his friends were ignorant.

The stomach of the dead woman and its contents were handed to Henry Letheby, lecturer in chemistry at the London Hospital. Then aged only twenty-eight, Letheby was to go on to a distinguished career in which he was appointed professor of chemistry, as well as medical officer of health and food analyst for the city of London. One of the new breed of English expert forensic examiners, he was reliable, knowledgeable and meticulous, giving his evidence with a detail and clarity that inspired confidence. The principal confirmations of the presence of prussic acid were the characteristic odour, and its reaction with certain metallic salts to produce a sediment.[12] Letheby performed a series of procedures on the turbid and strong-smelling fluid, all of which demonstrated that prussic acid was present, although he was unable to quantify how much had been taken. He also detected the presence of Epsom salts.

At the inquest, Letheby was asked what action he would take if he knew that a person had swallowed prussic acid; he replied that if death were not instantaneous, strong exertions by a practitioner would recover ninety cases out of a hundred. He would dash cold water in the face and, if the patient could swallow, give ammonia, and brandy and water. If these measures failed to rouse the patent, then artificial respiration, by pressure on the chest and abdomen, should be applied and continued for an hour, while a sponge moistened in chloride of lime should be held near the mouth so that the patient could inhale it. He would also send for a stomach pump, evacuate the contents of the stomach, and introduce stimulants into it. As a last resort, he would use galvanism. 'All these remedies should be known to professional men,'[13] he said, tellingly, making Belaney's warm water rub look decidedly suspicious.

Friends of the Belaneys told the court that husband and wife appeared to have been very fond of each other, but there was another side to the story that went unheard. Two ladies arrived intending to give evidence that Belaney had ill treated his wife, but they were never called as witnesses. Contrary to what Belaney had told Captain Clark and Mrs Heppingstall, Mrs Skelly's sister, Mrs Mary Stobbs, testified that Rachel's mother had enjoyed good health and had never suffered from fits.

The police, meanwhile, had sent an inspector to the north of England and had recovered some letters written by Belaney while in London. On 5 June, writing to Mr George Grey Bell, his agent for the lime-works, he stated that his wife was 'rather unwell' and referred to her delicate constitution and fatigue after the long journey. A second letter dated 8 June, which cannot have been posted until after Rachel Belaney's death, included a series of blatant lies, stating that she was dangerously ill with symptoms of premature labour and in the care of two medical attendants, one of whom had said that her heart was diseased.

The coroner was especially scathing about Belaney's failure both to apply 'various restoratives known to every medical man'[14] and to make any statement regarding prussic acid 'until after he had endeavoured to fish out of Mr Garratt what the post-mortem examination had suggested as the immediate cause of death'.[15] Two publications of Belaney's had been submitted to the court: a poem called 'The Steeple Chase', which the coroner described as 'ribald trash . . . evidently written in a reckless spirit of gross personal scandal, exhibiting the wild emotions of a disturbed and rambling brain . . . well calculated to cost a medical man whatever reputation he might happen to possess',[16] although a later volume on the subject of falconry 'showed a greatly improved spirit, and a much more refined taste'.[17]

The two letters written by Belaney came under minute scrutiny. The first, said the coroner, 'seemed as if purposely designed . . . to pave the way for the fatal result',[18] while the second 'contained a gross falsehood'.[19] The inquest jury returned a verdict of wilful murder against Belaney.

Belaney's trial opened at the Old Bailey on 21 August and lasted two days, during which the prisoner's many lies were exposed. He had claimed to have accidentally damaged the phial of prussic acid and spilt some of its contents, yet no broken glass was found in the room and no one had noticed the characteristic bitter almond aroma. The broken top of a phial had later been found in another room and broken glass on the front steps. The medical witnesses were scathing about Belaney's failure to apply well-known remedies to save his wife, and cast severe doubts on the suggestion that Rachel had been able to speak after she had swallowed prussic acid.

Henry Letheby said he had made numerous experiments with prussic acid on 'the lower animals'[20] and described the initial effect as a 'peculiar giddiness . . . a disposition to run round';[21] then the breathing becomes irregular, the animal falls and makes violent respiratory efforts, resulting in a loud shriek, followed by convulsions and foaming at the mouth. 'After the shriek or scream all sensibility and volition ceases . . . a person . . . would not be able to walk or converse.'[22] In his experiments on cats and horses, he had been able to restore the poisoned animals with ammonia and cold water. Ammonia, Letheby pointed out, existed in every house in the form of smelling salts. He and another doctor had also used artificial respiration with success. He had often tasted prussic acid himself and it was bitter, not hot.

For the defence, there were witnesses testifying to the happiness of the Belaney marriage, the good character of the accused and the fact that he had been taking prussic acid medicinally for some years. His failings were explained away as carelessness rather than as demonstrating criminal intent. It was also represented that, as he was already enjoying the proceeds of his wife's properties, he had no motive to commit murder.

In summing-up, Mr Justice Gurney said that the circumstances differed from almost every case of poisoning he had ever encountered, in which, 'generally speaking, there was a difficulty in ascertaining whether the death had been occasioned by poison, and . . . whether the poison came from the hands of the person charged with the crime'.[23] In this case, there were no such difficulties. The only question was whether the poison had been administered to the deceased by the accused, or purposely placed within her reach so she would take it herself; in either case, Belaney would be guilty of murder. If negligently left, as the accused described, then he would not be guilty of any crime. Gurney also drew the jury's attention to the fact that the medical witnesses had agreed that prussic acid was one of the easiest poisons to detect, which suggested that it was unlikely that a medical man contemplating murder would select it for the purpose. His final comment was that if the jurymen thought the case remained in doubt, then they must remember that 'it was better that many guilty men should escape than that one innocent man should perish'.[24] The jury was absent for an hour and delivered a verdict of not guilty.

It is rare to be given an insight into the workings of the jury room, but the Belaney case did unexpectedly provide an account of how the jurors came to their decision. Public outrage against the verdict stimulated considerable correspondence in the newspapers, and the story emerged that, initially, seven jurymen had been convinced that Belaney was guilty, but one man, who was violently opposed to capital punishment to the extent that he would refuse to condemn any prisoner in spite of being convinced of guilt, had persuaded two of them to change their vote. With one man still firmly holding out for a guilty verdict and the others wavering, all the points of the case were re-argued and Belaney was ultimately acquitted on the principle suggested by the judge. The Henry Fonda[25] figure in the jury later wrote to the newspapers claiming a victory for his principles, saying that, had the sentence for murder been transportation, Belaney would have been condemned.

A *Times* correspondent pointed out that the powerful odour of prussic acid would have prevented anyone from consuming it in the belief that it was water, adding, 'If there ever existed a case in which circumstantial evidence amounted to everything but demonstration, it is this case.'[26] Another correspondent demanded the exhumation of Rachel's mother, Mrs Skelly.

The *London Medical Gazette*[27] tried its own experiment, with Scheele's solution left in a tumbler for five minutes, which showed that had any been spilt, as Belaney claimed, no one entering the room could have failed to notice the unusual smell. The *Gazette* also took Gurney severely to task for his comments on prussic acid, which it believed had misled the jury. Although prussic acid was easily detectable at post-mortem, the *Gazette* said that this held true only if the examination was carried out soon after death. The speed of the acid's evaporation, said the writer, meant that it was unlikely to be detected even a few days after death. Belaney, if he had been studying poisons, might well have believed this. The *Gazette* may have been relying on the work of MM Leuret and Lassaigne described by Robert Christison in 1836.[28] They had found that when the body of an animal poisoned with prussic acid was left unburied for three days, the poison could no longer be detected. Even if buried within twenty-four hours after death, the poison, which was assumed to have volatilised or decomposed, could not be detected eight days later. Only a few years on, Alfred Swaine Taylor would have disagreed.[29]

Taylor was fast securing his position as the leading British forensic toxicologist of his day. His *Elements of Medical Jurisprudence* was published in 1842. This and his later works on poisons were to go through many editions, and became the standard texts on the subject. Taylor had established that prussic acid, even if mixed with organic matter, could be detected after long periods of time if not exposed to air. He had found it in porter after twelve months had elapsed, and when mixed with the decomposing stomach contents of a corpse three weeks after death.

Belaney would have selected prussic acid as his poison for a number of reasons: his history of medicinal use meant that he could be shown to have bought it for an innocent purpose; it was fast and certain if no measures were taken to recover the patient; and bystanders who did not know it had been used could easily be deceived into believing that what they saw was a stroke or the result of heart disease.

The Examiner suggested that Belaney's membership of the respectable class had worked in his favour, and many newspapers openly declared him guilty and were highly critical of the jury. *The Globe* was especially scathing: 'There is a screw loose somewhere in the public justice of this country ... If his horse had drunk prussic acid we cannot doubt his horse would have been cured ... So far as that jury have had power to affect the character of our criminal justice, we regard them as having employed it to decide that wife-murder goes free.'[30]

On the morning after the trial, Belaney, pausing only to purchase a case of champagne, left London and headed north, arriving at his destination on Saturday 24 August, but his ill fame preceded him. The critical letters in *The Times* had been reprinted at Alnwick, and widely circulated. His return, as well as that of some of the defence witnesses, was said to have been 'the reverse of welcome'.[31] One of Belaney's friends, a Mr Hall, was stoned as he neared his home.

Belaney took up residence not in his clifftop dwelling, which was as yet incomplete, but in a farmhouse he rented from the trustees of Bamburgh Castle. He must have hoped that time would overcome the general feeling against him, but handbills stirring up animosity continued to be circulated and shopkeepers refused to sell him even basic necessities. The people of Seahouses were angry and disgusted not only with Belaney, but also with the friends who had been witnesses for the defence, since they had a wholly different opinion of the character of the man and his behaviour towards his wife from that given in court. Many wanted to know more about Rachel's miscarriage in 1843, which had taken place five or six months after marriage, since a rumour had arisen that it had not been a miscarriage at all, but a birth that had been concealed, or even another murder.

On the Wednesday following his return, an effigy of Belaney was elevated on a pole, paraded around the neighbourhood by some hundreds of the population, and burned in front of the farmhouse. Belaney stayed put, but, on the Saturday, effigies of both him and his friend were publicly burned to the sound of much hissing and hooting. Still Belaney thought he could ride out the storm. The excitement continued into September and, on Monday 16 September, three effigies were carried about the streets by three men with their faces blackened, dressed in women's clothes and followed by a hooting crowd of about five hundred. One effigy represented Belaney, one was a defence witness and the third the 'arch-fiend'.'[32] These effigies were set alight at the farmhouse gate, and when Belaney saw the conflagration he ran out and fired a pistol at the crowd. No one was hurt, but the action sparked off a general rush towards him and he dropped the weapon and fled. The pistol was picked up and used to smash the windows and shutters of the house. The gate and the figures were then set alight.

Two days later, the crowds assembled to complete what they had begun. Three men entered the farmhouse, carrying some combustible material, went upstairs to the bedrooms and set fire to the beds, heaping furniture on top. Belaney, his clergyman brother Robert and some friends had been in the dining room and were obliged to make a hurried escape, running across cornfields to the house of a sympathetic innkeeper, where they remained incognito. A fire engine was sent for, but it proved useless – the whole house burned to the ground in a matter of hours. Belaney was eventually able to leave the district.

In October, Belaney's London solicitor was recognised at the Doncaster races, seized by a mob and ducked in a pond.[33] After a failed attempt by Belaney to prosecute George Grey Bell and another man for arson, the mob turned their attention to his clifftop retreat, smashing the windows and stealing lead from the roof.[34]

That December, it was reported that Belaney was staying with a friend in Stepney (not identified, but Captain Clark is a good candidate), 'not more than a stone-throw from the spot where his unfortunate partner met her sad fate'.[35] It is believed that, ultimately, he fled to the continent.[36]

History must judge Belaney guilty. As a medical man, he would have had no difficulty in persuading his wife to take a strange-smelling concoction of Epsom salts and prussic acid in water, perhaps offering her a biscuit to take away the taste. A young wife who might produce a large family was not a financial asset, and, had Belaney not been suspected of a second murder, he might well have cast about for another lady of property.

* * *

If Belaney was fortunate, two poisoning cases in 1850 show that there were other circumstances in which it was hard to obtain a conviction.

Eighteen-year-old Louisa Hartley lived in one room with her widowed father Joseph and, having no paid occupation, was entirely dependent on him. On 16 April 1850, Joseph found that the coffee Louisa had made for him tasted unpleasant and burned his lips and tongue. He spat it out, and accused Louisa of putting something in it, which she denied, saying that soda might have fallen in by mistake when she made it. He tasted her coffee and found it to be unadulterated. Joseph said he would find out what she had put in his coffee, but Louisa snatched it away, poured it into a wash-hand basin in which there was already a quantity of soapy water, and scoured the cup. Joseph was now certain that his daughter had tried to poison him and took the basin to Guy's Hospital to have the contents analysed.

The liquid was sent to Alfred Swaine Taylor, who found that it contained sulphuric acid. Louisa was arrested and charged with attempted murder. At the police court, Rebecca Yates, a dressmaker who roomed in the same house, revealed that Louisa and her father quarrelled a great deal and Louisa had twice said she would kill her father, once citing vitriol (sulphuric acid) as her chosen means, her other choice being a razor. Louisa had complained that her father beat her and showed Rebecca weals on her body. Joseph responded that his daughter was bad-tempered and disobedient.

Louisa was tried for attempted murder, still a hanging offence then, at the Old Bailey. Taylor told the court that the effects of sulphuric acid were dependent more on concentration than quantity.[37] The dilution in the coffee was such that, even if Hartley had swallowed it all, it would only have caused pain and vomiting and he would have recovered after a few hours, without medical assistance.

Joseph made a poor impression – he was described by the defence as 'a person devoid of all proper manly feeling'[38] – and was criticised for pursuing the case against Louisa in what was seen as a determined, vindictive and un-fatherly manner. It was even suggested that he had trumped up the charge so he would no longer have to support his daughter. The mood of the court was that Hartley was a rogue and a brute, and after a summing-up wholly sympathetic to the defendant, Louisa was acquitted and consigned to the care of the Ladies' Committee of the gaol.

There were no doubts about administration of poison in the case of Martha Sharp, a twenty-six-year-old widow passionately smitten by James Elphinstone, a comedian at London's Pavilion theatre. A married man with two children, he had declined Martha's many invitations to meet her and refused to answer her letters. Martha decided to bake him a very special jam tart; underneath the jam was a green layer composed of crushed cantharides beetles. The active compound produced by these beetles, cantharidin, is a highly irritant poison that, under the name 'Spanish Fly', gained a popular reputation as a stimulator of the passions, and Martha must have hoped it would act as a love philtre.

The tart was delivered to Elphinstone at the theatre, but Elphinstone gave it to his dresser, Thomas King, who took it home. King ate none himself, but his wife Charlotte ate most of it, fortunately leaving the centre portion where most of the curious green material was. She was taken violently ill, bringing up the tart together with phlegm and blood, but survived. Martha Sharp was tried at the Old Bailey, charged with administering cantharides to Charlotte and attempting to administer it to Elphinstone with intent to murder.[39] The judge, Mr Baron Platt, commented that he did not see how the prisoner could be convicted of intending to administer poison to someone she had never even seen. Whatever harm she had done, Martha had had no intention of killing anyone. She was clearly guilty of a lesser offence, such as wounding, but that was not the charge. She was acquitted.[40] It was not until the Offences Against the Person Act of 1861 that the law envisaged that poison might be administered without intent to murder, and it was henceforward a felony to administer poison with the intention of causing grievous bodily harm, and a misdemeanour if the intention was to 'injure, aggrieve or annoy'.[41] Charlotte King must have felt all three.

CHAPTER NINE

Alarm and Indignation

Throughout the nineteenth century, the public's perception of the incidence of murder by poison was largely determined by dramatic trials given extensive coverage in newspapers and made the subject of widespread gossip and rumour. Until the compulsory registration of deaths, which commenced in the last quarter of 1837, the best statistics were provided by the annual publication of the Bills of Mortality, compiled from weekly returns of parish clerks, in which poisoning in general did not feature strongly as a cause of death. The return for the City of London for 1831 is fairly typical of the period, deaths being divided between disease and casualties. Out of 411 casualties, there are seven simply referred to as 'poisoned'.[1] Even after the appointment of the Registrar General, there was still no legal requirement for a certificate to be provided by a qualified doctor who had actually attended the deceased. Many deaths were registered with no certificate at all.[2] There was understandable concern that, hidden amongst deaths from 'unknown causes' or 'by the visitation of God' or described vaguely as due to 'inflammation' or 'fever', were numerous murders that had gone undetected and unpunished.

In 1829, the *Morning Chronicle*, reporting a poison murder in France, commented, 'In England, on the contrary, until within the last few years, a case of poisoning was almost unheard of; or if such a thing did occur, it excited the utmost alarm and indignation, and produced at once that diligent and prompt enquiry which vindicated the efficiency of our laws...' Now, it seemed to the *Chronicle*'s editor, scarcely a day could pass without the paper having to report a death by poison. Was it because the law was less strictly administered, or science had produced refinements in the art of killing that the usual tests could not detect, or morals had declined? He could not say. The possibility that it was the chance of detection of the poison, rather than the incidence of crime, that had increased was not considered. The *Chronicle*'s conclusion was that the ease with which people could get poison was 'an evil, a crying grievance, which demands the immediate attention of the legislature'.[3]

In the same year, the *Royal Cornwall Gazette* commented:

> Every reader of the criminal proceedings which are recorded in the newspapers, must have observed the increase in the administration of poisons in this country, in cases of murder as well as suicide . . . There have been several mysterious cases in which death has ensued from poison without any means of discovering who administered it, so great is the facility with which this treacherous crime may be committed . . . It has always appeared to us that the sale of poisons from the druggists has been infinitely too lax . . . The only precautions now taken to prevent the evil use of poisons is the asking of some questions, which a guilty party goes prepared to answer falsely, and the wrapping up in paper marked 'poison'.[4]

Parliament, however, continued to remain unmoved by isolated protests, occasional petitions and the recommendations of medical societies.

Public demand for legislation to control the sale of poisons did not develop into a serious movement until the 1840s, when it was stimulated by the outrage that followed several trials involving blatant and callous killings. The impression that emerged from these cases and took firm hold of the public imagination was that the typical poison murder was perpetrated in a home of the labouring class, and the culprit was almost always a female who administered arsenic to members of her family in the food she prepared. The idea of poison as predominantly a woman's weapon persists to this day. While the majority of murderers are male, when one looks only at poison murders, the perpetrators are more equally divided between male and female.

In 1842, Little Bolton, a township north-west of Manchester, found itself at the heart of a murder case that rapidly became deeply disturbing and attracted nationwide attention. When a coroner's jury was summoned to the Crown Inn concerning William Eccles, who had died on 26 September, aged nearly fifteen, they at first believed that it was 'an ordinary case of sudden death'.[5] This is not an assumption that would be made today, and it demonstrates that, in 1842, the sudden unexplained death of a teenage boy was not considered to be anything extraordinary. The facts produced at the hearing were, however, suspicious enough to warrant an order for a post-mortem examination. This was carried out on 28 September by local surgeon Joseph Denham, who found the bladder contracted and hard, and the internal mucous membranes of the stomach and intestines in a highly inflamed state. Sticking to the membranes was a quantity of white powder, which Denham was able to scrape off, and this, with some dark dirty-coloured liquid found in the intestines, was passed to Henry Hough Watson, a chemist of Little Bolton. Watson 'employed the usual and most approved tests';[6] both Marsh's and heating the samples with carbon. The white powder was arsenic, and the dark liquid, which was mainly blood and mucus, contained arsenic in solution. There was only one possible suspect: the boy's stepmother.

William's father, Henry, was a respectable and hard-working carter whose first wife Cisley had died in April 1840, aged forty-one. Later that year, Henry met Betty Haslam (formerly Heywood), a thirty-nine-year-old widow with three daughters: ten-year-old Alice; Nancy (five);[7] and Hannah (two).[8] Within a few weeks of their first meeting, they were courting, but the relationship may have cooled, since Henry did not visit Betty for a few weeks. The next time he saw Betty, he learned that Nancy and Hannah were dead. Hannah had been buried in October, and Nancy two months later. Both, said Betty, had died of measles. Soon after Nancy's death, a neighbour, on being told that Betty was beating her surviving daughter, peered through a window of Betty's home and saw her in bed, with Alice lying on the floor almost naked and crying. The neighbour angrily took Betty to task for this shameful treatment, but Betty simply retorted that she could do as she liked with her own children.

On 3 January 1841, Henry and Betty were married. Henry never complained, but he must have regretted his decision. That August, he was discharged from his employment in nearby Firwood because of what the newspapers called 'some *faux pas* on the part of his wife'.[9] The nature of the *faux pas* is suggested by the fact that Betty was later described as 'an abandoned woman given to drinking'.[10] Henry made the best of things and found new employment, but, since it was in Manchester, he was obliged to go out early every Monday morning and return home late on Saturday night. Perhaps spending the week away from his new wife was part of the attraction.

In February 1842, Betty was employed to nurse William Hatton Heywood, the ten-month-old son of James,[11] a rag dealer, for which she was paid 3s. 6d. a week. In March, Betty asked James to let her have two flock beds as she wanted to take in lodgers. Two days later, he supplied them, saying she could pay for them out of what she received for nursing his child. She never made the payment and the child died the following day. Margaret Benison, who was employed to make the funeral suit for little William Heywood, noticed when she dressed the corpse that the child had been neglected and was dirty, with vomit about the mouth. All that Henry Eccles knew about the matter was that he had gone to work on the Monday leaving the child well, and had come home on Saturday to find him dead. It was a pattern that would be repeated.

When Henry went to work in Manchester on 5 September, his step-daughter Alice was well. He returned on Saturday to find that she had died that afternoon. Betty told him that her daughter had been suffering from fits. Richard, William Eccles's younger brother, was later to recall that both Alice and William Heywood had died after similar kinds of fit.

On the morning of Monday 26 September, Henry went to Manchester, and William, who worked in the finishing room of a nearby bleach-works, walked to his work a mile away. He was in his normal good health and spirits when he came home for dinner and ate a damson pudding prepared by Betty, but, on his return at a quarter to two, it was seen that the boy had been crying and he said he was very ill with stomach pains. He then started vomiting a greenish fluid. These attacks continued throughout the day and, at three o'clock, William decided to go home. An hour later, one of his fellow workmen, Thomas Davenport, was walking home when he found William lying in a ditch in great pain, vomiting copiously. William said that he was too ill to go on any further, so Davenport lifted him up and helped him walk towards his home, William continuing to heave up a watery liquid as he went. On the way, they met Betty, and Davenport asked her to take William home and give him some hot tea. William died later the same day.

Mary Anne Hopkinson, a joiner's wife, was called to the house to take an order for a coffin, but when Mary Eccles, Henry's eight-year-old daughter, started to talk about her brother's death, she was quickly ordered by Betty to hold her tongue. Betty claimed that she had gone to get a doctor but had not found him in; in any case, she added, a doctor would have been of no use, as William 'would not get better of what he ailed'.[12]

The following day, a neighbour suggested that Betty should have the body opened in case William had died of some infectious disease, but she responded angrily that she had trouble enough without having more. Betty was not, however, too troubled to go to the bleach factory to demand the money payable on the death of an employee. Only seventeen days earlier, Betty had tried to make a claim on Alice's death, and had then been told that payment could only be made for an employee or the child of an employee. The bookkeeper, who had seen William in good health the previous morning, became suspicious and asked Betty to call again the next day. In the light of William's unexpected death, it was now suspected that Betty's daughters had also been poisoned and, as the rumour mill began to grind, it was further whispered that Betty had had ten children, of whom eight had died suddenly, as had her first husband, and 'it is now hinted . . . that all her family, excepting those alive, have received a helping hand'.[13] Concern was also expressed at the death of William Hatton Heywood, and it was alleged that Betty had poisoned the child to avoid paying for the beds.

The bodies of Betty's daughters and William Hatton Heywood were exhumed. The intestines of Alice Haslam who had been buried for just a few weeks were red and inflamed, and the interior of the stomach was coated in a white powder. Nancy Haslam had been buried for more than a year, and her face was decomposed, but the intestines were well preserved and in a similar state to Alice's. Hannah's body, however, which had been buried two months longer than Nancy's, was reduced to a skeleton, and the coffin was broken and full of earth. Denham was able to collect earth for analysis only from where the stomach had been. He was able to remove the stomach and duodenum of baby William.

All the samples were passed to Henry Hough Watson, who identified arsenic in the remains of Alice and Nancy, but not in the samples from Hannah and William Hatton Heywood.

The only possible suspect was Betty. The deaths had occurred when Henry was away and she had prepared all the family meals. The police made a careful search of the Eccles home but no trace of any poison was found. Betty said she 'never had a bit of poison in her life',[14] but this was a lie. In August 1842, she had gone to local druggist, Mr Moscrop, asking for a pennyworth of arsenic to destroy mice. Moscrop, with what a newspaper described as 'laudable caution',[15] had refused to sell her any unless another person was there as a witness. Betty left the shop and returned soon afterwards with another woman whom Mr Moscrop did not know, and the druggist, all requirements now fully satisfied, supplied the arsenic wrapped in two papers, clearly writing 'POISON' on each one. On 29 September, Betty was arrested.

By the time the inquest was resumed on 4 October, lively public interest had demanded that it should be transferred to the larger facilities of Little Bolton town hall. Feelings were running high against the 'unnatural parent',[16] and many thousands of people assembled in front of the building, while those who could not get near enough crowded the adjacent streets. When Betty arrived in a chaise guarded by police, there were yells and hoots and it was all the police could do to prevent the mob from breaking into the inquest room and inflicting summary punishment. The inquiry was now considered of sufficient importance for *The Times* to no longer rely on reprinting reports from the *Bolton Chronicle*; instead, it sent its own correspondent, and *The Spectator* described the case as 'one of the most atrocious stories of crime on record'.[17] Even before the inquest had closed, the public verdict was settled. 'We should suppose that there never was, in this country, such a wanton and cold-blooded number of murders without cause. She was not in distress; had not quarrelled with them; and, what is more, could not have done it for the sake of money, as they did not belong to any burial society.'[18] The sum of forty-five (in some papers reported as fifty) shillings was payable on William's death, by his employer, although most of this would have been needed to pay for the funeral, and it appears that Betty received nothing on the death of her younger daughters.

At the Liverpool Assizes in April 1843, Betty was found guilty of the murder of William Eccles. She was hanged on 6 May. It is impossible to determine how many people Betty Eccles murdered, and the motives may seem slight, or even mysterious. For a serial poisoner like Betty, early success breeds repetition, but it is hard to see what she gained from killing a boy who was bringing wages into the house. The deaths of Nancy and Hannah in 1840 may well have been motivated by Henry's cooling off towards Betty. In the annals of murder, there are many cases of single women with children murdering their offspring when they see the chance of a permanent relationship with a man, the murderess believing rightly or wrongly that he is hesitant about supporting another man's children.

The ease with which Betty Eccles obtained pure arsenic did not arouse comment at the time, but more shocking was the 1844 case of Mary Ann Johnson. Mary Ann was born in 1831, the illegitimate daughter of a servant. In 1834, her mother Mary married agricultural labourer Christopher Farr of Benington in Lincolnshire. Two sons, Daniel and William, were born in 1835 and 1838. A daughter, Martha, was baptised in July 1842 but died soon afterwards. In December 1843, Mary Farr died aged thirty-four. The household now consisted of Christopher, his two sons, stepdaughter

Mary Ann and a housekeeper, thirty-year-old Elizabeth Johnson. There were only two beds in the house, and Christopher and the boys shared one, while Mary Ann and Elizabeth slept in the other.

After the family evening meal on 6 February 1844, Christopher Farr and his housekeeper went to a ranting chapel, leaving twelve-year-old Mary Ann looking after her half-brothers. On their return, they found that the boys had been taken ill with vomiting and diarrhoea and were at the home of a neighbour, Mrs Chevins. Farr went to fetch a surgeon, Richard Cummack, but, soon after his arrival, William died. Daniel died two hours later.

The following day, Christopher, discovering that Mary Ann had obtained arsenic from a grocery shop, took her there and confronted the manager William Overton, demanding to know if his stepdaughter had bought poison from him. Told that she had, Christopher asked Mary Ann what she had done with it, and she said that she had bought it for a woman she had met on the road, a woman she had never seen before or since.

Richard Cummack found arsenic in the stomachs of both victims and, at the inquest held on 19 February, there was some very close questioning of all the witnesses. Mr Overton said that Mary Ann had told him she wanted arsenic to kill mice, and, after warning her that it was highly dangerous and she should be careful where she put it, he sold her a pennyworth, a quarter of a pound, taking care to label it 'POISON'. Mary Ann described the woman she said had sent her for the arsenic as wearing a brown cloak and straw bonnet, but another witness had seen both Mary Ann and a woman of that description on the road on the day the arsenic had been purchased and said they had not communicated in any way. No information incriminated anyone other than the twelve-year-old girl, and the coroner undertook the 'painful duty'[19] of committing Mary Ann to prison in Lincoln Castle to await trial. The only person in court who appeared unmoved was Mary Ann. Since it now seemed that the woman in the brown cloak was not involved, the only unanswered question was where Mary had got the penny to buy the arsenic, as her father said he never gave her any money.

The day after, Mary Ann made a full confession to the prison chaplain, but said that she had not acted alone. On 9 March, at the Lincolnshire Assizes, there was a sudden turnaround in the case. The coroner's order was quashed on a technicality – the word 'feloniously' had been omitted – and the prisoner was discharged. In July, the Farrs' housekeeper Elizabeth Johnson was tried for the murder of the two children, and the principal witness against her was Mary Ann. It was Elizabeth, said Mary Ann, who had sent her for the arsenic, mixed it with water in a mug and told her to give it to the boys. When Mary had asked if the drink would harm the children, Elizabeth said it would not, as it would only poison mice. Mary did as she was directed and, afterwards, Elizabeth told her to say she had bought the arsenic for a woman in a brown cloak. After the death of the children, Elizabeth and Christopher had slept in the same bed. The only other evidence was that of Sarah Johnson, the sister of Farr's late wife, who said that Elizabeth had wanted a home of her own and, before she went to work for Farr, had gone to a fortune teller to see if she would marry him.

Had Elizabeth really poisoned two small children to gain a place in their father's bed? Murder has been committed for even flimsier reasons, and if Mary Ann did not have any money, as Farr said, then the most probable source of the fatal penny must have been the housekeeper.

Mr Macaulay, defending Elizabeth, accused Mary Ann of being the guilty party, arguing that 'the intelligence and care the child had taken to conceal what she now stated rendered her responsible as a principal felon'.[20] The jurymen must have been in no doubt that this was a case of murder, and that either the accused or the accuser was guilty, but they were unable to decide which, and, after five minutes of deliberation, acquitted the prisoner. The murders remain unsolved. No one seems to have commented publicly on the fact that a quarter of a pound of arsenic had been sold to an unaccompanied twelve-year-old girl.

Another case that must have raised alarm was the Edinburgh trial in January 1844 of a Mrs Gilmour for the murder of her husband by arsenic, just six weeks after their wedding. In the absence of an obvious motive, the Scottish verdict of 'not proven' was delivered. Alexander Wylie, the druggist from whom Mrs Gilmour had bought the arsenic, said that, since 1832, he had recorded the names of all persons to whom he sold it. Mrs Gilmour, whom he identified in court, had told him her name was Roberts and that she was buying the arsenic for a farmer, John Ferguson, who wanted it to kill rats, although she was unable to recall the name of his farm. The druggist had checked with a neighbour if he knew of a farmer called John Ferguson, but he did not. Despite this, he sold his customer twopence-worth of arsenic.

The trial of Eliza Joyce in July (Chapter 7) attracted national attention, to be followed in September by the trial of Mary Sheming of Woodbridge, Suffolk. Fifty-one-year-old Mary, mother of six children, finding herself faced with the prospect of caring for her daughter's illegitimate ten-week-old son, bought a pennyworth of arsenic and notified the local carpenter that a child-sized coffin would soon be wanted. A few days later, she was minding the baby when it was taken ill and died in a few hours. The post-mortem left no doubt that it had died from arsenic. Mary was tried for murder at Bury St Edmunds and attempted to evade the consequences of what *The Times* called 'one of the most revolting and disgusting cases ever known to transpire in a court of justice in this part of the kingdom'[21] by accusing two of her daughters of the crime. She was found guilty and hanged at Ipswich on 11 January 1845.

In none of these cases were any of the druggists thought to be at fault for supplying arsenic and it must have been increasingly apparent that the existing law and practice on the sale of poison was wholly inadequate.

On 14 January 1845, *The Times* published a letter from one of its regular correspondents, Senex, who suggested that, since arsenic was so frequently used for murder and suicide after being sold for the alleged purpose of destroying rats, a law should be enacted to prevent chemists from supplying it for that purpose. Another correspondent, Vigilans, wrote in agreement. 'The frightful frequency of cases of murder by poison . . . renders it imperative for the legislature to interfere.' Vigilans suggested that it should be illegal to sell poison to any person not a medical practitioner, without the production of a certificate signed by two householders stating that they believed it could be safely supplied. 'The very poor can never want poison for any innocent purpose,' he went on, 'and the rest of the community can procure the requisite certificate.'[22] Where drugs that were taken medicinally were only poisonous in excess, the law should provide that no more than a medicinal dose should be sold without a certificate. Vigilans may have been unaware that the Victorian chemist's shop was an Aladdin's cave of poison to which anyone might freely resort, and that acids, corrosives and metallic poisons were widely used for domestic cleaning tasks and in many trades. Another correspondent, signing himself 'A Chymist', advised Vigilans that 'the poorer classes are, from the very nature of things, the principal purchasers of such articles',[23] and if such a law were passed, it was doubtful that anyone would comply when nineteen out of twenty substances sold by druggists were poison and were in hourly demand by all classes of society. A demand for any drug that was dangerous to life and not in habitual use for domestic and medicinal purposes would normally be met with refusal or a strict enquiry. His only concern was that: 'In the country, where arsenic is extensively used as a dressing for sheep, the same degree of precaution cannot be used. . .'[24]

A retired chemist had told another *Times* correspondent that it was not unusual, after refusing to supply a poison, to be called upon within a few hours to help counteract the poison that had been obtained elsewhere. He had avoided being an accessory to crime by requiring all applications for poison to be made by two people, both of whom knew the use to which it was to be put, thus 'placing a not easily surmountable difficulty in the way of the felonious applicant'.[25] The retired chemist may have been fortunate, for such a difficulty was very easily surmounted. For the time being, even this elementary precaution was being left to the discretion of the chemist. Concern was also being expressed about the time-honoured practice of parents drugging their children with narcotic mixtures – such as Godfrey's Cordial, a solution of opium and treacle – which were available not only from chemists, but also from grocers and chandlers.

The *Pharmaceutical Journal,* its eye very firmly on the status of the profession, pointed out that prohibitions on the sale of poisons 'even if practicable', would result in 'very serious commercial inconvenience' due to the many tons of arsenic and sulphate of copper used annually in agriculture. It had been suggested that in every shop where medicines were dispensed, poisons should be kept under lock and key. This was the practice in Germany, but in that country the medical profession and sale of drugs formed 'a monopoly under the immediate control of the government, a system which would not suit the prejudices and notions of freedom prevalent in Great Britain'.[26] Addressing proposals that a list of poisons should be published annually and that it should be illegal for any person except a medical practitioner to prescribe or administer them, the journal objected that 'this would put a stop to the domestic use of some of the most valuable remedies, which, although undoubtedly poisons, are nevertheless frequently resorted to by persons unconnected with the medical profession, who would be indignant if they were told that they must not take a calomel pill or a dose of Jameses [sic] powder without consulting a doctor'.[27]

The journal felt that proper labelling, refusal to sell poison to anyone who appeared to be agitated, and only selling small quantities to strangers were more than adequate precautions, and all that was required was improved professional education. No chemist wanted to be named in the newspapers in connection with a case of poisoning, 'and the caution arising from this dread of exposure is as effectual a check on the sale of poisons for criminal purposes as a legal prohibition'.[28]

In 1844, the bastardy clause, which had caused so much distress to mothers of illegitimate children, was effectively repealed by the Poor Law

Amendment Act,[29] which empowered an unmarried mother to apply to the petty sessions for an affiliation order against the father for maintenance of herself and her child. The financial motivation to murder illegitimate infants was therefore transferred from the mother of the child to the father.[30] This was to play a part in one of a series of murders that, for a time, made a small Essex village appear to be the poisoning capital of Britain and finally led to the long-needed change in the law controlling the sale of poisons.

Clavering was an agricultural village with a population of about a thousand, most of whom were labourers and their dependants. The Newport family of Curles Farm were landowners, employing fifteen men and boys and occupying a substantial house staffed by servants. The eldest son, Thomas, born around 1818, seems to have been a young man who felt entitled by his elevated position in the small community to take his pleasure as he liked and then use money and influence to wriggle out of the consequences. In the spring of 1845, maidservant Lydia Taylor was dismissed from her employment at Curles Farm because she was pregnant. Thomas, who was the father, had an easy solution to the problem: he offered Lydia some medicine, which he said would take everything away. Lydia angrily refused. The nature and source of the medicine was never discovered. Lydia returned to nearby Manuden, where a clean but crowded hovel was home to her parents and their seven other children. Here she gave birth to a healthy son, Solomon, in December. Thomas showed no interest in the baby, nor did he have any intention of supporting it, so Lydia's mother went to see him and he grudgingly handed her a few shillings. It must have been apparent to Thomas that this was only the first of many such requests and he would be expected to pay for the Soloman's upkeep until the child reached adulthood. Thomas had a better alternative – murder – and for that he employed the services of a woman who had already committed murder and got away with it.

Sarah Chesham was born in 1809 and was said to have been a 'masculine-looking woman'.[31] As there are no pictures of her, she may be imagined as a sturdy countrywoman with strong features. She and her labourer husband Richard had six children. In January 1845, two of Sarah's sons, ten-year-old Joseph and James, eight, became desperately ill. Painter Thomas Deards, who lived in the same cottage as the Cheshams, the living spaces of the two households being divided by a thin partition wall, had special reason to remember the events. On 18 January, Sarah told him that Joseph had been taken ill after being struck by Thomas Newport with a stick. Deards told her to call a doctor and she said she would, but no doctor was called. When Richard returned home that night he found both boys ill in bed, vomiting. Richard went to local surgeon Mr Hawkes, who didn't think the matter was serious enough to warrant a call, and sent a bottle of laxative mixture.

The upper room in which the boys slept projected several feet over a room occupied by Deards and, that night, he heard the two children groaning in pain. Next morning, as he sat at breakfast, the boys' vomit poured between the cracks in the floorboards and on to his table and the floor. He knocked on the Cheshams' door, but there was no answer. Later that day, he was astonished to see Sarah Chesham in the street. 'Mrs Chesham, are you aware how bad your children are?' he exclaimed, adding, 'We can scarcely live in the house!' Her reply was, 'I will go home and alter it.'[32]

Back home that afternoon, Deards could still hear the boys being sick, first on one side of the bed, then the other, and their vomit continued to pour through the floorboards. Unable to stay in the house, he went out and again found Sarah Chesham in the street. He marched up to her, and told her that her children were very ill and there was no one in the house to look after them. He suggested she call Mr Hawkes and she promised she would. Hawkes was not called, but, at nine o'clock that night, Sarah knocked on Deards's door and called him to see her son, who was dying. When Deards arrived, he found that Joseph had just died. Next morning, Richard went to tell Hawkes that Joseph was dead, and, this time, the surgeon called and found James very weak and in great pain. He prescribed opium and a laxative. At no time did Hawkes ever suggest that the Chesham boys had been poisoned. On 22 January, James died. Hawkes wanted to carry out a post-mortem but Sarah was reluctant and he didn't press the matter. He certified both deaths as due to cholera and the boys were buried in Clavering churchyard.

Sarah Chesham, as was to become clear, had murdered her sons with arsenic, and it was later alleged, though not proven, that she had also made an attempt on the life of two other sons. When a woman cold-bloodedly attempts to eliminate her children, she usually has her eye on a man, and that man must have been Thomas Newport, a prime catch compared with Sarah's labourer husband, and the person who might well have provided the arsenic. Whether or not the two were lovers, as has sometimes been alleged, is unknown. Early in 1846, Sarah, either at Newport's instigation or to gain his favour, decided to take care of Lydia and Solomon Taylor. Unlike the typical secret poisoner, her methods were shockingly open, crude, callous and blatant, to the extent that it is hard to imagine how she thought she would get away with it. Sarah was far from stupid, and not indifferent to whether or not she suffered for her murders, so how she went about her crimes suggests that she was either arrogant enough to believe that she was immune, or else she had a powerful backer: Thomas Newport.

Although Sarah barely knew Lydia, she called upon her to admire the baby, bringing gifts of a rice pudding, apple turnover, butter, tea and sugar, but, as she held the baby in her arms, he was very suddenly taken ill. Lydia saw something white and slimy on her child's lips and demanded to know what Sarah had given him, but Sarah protested that it was only sugar. After Sarah left, Lydia threw away all the food she had brought.

Solomon recovered, but Lydia was bombarded with requests by Sarah to be allowed to call again. Lydia finally relented, but made sure to hold on to her baby. Sarah waited for an opportunity, then suddenly snatched the child and ran out of the house. The terrified mother gave chase and, when she caught up with Sarah, saw her wiping her fingers on her gown. The baby had been sick and there was something like pink ointment in its mouth. Solomon was ill for three weeks.

Once Solomon was weaned, Lydia went into service, leaving the baby with her mother. The next day, Sarah Chesham was back. Mrs Taylor tried to avoid the unwelcome visitor by going out, but, once again, Sarah was able to snatch the baby and run around a corner. By the time Mrs Taylor had caught up with her, Sarah had put something in the baby's mouth. From that moment, Solomon was a very sick child and began to waste away.

Thomas Newport had failed to support his son, so, in August, the Taylor family had him summoned to the Saffron Walden petty sessions, where they obtained an order for him to pay 2s. 6d. a week. They also handed in a solicitor's letter accusing Sarah Chesham of being Thomas Newport's agent in an attempt to murder Solomon. Sarah was arrested and, when tongues started to wag about the events of January 1845, an order was made to exhume James and Joseph Chesham. Their stomachs were taken to London and examined by Alfred Swaine Taylor. On 4 September, he told an inquest jury that both boys had taken a quantity of arsenic that would have been fatal to an adult and that this was undoubtedly the cause of death. Two villagers who used arsenic for pest control told the court that Sarah Chesham had asked them for some, but both denied having supplied her. Later that month, little Solomon died; not from the immediate effects of arsenic, but from the damage done to his organs, which had prevented him from absorbing nourishment. Over the weeks, all traces of arsenic had disappeared from his body, and there was insufficient evidence for Taylor to determine for certain if he had died as a result of being poisoned.

While Sarah waited in gaol for her trial, she wrote a letter to Thomas Newport, which she must have guessed would be seized by the authorities, in which she claimed that she was innocent and being made to suffer for his crimes. Sarah Chesham stood trial for the murder of her sons in March 1847, and Thomas Newport was charged with aiding and abetting her. Sarah's defence was that there was no proof she had ever given her sons arsenic. Despite careful searches of her house, no arsenic was found or traced to her possession, and no shop had a record of selling her any. Witnesses testified that she had always been an affectionate mother. Sarah was acquitted; however, there was still the charge of murdering baby Solomon, since she had been seen giving him something harmful on three separate occasions. When Professor Taylor said that he was unable to prove the cause of death, the prosecution was withdrawn, and Sarah was once again found not guilty.

The charges against Thomas Newport were dropped. Sarah returned to Clavering, and legend has it that she became a celebrity, having earned the nickname 'Sally arsenic'. It was rumoured that she had freely admitted carrying out the murders and that she was now available to give advice to others on how to commit murder and get away with it.

CHAPTER TEN

The Essex Factor

In 1842, a new test for arsenic was proposed by German chemist Hugo Reinsch. He boiled suspect matter with hydrochloric acid then put a fine copper mesh in the liquid. If arsenic were present, it would precipitate on to the copper as a steel-grey coating. The Reinsch test, which eliminated the possibility of contamination by impure zinc, soon joined the battery of methods used to detect arsenic. The Marsh and Reinsch tests were not, however, as had been hoped, the ultimate answer, since their delicacy made them a snare for the inexperienced and, sometimes, as was later to emerge, even for the very experienced.

The late 1840s was a time of increasing public concern about poison, especially about the widespread use of arsenic in rural areas, where, purchased by the pound, it was in daily use in making up sheep dip, steeping seed grain and poisoning rats. A crop failure in 1845 led to great poverty and distress, exacerbated by the high prices of basic foodstuffs under the Corn Laws, and the period became known as the 'hungry forties'. Long hours of physical labour on starvation rations, or the threat of homelessness due to lack of work, encouraged begging, stealing and, occasionally, murder.[1]

On 25 September 1847, *The Times* devoted a leader to the subject of 'the alarming increase of deaths through the criminal or accidental administering of arsenic' and the 'ease with which arsenic may be administered without exciting suspicion',[2] calling upon the government to determine some means by which its presence should be obvious when mixed with other substances. 'We hope to see our advice taken without delay, for, of all the horrors with which our criminal annals have been recently stained, none is so revolting or so frequent as the crime of murder in the domestic circle by means of arsenic.'[3] The following month, the *London Medical*

Gazette reported that the Council of Salubrity of Paris had proposed that all persons who kept arsenic on their premises should mix it with Prussian blue to colour it and nux vomica to give it a bitter taste, recommending this as a good example for the UK government to follow.[4]

Not all poisonings involved arsenic or excited such revulsion. In 1848, Hannah Leath of Marylebone,[5] whose husband had been out of work for a year, terrified that a broker would remove their few items of property to pay for arrears of rent and turn her family out on to the street, attempted to poison her three children with lead acetate, known as sugar of lead. The children recovered, and Hannah was tried and found guilty of attempted murder. A sympathetic jury recommended mercy, and her mandatory death sentence was commuted to transportation for seven years.

In 1848, with England suffering its second epidemic of Asiatic cholera,[6] there was some anxiety in the medical profession that its prevalence would mask unsuspected cases of poisoning, and Henry Letheby published a paper showing how the symptoms of the two conditions could be distinguished.[7]

Arsenic continued to dominate poison crime. A law to restrict its sale remained frustratingly out of reach, and it must have seemed that only a shocking murder that created national outrage would galvanise parliament into action. When the catalyst finally came, it emerged not from the crowded cities, but the sleepy rural villages of north-east Essex.

Twenty-eight-year-old Mary May, described as 'a repulsive-looking woman',[8] kept a small grocer's shop in the village of Wix near Harwich, where she lived with her husband, two children, and two lodgers, one of whom was her half-brother, forty-five-year-old William. A pedlar and odd-job man, he went by the nickname Spratty Watts. On Thursday 8 June 1848, Watts came home from his work as usual and Mary warmed up porter in a saucepan, to which she was seen to add some powder from a screw of paper she took from a drawer. Soon after drinking the porter, Watts was taken very ill with vomiting and diarrhoea.

From the very start of her half-brother's illness, Mary told her friends and neighbours that she expected him to die, and that, on his death, she would receive some £9 or £10. She had already planned what to do with the money. Mary had insured Watt's life with a burial club in Harwich, giving his name as William Constable, and stating his age as thirty-eight, when he was actually at an age that would have disqualified him from membership. There were two burial clubs nearer to her home, but she most probably used the more distant one to prevent exposure of the fraud.

On the Friday morning, with Watts still very unwell, Mary's husband Robert called on local surgeon Mr Thompson, whose assistant supplied a laxative dose of rhubarb and a sedative draught. Thompson came to see his patient on Saturday afternoon, when Watts said he was much better. Thompson told Mary to send to him next morning for more medicine and that if Watts became any worse, he would come again to see him. Watts did become worse, but Thompson heard no more until the Monday when he learned that Watts had died the previous day.

Mary arranged for a speedy funeral, and got a neighbour to write to the burial society to inform them of the death and ask them to send the money. Her deception was soon discovered, however, and the insurance was never paid. There was already a stir of rumours about the village that Watts had been poisoned, and Mary made a number of strange and often conflicting statements to explain his death. Initially she denied that she had ever had arsenic, but said that Watts was in the habit of using it about sheep, and so, if he had been poisoned, then he had taken it himself. She claimed that he had once tried to hang himself and she had prevented him. Mary then alleged that Watts had been ill ever since he had drunk something from a bottle he had found in a field, and asked a number of people in Wix to support this unlikely story. Later, she admitted that she had bought some arsenic from a chemist in Manningtree, and put it on bread and butter to poison a particularly persistent rat that used to pop up from a hole in a room where her son would play with it. Finally, she changed her story again and said it was not arsenic she had bought.

The funeral went ahead, but Watts's body was exhumed on 30 June, and the stomach removed and submitted to Professor Alfred Swaine Taylor, who found eight grains of arsenic. The following July, Mary May was tried for murder at the Chelmsford Assizes. There was little that could be said in her defence. The chemist of Manningtree could not remember having sold her arsenic, but said that his rat poison was composed of arsenic and, had she come to him for some, he would have sold it to her. Her counsel made an earnest appeal for his client to be given the benefit of the doubt, but the jury had no doubts and she was convicted and sentenced to death. She was hanged at Chelmsford on 14 August.

Commonplace as her crime might have appeared, Mary May's twitching body became a symbol of all that was most loathed and feared about the secret poisoner. It was rumoured that she was responsible for many other similar murders, including the death of a former husband and no fewer than fourteen children, and that she had advised other women to follow her example. The result, it was believed, was an outbreak of poison murder concentrated in one small part of Essex. A torrent of information poured into police stations, alleging that atrocities similar to those of Mary May had been committed in the surrounding villages of Bradfield, Ramsey, Dovercourt and Tendring. In every case, it was said, a husband had been poisoned by his wife.

On 23 August, the body of Thomas Ham, a blacksmith of Tendring, was exhumed. He had died aged twenty-eight, on 27 April 1847, after two months of persistent ill health, during which he had suffered burning pains in his throat and vomited yellow and green matter streaked with blood.

The principal witness at the inquest that followed was Ham's servant Phoebe Read, who reported in considerable detail the quarrels she had overheard between her late master and his wife. Hannah Ham often spent up to a fortnight at a time away from the marital home, in the close company of farm worker John Southgate, and made no secret of the fact that she wanted him, saying that she liked his little finger better than her husband's whole body. 'Well, poor fellow, he's gone,' Hannah was alleged to have said just hours after her husband's death, 'and I'm glad of it, for we never lived happy together. I never liked him, and I wish he'd died before.'[9] Hannah and John were married soon afterwards.

There had been arsenic in abundance in the blacksmith's house, sprinkled on pieces of bread and butter that were distributed to kill rats, and, more to the point, there had been visits from Hannah's friend Mary May. The two women were said to have discussed Hannah's quarrels with Thomas, and Mary had said, 'If he was my husband, I would give him a pill.'[10]

Thomas had taken more than a pill, for Professor Taylor found fifteen grains of arsenic in the corpse's stomach. Hannah was arrested and there were shocked exclamations at the inquest when several witnesses testified that she had threatened to kill her husband. More shocking still were the rumours that application would be made to the Secretary of State for the Home Office for the exhumation of more bodies, including those of Mary May's former husband and children. The inquest jury returned a verdict of murder against Hannah Southgate and she was committed for trial at the Chelmsford Assizes.

The issue of arsenic poison did not, however, remain confined to rural Essex, but attracted nationwide concern, amounting to a phenomenon sometimes termed 'moral panic', in which there is widespread acceptance of the existence of a threat to society out of all proportion to its actual occurrence. Fear of an increase in crime, especially violent crime, often lies at the heart of such panics.

On 1 September 1848, *The Times* devoted a leader to the inexplicable phenomenon of 'epidemic crime' in Essex. 'Why should the practice of secret poisoning obtain rather in the marshes of Essex than elsewhere? In other counties as well as Essex, wives are weary of their husbands and would gladly follow them to the grave ... How is it then in Essex alone that inclination should become an act, and result in murder?'[11]

How much of the Essex poison furore was truth and how much was fear and gossip is highly debatable,[12] but, if nothing else, it had drawn attention once more to the need for regulation of the sale of poisons, especially arsenic. What must have been seen as a flood of uncontrollable poisonings stimulated debate in the national newspapers, escalating to a public outcry at the continued unrestricted sale of arsenic; although, as late as October 1848, *The Times* observed, 'It seems that considerable practical difficulty exists in forbidding the sale of arsenic, as it enters so largely into processes of trade.'[13]

At Hannah Southgate's trial in March 1849, the defence alleged that the witness Phoebe Read was unreliable, and urged the jury to ignore her testimony, suggesting that the deceased had taken poison by accident. Hannah was acquitted and she and 'her Johnny',[14] little finger and all, were reunited and went on to raise a substantial brood of children.

The Essex rumour-mongers quietened down; although, in December 1849, the *Chelmsford Chronicle* reported that most of the burial clubs in the county had been broken up, after the proceedings in the Tendring Hundred showed what crimes the clubs might instigate, and all the funds paid into them were lost to the contributors.[15]

The casual sale of arsenic continued, and it was still left to the individual discretion of chemists as to whether they asked the buyer to provide a witness or a signature. A Mr Sturton wrote to *The Times* in 1849 saying that he kept a register of the sale of arsenic, which he believed 'must act as a most salutary (perhaps efficient) check on its felonious administration';[16] he recommended that the law should make such a register imperative, a monthly copy being sent to the coroner of the district. No one, he suggested, should be allowed to sign by marking with an X, a stipulation that would inevitably exclude many of the poorly educated labouring classes. The seller should know the buyer and only chemists should be allowed to sell arsenic. The law did not respond.

In the same year, *Punch* satirised the situation with a cartoon entitled *Fatal Facility; or Poisons for the Asking*,[17] showing a child with a pinched bleak face asking an obliging chemist for a refill of her mother's laudanum bottle, and another pound and a half of arsenic for the rats. The cartoon not only showed the ease with which poison was sold even to children, but also reinforced the idea of poison murder as an issue relating to women of the poorer classes. During the intense debates that followed, it was impossible to separate the problems associated with unrestricted purchase of poison from issues of gender and class.

Campaigners for the regulation of the sale of arsenic were assisted by statistics quoted by Alfred Swaine Taylor in 1848,[18] giving the number of deaths attributable to arsenic in coroners' inquisitions in 1837 and 1838. Out of 541 poisoning cases, in fourteen of which the poison was unknown, opiates accounted for 196 deaths, most of which were held to be either suicide or accident, and arsenic 185, the greater number, according to Taylor, being criminal poisoning. Taylor also referred to the Registrar General's report that recorded 349 cases of death by poisoning, of which seventy-five were due to opium, thirty-two to arsenic, and the remaining 242, to his great regret, were not analysed. Of the opium deaths, forty-two occurred in children under the age of five.

A leading factor in the call for legal reform was the growth of professional associations. The Pharmaceutical Society (founded in 1841) and the Provincial Medical and Surgical Association (founded in 1832; it became the British Medical Association in 1856) were both seeking better regulation of their professions, and a recognised system of education and registration, in order to protect the public from unqualified practitioners. Hand in hand with the call for better regulation came demands for the sale of poisons to be placed under more stringent professional control.

Another trial that provoked a demand for reform was that of twenty-eight-year-old Sarah Freeman in 1845. That January, when her brother Charles Dimond died from poison in Shapwick, near Bridgwater in Somerset, it was recalled that Sarah's mother, her husband Henry and her illegitimate son, seven-year-old James, had all died in similar circumstances, their deaths having been attributed to natural causes. Sarah had bought arsenic at a chemist's shop, however. The assistant had initially refused to sell it to her as it was a dangerous poison, but it was only necessary for Sarah to lie that she was the sister of a local carrier well known to him, and he sold her half an ounce. An order was made for the exhumation of Sarah's mother, husband and son, and arsenic was found in all the bodies.

Sarah Freeman was found guilty of the murder of her brother, and hanged on 23 April 1845. Somerset doctor Jonathan Toogood was deeply shocked at the careless manner in which suspicious deaths were investigated, and wrote to *The Times*. With the case still weighing heavily on his mind, he took up the cause of the restriction of sale of poisons with considerable energy. Toogood successfully urged the town council of Bridgwater to present a petition to parliament, and personally wrote to Prime Minister Sir Robert Peel, Home Secretary Sir James Graham, and Graham's successor Sir George Grey. All his efforts were unsuccessful and, in 1849, he decided to harness the influence of the Provincial Medical and Surgical Association, and wrote to its president, Charles Hastings. His letter, entitled 'On the Crime of Secret Poisoning', was published in the association's journal. Toogood claimed that secret poisoning had become so common 'that scarcely a week passes in which some instance is not brought before the public'.[19] Confining his efforts to arsenic, which he said was 'seldom required for legitimate uses' and could be bought 'under the most frivolous pretences', he suggested that a petition from the PMSA 'would rouse the public mind, and command that attention which an humble individual can hardly expect'.[20] His object was 'to check the indiscriminate sale of poisons to "illiterate persons"'.[21] The issue was discussed at the subsequent AGM,[22] and the association formed a committee to gather facts, and present a petition to parliament, its 'philanthropic object the prevention of the indiscriminate sale of arsenic, and the crime of secret murder ... and thus enable our association to claim the honour of having directed the public attention to the greatest blot upon the civilisation of the nineteenth century'.[23]

It had not escaped the association that, while undertaking the commendable task of fighting an evil for the public good, there was much to be gained professionally from the campaign, both in enhanced status and by eliminating unqualified rivals. In November, the PMSA's arsenic committee joined forces with a similar committee formed by the Pharmaceutical Society. In their united eagerness, their report may well have over-egged the poisoned pudding, making arsenic murder appear far more common than it actually was. Parliament was presented with three recommendations for reform: that arsenic could only be sold by chemists, druggists and apothecaries (while drawing attention to the fact that this would only be effective if accompanied by a provision for regulating the appropriate qualifications); that a record of sales be kept; and – probably the only petition to suggest this – that only adult males should be eligible to make purchases.

Parliament was now taking the issue seriously enough that, by May 1850, a government bill was in preparation.

Meanwhile, that doyenne of Essex poisoning, the seemingly unstoppable and untouchable Sarah Chesham, whose exploits may well have inspired Mary May, had not been idle. She had realised, especially after Mary's execution, that a large dose of poison leading to sudden death might excite suspicion, and decided to try a new approach.

Ever since the harvest of 1849, her trusting husband Richard had complained of stomach trouble, and, in December that year, he was taken very ill. On 11 February, Surgeon Hawkes was asked to attend the sick man and, despite everything that had gone before, declared himself unable to form an opinion of what was wrong. Hawkes's behaviour during the next few weeks is mystifying to the point of being downright suspicious. It was as if Sarah's trials for the murder by arsenic of three children had never occurred. He later claimed that he had advised Sarah to call in another man for a second opinion as her husband's disease was obscure, but that she had declared herself satisfied with his services. Richard Chesham died the following May, and, despite the fact that he had been suffering from sickness and diarrhoea, Hawkes, who officiated at the post-mortem, said that the patient had died from a disease of the lungs. Another doctor who was present, however, saw the obvious signs of an irritant poison, and the stomach and its contents were sent to Alfred Swaine Taylor for analysis.

Sarah seemed content to have her cottage searched again for evidence of poison, until it was proposed to remove a bag of rice she had used to make milk puddings for her husband, though she repeatedly begged that it should not be taken away. The rice was sent to Professor Taylor, and there, nestling amongst the grains, were crystals of white arsenic. At the inquest, however, Taylor was unable to say for certain that Richard Chesham had died from the effects of arsenic. There had been traces of consumption and it was possible that repeated small doses of arsenic had weakened him and hastened death from that cause. The coroner's jurors, who had been expected to make a decision as to whether or not Sarah had killed her husband, were in a quandary until the coroner made a suggestion. He asked if they would be willing to proceed against her on a charge of attempted murder. They agreed.

When enquiries were launched, an important new witness came forward. Hannah Phillips was the wife of an agricultural labourer who worked for Thomas Newport, the landowner who had been implicated in Sarah's poisoning of his illegitimate son Solomon Taylor. Hannah said she had previously been intimidated into keeping quiet but was now prepared to tell all she knew. Sarah had told her that Newport had given her poison to kill both Lydia Taylor and her child. She had hidden the poison under a 'stub' (a tree root), and recovered it on her release. On learning that Hannah was ill treated by her husband, Sarah had said she knew how to make a poisoned pie, it being no sin to bury such a husband.

Sarah Chesham was arrested and, in September 1850, was committed to be tried for attempted murder at the Chelmsford Spring Assizes. She blamed Hawkes for Richard's death; however, her own family members were able to confirm that it was she and only she who had made the rice pudding and fed it to her sick husband.

'She seems to have acted as a regular poisoner,' said Professor Alfred Swaine Taylor, some years later. 'The coroners of Essex told me that they were almost afraid to look back to the extent to which, it was suspected, she had been guilty . . . She was able to use [arsenic] in any way that she pleased; and there was no possibility . . . of tracing when or where, or how she got it.'[24]

The Times, recalling her earlier acquittals, declared: 'The woman has thus led a notorious and almost public career for upwards of four years,' and she 'was regarded as a professed murderess in her own neighbourhood . . . mothers locked their children up when she was seen about the premises'.[25] The case was 'nothing less than wholesale indiscriminate and almost gratuitous assassination'. Citing a spate of poisoning cases in other counties, which 'possessed certain common features of atrocity and terror', *The Times* observed that the scene was always a remote village where the inhabitants had 'become perfectly familiarised with the idea of murder by poison. Certain parties were currently understood to be "poisoners" without incurring thereby any greater disrepute than if they had been poachers or smugglers.'[26] Arsenic was 'a weapon in the hands of the weaker vessel' and one aspect of Sarah's career was

> too remarkable to be passed over, for we doubt whether we will find a parallel in the truths or fictions of even French life. On her first trial, a medical witness detailed at some length the deleterious properties of arsenic, and its effects when administered under given conditions and circumstances. The woman, then in peril of her life, stood quietly at the bar, listened, and learnt. No sooner was she discharged than she availed herself of the lesson.

She had previously poisoned in a 'coarse and unscientific fashion', but her husband had been poisoned in small doses, and expired after six months of slow torture, so very little arsenic remained to be found.

The 'surest means of precaution' was the druggist's shop, and, fortunately, 'the dabblers in death confine themselves to the employment of arsenic almost exclusively'.[27] The editor could only look to parliament for 'any measure which shall restrict or qualify the sale of arsenic'.[28]

November's issue of *Household Words*, the weekly magazine edited by Charles Dickens, in an article titled 'POISON SOLD HERE!!', declared that poisoning, once a science, 'may be safely practised by the meanest capacity. The exciting extent to which murder has been recently done by poison fills a column of every newspaper, and furnishes a topic for every conversation.'[29] The article suggests that the sale be confined to 'those qualified by education to exercise wholesome care and to use a sound judgement in dispensing it'.

The Chesham case was the final straw, provoking sufficient outrage for the legislature to finally bow to public opinion.

On 17 February 1851, it was announced that parliament would introduce a measure to restrict the sale of arsenic. The bill, as instigated by Lord Carlisle, began with the words, 'Whereas the unrestricted sale of arsenic facilitates the commission of crime . . .'[30] and provided only that the sale be registered.

At the second reading of the bill, on 13 March, Carlisle said that while he could have urged the necessity of the measure by prefacing his speech 'with some striking details . . . with respect to the crime of poisoning . . . he felt there was a degree of mysterious horror attached to the use of poison, which seemed to attract and fascinate a certain class of minds more than any other crime', and therefore said only what was strictly necessary. He admitted that there were deficiencies in the bill. It was limited only to the sale of arsenic; however, arsenic 'afforded great facilities for the commission of the crime of poisoning' and, 'among those classes where the crime was rife, "arsenic" and "poison" were looked on as synonymous. We could not debar the use of poison from those whose knowledge, opportunity or instruction might indicate to them other poisons for their purpose, besides arsenic, but we could debar the ignorant from the use of arsenic.' Similarly, he did not wish to draw up a schedule of poisons, as that would be advertising what substances were 'destructive to life', neither did he wish to specify a minimum amount to be sold, in case people requiring it for purposes such as sheep dip bought more than they needed and left the residue lying around. Carlisle realised that the bill was only a part-measure, and thought that, however legislation tried to deal with the evil, 'it could only be successfully combatted by teaching our people the true spirit of Christianity'.[31] Carlisle received numerous communications from the PMSA and the Pharmaceutical Society suggesting improvements to make the bill more effective, and took their advice on board. At the third reading, on 24 March, there were additional proposals, including the mixing of arsenic with colouring matter to avoid mistakes, and also 'that arsenic should be sold to none but male adults, as several deplorable accidents had occurred from young children and female servants having been sent to purchase it'.[32] This last measure was later dropped, 'owing to the indignant remonstrances of ladies'.[33] Had it not been, the anomalous result would have been that female grocers would have been allowed to sell arsenic but not purchase it.

On 25 March, the day after the third reading of the Sale of Arsenic Bill, Sarah Chesham, having been convicted at the Essex Spring Assizes, went to the gallows at Springfield Gaol, Chelmsford. She was the last woman to be hanged for attempted murder.

The Sale of Arsenic Regulation Act, which came into force when it received royal assent on 5 June 1851, placed no restrictions on who could sell arsenic, but stipulated that, in the case of each sale, the seller should keep a record of the quantity sold, the purpose for which it was required, the date of sale, the occupation of the purchaser, and a signature, unless the purchaser was unable to write. Arsenic should not be sold to a person who was unknown to the seller, unless in the presence of a witness who was known to both the purchaser and the seller, that witness being obliged to provide both name and address. Arsenic was not to be sold to any person not of full age, i.e. twenty-one. The arsenic sold was to be mixed with soot or indigo in the proportion of at least one ounce of soot or half an ounce of indigo to a pound of arsenic, unless it was required for a purpose where the colouring would make it unfit for use. If sold without such a mixture, it was to be sold in a quantity of not less than ten pounds at a time. The penalty for failing to comply was a fine not exceeding £20. The act did not apply to wholesalers or where arsenic was part of a medical prescription.

It was progress of a sort, but very little, since there were still gaping holes in the legislation. The act did not provide that arsenic could only be supplied by qualified chemists and druggists – it could not, as they still had no legal definition. The PMSA, however, welcomed 'an enactment in the right direction and, whatever may be the defects of the law, it cannot fail to produce good results'.[34] In 1852, the Pharmacy Act established a register of pharmaceutical chemists, who had taken the society's examinations, and only those registered with the society could use that title, though it still did not restrict the practice of pharmacy to qualified people.

It took some time for the new provisions to be implemented and, in the meantime, white arsenic was frequently sold in the same way, while old supplies still lurked in homes. It is impossible to guess to what extent the new law was complied with, and what, if any, effect it had on the purchase of arsenic and its use for murder.

In December 1851, *Household Words*, almost certainly responding to the Chesham case, declared that 'the crime of murder by means of poison – and more particularly of slow poison, or poison administered in very small doses from time to time – admits more readily of a fiendish sophistication in the mind of the perpetrator than any other form by which murder is committed'.[35] The crime loomed so large in the writer's fertile imagination that he wrote incorrectly of the 'great numerical preponderance of murders by means of poison over every other means of destruction',[36] saying that the 'dreadful frequency' of recent cases 'ought to awaken the public to a demand for the absolute enforcement of legislative regulations for the sale of all such drugs and deadly ingredients'.[37]

Any examination of trial statistics shows that the great majority of murders and attempted murders that came to court involve physical violence,[38] but *Household Words* must have reflected public opinion. The writer believed that the provisions of the new act were being largely ignored, and cited cases in which poison had been purchased on behalf of the murderer by innocent third parties or had been carelessly sold with no real understanding of the new law.

Seventy-nine-year-old William Rowlinson,[39] of Great Thurlow in Suffolk, who was charged in October 1851 with the attempted murder of his daughter-in-law Mary and the murder of her sister Ann Cornell, had been such a regular purchaser of arsenic that he had no difficulty in obtaining what he wanted even after the act became law. Ann had been looking after Mary, who had been ill for some time, and had died after eating some broth Mary had made for herself. The death was initially attributed to cholera, but when the family, as well as a cat and dog, all became ill after eating dumplings, suspicions were aroused, and tests showed that the dumplings, and the flour they were made from, contained white arsenic. The accused had last been sold white arsenic in August 1851, and the only precaution that had been taken by the druggist was to write the word 'POISON' on the paper. Rowlinson was found guilty, but the death sentence was commuted to transportation to avoid the distasteful spectacle of the hanging of a man aged nearly eighty.

More recent studies confirm that, in the nineteenth century, the principal danger of poison came from accidental exposure, and not homicide,[40] but the clandestine poisoner who had gripped the public imagination, still usually thought of as female, became and remained an iconic figure, akin to the cloaked and top-hatted Jack the Ripper. Sensational trials thrilled the public with horror and took up newspaper space wholly out of proportion to their actual frequency, leading to the belief that an epidemic of secret poisoning was sweeping the land.

The 1851 act did not put an end to arsenic murder, but it was the first demonstration that the sale of poisons could be governed by legislation, and it opened the door to further reforms and better regulation of both the medical and the pharmaceutical professions.

CHAPTER ELEVEN

Tobacco Kills

By the time Alfred Swaine Taylor published *On Poisons in Relation to Medical Jurisprudence and Medicine* in 1848, organs such as the stomach or spleen could be made to give up mineral poisons locked in their tissues by a variety of procedures, the main objective of which was to destroy the organic material with acid, and work was continuing to further refine the process. The really obstinate problem faced by toxicologists was extracting poisons of vegetable origin.

'In the present state of our knowledge,' wrote Taylor, '[t]he greater number of vegetable poisons are beyond the reach of chemical analysis.'[1]

If a vegetable poison had been swallowed in the form of leaves, roots or seeds, then botanical residue would give a toxicologist important clues as to its presence, but when the poison had been taken as a juice or infusion, chemical analysis was useless.

The same was true of the powerful alkaloids that chemists were learning to extract from plant material. In their pure form, such as morphine or strychnine, the addition of reagents would produce characteristic clouding, precipitates or colour changes, which was all very well on the laboratory bench, but not when the chemist was handed the organs of a possible poison victim. Any procedure that destroyed the organic material would also destroy the poison. Opium tended to be so quickly absorbed that it was rarely detectable in the stomach of the deceased; even tests to identify it in stomach contents that had been vomited or removed by pump were 'open to objections'.[2]

The only possible test in cases where the vehicle in which the poison had been given was still available was to feed it to an animal, and observe the results. In 1838, Mary Sherrington was tried at the Lancashire Assizes for the attempted murder of Mary Byres. Sherrington had asked two children to take a poisoned pudding to her intended victim, but, on the way, the children tasted the pudding and found it unpleasant. It was sent to a surgeon to be analysed. He was unable to detect any poison, but, suspecting a vegetable narcotic, he gave a piece of the pudding the size of an egg to a dog. In twenty minutes, the dog was sick, and, in another twenty minutes, it had lost the use of its legs. Three hours after eating the pudding, the dog was dead. Mary Sherrington was convicted of attempted murder and transported for life.

* * *

As the 1840s drew to a close, Alfred Swaine Taylor and Mathieu Orfila, the two giants of forensic toxicology, had still failed to solve the problem posed by vegetable poisons, and no one knew this better than Count Hippolyte Visart de Bocarmé.

De Bocarmé, born in 1818, was descended from Belgian nobility. His father Julien was vice-governor of Java, his mother Ida, the granddaughter of the Marquis de Chasteler. In 1836, Julien decided to settle in Arkansas, where he purchased farmland and remained until his death in 1851.[3] De Bocarmé lived with his father for a time, but later returned to Belgium, where he occupied the family seat, the Château de Bitremont, in the municipality of Bury, near Mons.

De Bocarmé later claimed that he had been neglected by his father and allowed to do as he pleased, which, even if true, is hardly an excuse for his subsequent behaviour. Nevertheless, an undisciplined young man, conscious of his noble status but finding himself in need of money, will often look for easy methods of making the fortune he feels he deserves. De Bocarmé's annual income of 2,400 francs[4] was too small to maintain the luxurious style he craved. He decided to marry for money, and sought out a wealthy *bourgeoise* who was looking for a title. In 1843, he married twenty-five-year-old Lydie Fougnies, the daughter of a retired grocer, but, after the wedding, the hopeful bridegroom discovered that he had not acquired the anticipated fortune; Lydie was entitled only to an annual income of 2,000 francs paid by her father. The couple were incapable of living within their means; rather they were determined to enjoy a recklessly extravagant lifestyle, and, in addition, de Bocarmé kept a mistress. It was a turbulent marriage in which quarrels that erupted into violence alternated with assionate reconciliations. To support their profligacy, they borrowed 40,000 francs from a notary, which they were unable to repay.

The death of Lydie's father in 1845 increased her annual income by another 5,000 francs, but the couple's expenses only increased to match, and they were obliged to sell the properties that produced her income. By 1849, they were supporting three children and employing several servants. They had sold as much land as they could and, still heavily in debt, were reduced to pawning jewellery. There was only one hope. Lydie's unmarried brother Gustave, some four or five years her junior, had inherited the greater part of their father's fortune, and, if he died, it would pass to his sister. Better still, from de Bocarmé's point of view, Gustave was in poor health. He had been very ill since having a leg amputated and could walk only with the aid of crutches. His early death would not be a surprise. De Bocarmé, hoping that Gustave would die a natural death if he waited long enough, consulted a physician about that very possibility. In November 1850, however, Gustave announced his intention of becoming engaged to a Mlle de Dudzeele. If the marriage took place, all de Bocarmé's hopes of inheritance were gone. The count and countess did their best to dissuade Gustave from the marriage, and the countess even wrote two letters to her brother vilifying his intended bride's character, but to no avail. Gustave had to die.

The plan, which unfolded on 20 November, was that Gustave should appear to have died from natural causes. The execution of this scheme must count as one of the most spectacular failures in criminal history. Gustave arrived at the chateau at 10 a.m. and breakfasted with his sister. As preparations were made for dinner, which was to start at 3 p.m., the scene was being set for murder. The governess took charge of the children, and the countess ordered her maid, Emerance, to retire after she had served the second course, attending to the rest of the service herself. The meal was over by half past four, but the diners remained by the fireside talking. As dusk fell, Emerance, as was usual, went into the room to ask if lights were needed, and both the count and countess quickly said that they were not. The dining room was, therefore, occupied by only three people: the murderers and their victim. The excuse later given for the unusual arrangements was that they were discussing business. Gustave decided that it was time for him to leave, and de Bocarmé went to tell his coachman, Gilles, to prepare a carriage; he then returned to the dining room.

Soon afterwards, Justine, the nursemaid was near the dining room door when she heard Gustave cry out in a muffled voice. The other servants had heard it too, from further away, but assumed that it was from the pain of having his leg dressed. Justine, terrified by what she had heard, ran to the others and insisted that it was the voice of someone being suffocated. There was later some dispute as to whether she had heard, 'Aie! Aie! Hippolyte! Pardon!'[5] or actual cries for help. The servants were by now all too frightened to do anything other than wait for orders. After a few minutes, the countess emerged from the dining room, shutting the door behind her, went into the kitchen and washed her hands well with black soap, a heavy-duty washing type.

When Gilles returned to say that the carriage was ready, he found the house in uproar, his employers wailing that there had been a terrible misfortune. Loud as their cries were, however, neither of them shed a tear. Gilles entered the dining room and saw Gustave lying on the floor, dead, the count, breathless, pale and trembling, his hair dishevelled, sponging the face of the corpse with a cloth soaked in vinegar. Emerance, at last, dared to enter, and was appalled to see the count trying to force the cloth into Gustave's mouth. She made him stop, then started to rub Gustave's temples with cologne, but, on bringing a lamp to see better, she cried out in horror. Gustave's tongue was black as coal and his lips were burnt. The servants were told that Gustave had suffered a fit of apoplexy, and the countess ordered Gilles to carry the body upstairs, undress it, wash it with hot water and vinegar, and pour vinegar into the mouth. Gilles laid the body on the bed in Emerance's room and did as he was told, apart from pouring vinegar into the mouth, which he thought a useless thing to do to a dead man. Gustave's mouth, he noticed, was blackened as if with a corrosive, and there was blood on his cheeks.

The countess, who was far calmer than her husband, ordered the cook to bring hot water and scrub the floor of the dining room. As the girl did so, she noticed that the pervading smell in the room was that of vinegar. The countess also gave orders that 'those two *coquines* [hussies], Mme Dudzeele and her daughter, should be informed that her brother had died suddenly of apoplexy'.[6]

Gilles was next ordered to clean the floor with a brush, and, once it had been scoured, de Bocarmé scraped the boards with a knife. The countess told Gilles to burn Gustave's cravat and waistcoat. The rest of the deceased's clothes, together with de Bocarmé's jacket, were taken to the laundry, soaked in cold water and then washed in boiling soapsuds. The countess then ordered Gilles to scrub Gustave's crutches with hot water, but later decided that they too should be burned.

A doctor was summoned, and conducted to the bedroom where the body lay in darkness. Having no reason to doubt the de Bocarmés' version of events, he pronounced Gustave dead, then departed.

The count and countess spent the next few hours burning papers in the fireplace. Finally, exhausted but satisfied that they had got away with murder, they went to bed, although de Bocarmé was so shaken that he spent much of the night vomiting. The following day, the terrified servants met to discuss what should be done, and, that evening, Emerance, Gilles and two nursemaids called on the parish priest of Bury to describe what had occurred. News of the tragic death soon reached M Heughebaert, the examining magistrate of Tournai. Doubting such a sensational story, he was nevertheless obliged to investigate, and, on 22 November, he arrived in Bury and proceeded to the chateau, accompanied by the town clerk and three doctors. There he found the countess eating her breakfast. The dining room fireplace was choked with the ashes of burnt papers, the scraped floor was still littered with wood shavings, and the count was curiously unwilling to appear.

The magistrate was starting to entertain suspicions, and, when the countess informed him that Gustave had died from a congestion of the brain, he insisted on seeing the body. Eventually, and with great reluctance, the count arrived and conducted Heughebaert to where the body lay in the darkened bedroom. When the magistrate pulled the bed near to the window to get a better look, he was appalled at Gustave's appearance – the scratched and bruised cheeks, and the swollen blackened lips and tongue. He ordered that the body should be carried to a coach house for examination. Two hours later, the doctors reported that Gustave had not died from natural causes, but from the effects of drinking a corrosive liquid, which they suggested might be sulphuric acid. Gustave's tongue, gullet, stomach, intestines, liver and lungs were placed in jars containing pure alcohol to preserve them, and removed for further examination. When Heughebaert started to make detailed enquiries, de Bocarmé approached him with tears in his eyes and confided that Gustave had poisoned himself and that he had had to prise the phial from his brother-in-law's hands. The magistrate promptly rejected this statement as a lie and an absurdity.

The count and countess were arrested. When the count was examined, it was found that one of his fingers had been badly bitten and there were traces of what looked like blood under his fingernails.

Had the death of Gustave Fougnies occurred in France, the remains might have been sent to the Faculty of Medicine in Paris, but, instead, Heughebaert took the specimens to Brussels. The man about to make medical history was thirty-seven-year-old Jean Servais Stas, professor of chemistry at the École Royale Militaire. Stas had studied in his native

1 An execution outside Debtors' Door, Newgate Prison.

2 Interior of the condemned cell, Newgate Prison.

CONDEMNED CELL IN NEWGATE.

3 Eliza Fenning in the condemned cell, Newgate Prison.

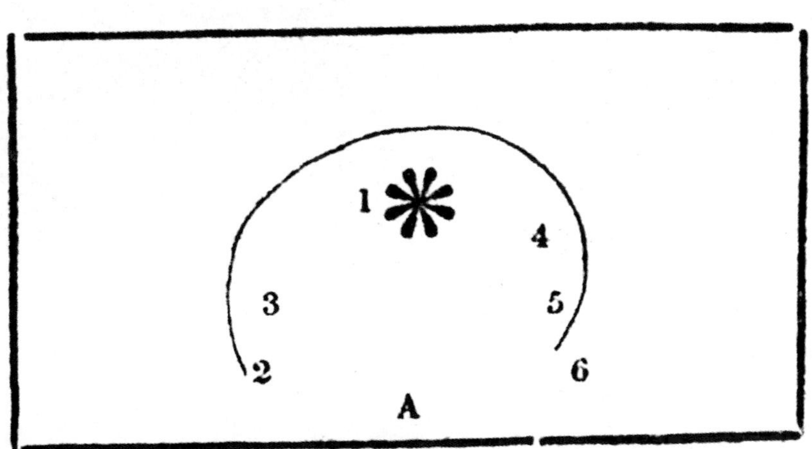

* The Tea Table. A. The Fire Place.
1 Mrs. Donall, at the Table.
2 The Prisoner.
3 Mr. Samuel Downing.
4 Mrs. Jordan.
5 Mr. Edward Downing.
6 The Deceased.

4 Contemporary sketch of the Donnall parlour showing the location of those present and Robert Sawle Donnall's unusual route around the table.

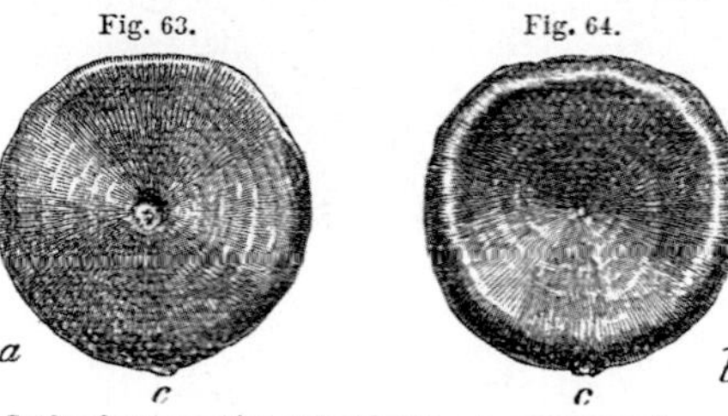

5 The nux vomica bean as illustrated in Taylor's *Principles and Practice of Medical Jurisprudence*. He described it as being the size of a shilling.

6 Mathieu Joseph Bonaventure Orfila in his robes as Dean of the Faculty of Medicine, Paris.

7 Edme Samuel Castaing and his victims, Auguste and Hippolyte Ballet.

8 William Herapath.

9 Marie-Fortunée Lafarge.

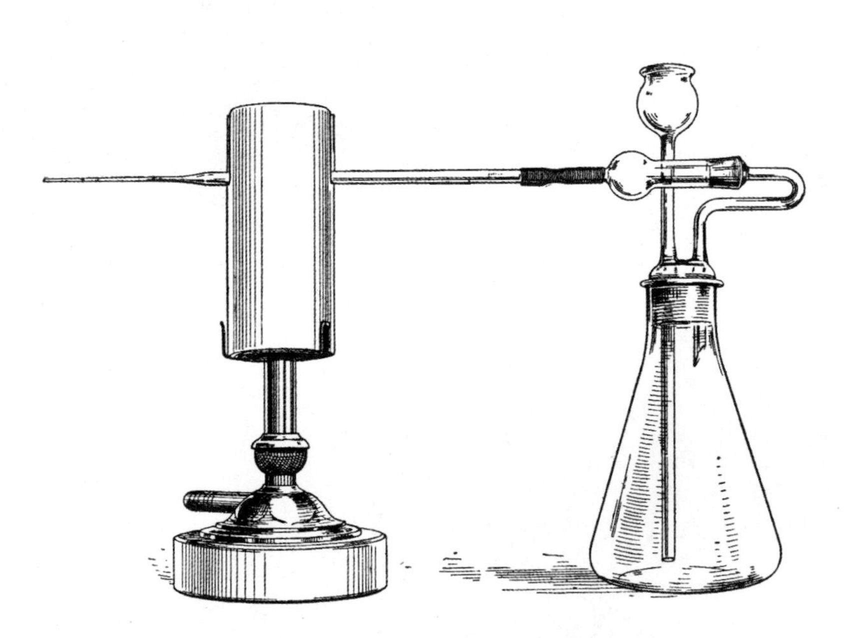

10 James Marsh's apparatus for the separation of arsenic.

11 François-Vincent Raspail.

12 James Cockburn Belaney.

13 Henry Letheby.

14 *Punch* cartoon, 8 September 1849.

15 Château de Bitremont, family seat of the de Bocarmé family.

16 Hippolyte Visart de Bocarmé.

17 Jean Servais Stas.

18 Sir Robert Christison.

19 Rugeley High Street, showing the Talbot Arms (left) and Palmer's house, opposite.

20 Professor Alfred Swaine Taylor (left) and Dr George Owen Rees analysing the remains in the Palmer case.

21 Thomas Smethurst.

22 Aconite root compared with horseradish from Taylor's *Principles and Practice of Medical Jurisprudence*.

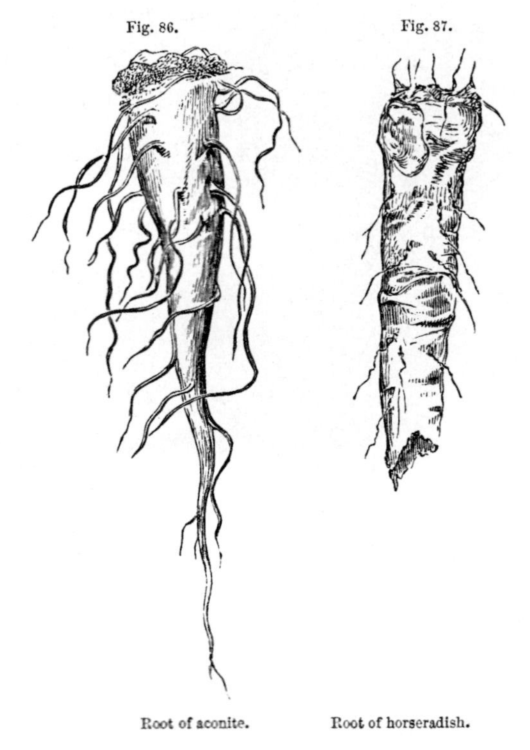

23 Edmond-Désiré Couty de la Pommerais.

24 Dr Pritchard and his family. The two seated ladies are Mrs Pritchard (centre) and Mrs Taylor (far right).

Glasgow March 18th 1865

Sir

Dr Pritchard's Mother in law died suddenly and unexpectedly about three weeks ago in his house Sauchiehall Street Glasgow under circumstances at least very suspicious. His wife died to-day also suddenly and unexpectedly and under circumstances equally suspicious. We think it right to draw your attention to the above as the proper person to take action in the matter and see justice done.

Yours &
Amor Justitiae

[illegible] Hart Esq

25 Anonymous letter drawing attention to the suspicious deaths of Dr Pritchard's wife and mother-in-law.

26 Auguste Ambroise Tardieu.

27 Eugène Chantrelle.

28 Christiana Edmunds.

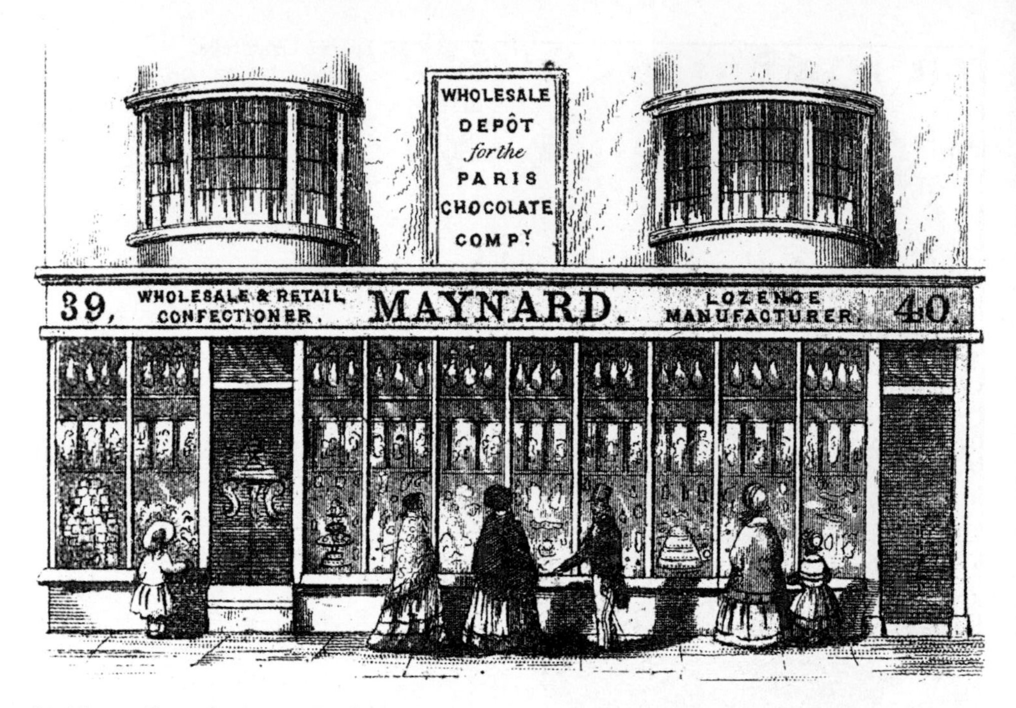

29 Maynard's confectioners, Brighton.

30 Dr Henry Lamson.

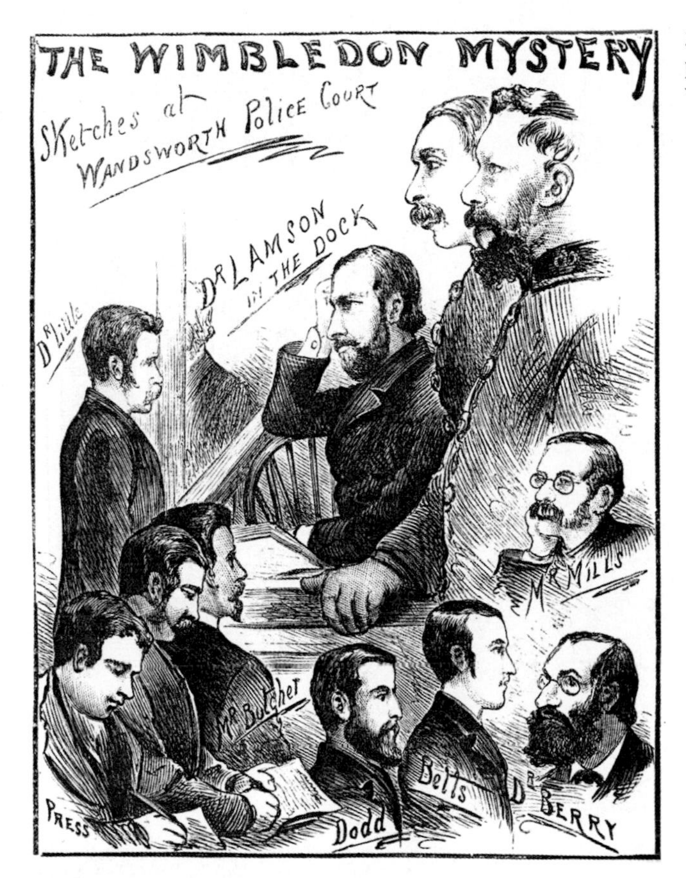

31 Dr Lamson in Wandsworth police court with police, witnesses and press.

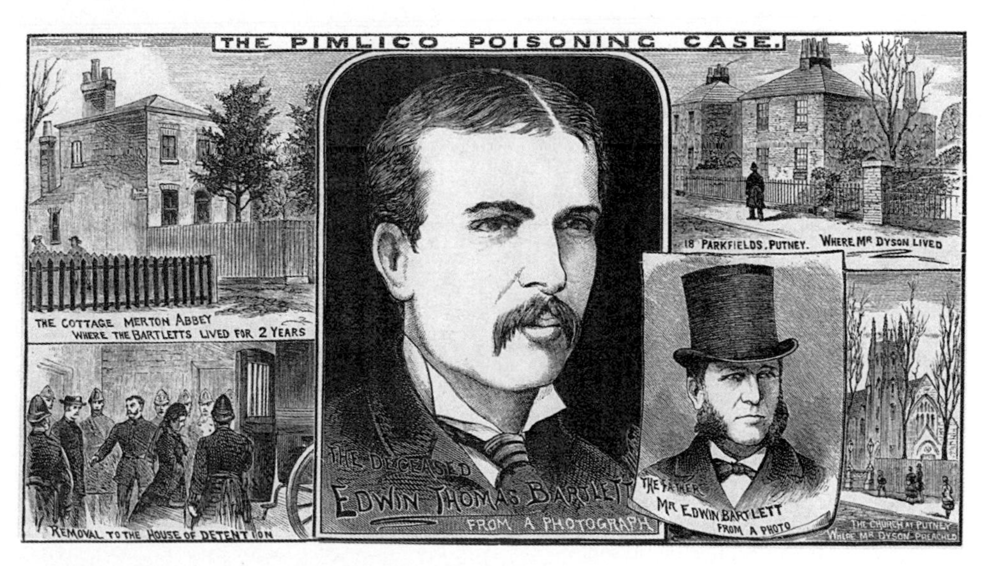

32 Scenes and personalities in the Pimlico poisoning case.

33 Carlyle Harris.

34 Advertisement for Rodine rat poison.

Louvain and also spent time in Paris, attending the lectures of Professor Orfila. He was known for his painstaking methodology and groundbreaking work on the determination of atomic weights. Stas had attracted the attention of the great chemist Liebig, who had wanted him to work in Germany, since the Belgian École was not well equipped, though Stas preferred to stay in Belgium using his own resources, which he could ill afford, to create his own laboratory in Ixelles.

Told that sulphuric acid was suspected, Stas quickly ruled it out, but at once noticed the pronounced smell of vinegar. When he learned of the repeated washings of the body, he wondered if this had been done to mask something else. He set about filtering and distilling the contents of the stomach, intestines and bladder, setting aside many samples for a series of experiments. The odour of vinegar continued to cling to what was now a dark syrup but, when Stas dissolved a portion in a powerful alkali, caustic potash, he noticed another smell, similar to mouse urine, which chemists associated with coniine, an alkaloid extracted from hemlock. This important clue suggested that he was dealing with a vegetable poison. Stas worked long hours, repeatedly filtering and refining his samples, before once again adding caustic potash. This time, there was a sharper smell, more closely resembling another alkaloid: nicotine.

Nicotine is specific to the genus *Nicotiana*, of which the most commonly cultivated variety is *N. tabacum*, the tobacco plant. It was first extracted in an impure form in 1809 and isolated in 1828. Nicotine is soluble in both water and alcohol and has a pungent taste and characteristic smell. It is also a powerful toxin; more so than arsenic or cyanide. In 1848, Taylor called the extract 'nicotina' and reported that deaths from tobacco poisoning were rare. Usually they occurred not by ingestion, but when tobacco was brought into contact with diseased or delicate membranes. In one case, a decoction of tobacco (an extract made by boiling the material) used as an enema to treat a strangulated hernia proved fatal. Its action can be very rapid. In another similar case, death had ensued in just eighteen minutes. Taylor also mentioned two instances in which a large quantity of tobacco consumed by smoking led to death.[7] There had been few cases affording examination of the body of someone poisoned by tobacco, but, about ten years before the death of Gustave Fougnies, Orfila had conducted experiments with dogs. A single drop of nicotine in the eye or five drops on the tongue resulted in rapid collapse, violent convulsions and death within minutes.

Stas dissolved some of his sample in ether, then allowed the ether to evaporate in a dish. What remained was a brown ring smelling of tobacco.

Stas touched his tongue to it, and the burning taste of tobacco filled his mouth. It was only after days of further extensive testing that Stas was able to confirm that he had extracted nicotine from the remains of Gustave Fougnies, and he wrote to Heughebaert asking if the count and countess had ever had nicotine. The chateau was searched again without result and, once again, the servants were questioned. This time, the gardener revealed something that showed that de Bocarmé had had murder on his mind for some time. In the summer and autumn of that year, the gardener had assisted his employer in the manufacture of what he was told was eau de cologne in a laboratory at the castle. The chief ingredient was tobacco leaves. The work had been completed on 10 November, two phials of the extract being placed in a cupboard in the dining room. The following day, all trace of the equipment used in the work had been removed. The coachman also revealed that, as long ago as February, de Bocarmé had gone to Ghent to see a professor of chemistry. Heughebaert went to Ghent and traced a Professor Loppens, who said that he had been visited by a gentleman from Bury with whom he had later engaged in a lengthy correspondence about the extraction of nicotine from tobacco leaves. The gentleman had given the name Bérant, saying he came from America where his relatives were being attacked by Indians [*sic*] who dipped their arrows in vegetable poisons, and he wanted to learn about these substances. In particular, he wanted to know if a vegetable poison left no trace in the poisoned person. Loppens said that was the case. When 'Bérant' next visited, he said he wanted to produce an extract of nicotine in order to study its effects, and the obliging Loppens told him how to do this, and recommended a coppersmith and apothecary where he might obtain everything he needed. The apparatus was delivered to Bury and, by October, de Bocarmé had been able to produce pure nicotine.

Stas, meanwhile, had discovered why his work had led to such an important result. It was the chance combination of the alcohol in which the remains had been preserved and the vinegar with which the body had been washed, which had effectively digested the material. Most of the matter comprising the human body is soluble in either water or alcohol, or neither – but never both. The vegetable poisons, on the other hand, are soluble in both. When the samples were filtered, the alcohol that ran off took with it those parts of the body matter that were soluble in alcohol, together with the alkaloid. The filtrate was then mixed with water and filtered again. This time, the organic material, which was not soluble in water, stayed behind, the water-soluble alkaloids passing through. The addition of an alkali and then a solvent separated the alkaloid from the solution. What Stas could not then have known was that his method was not specific to nicotine, but would, in time, revolutionise forensic toxicology by enabling chemists to identify other alkaloids in the bodies of murder victims.

Now that Stas knew what he was looking for, it seemed that nicotine was almost everywhere; not on Gustave's clothing, which had been so thoroughly washed, but absorbed into the floorboards of the chateau's dining room, on the gardener's trousers, and in the bodies of the cats and ducks used by de Bocarmé in his experiments, which were found buried in his garden. Of the two phials placed in the dining room cupboard, there was no sign.

The one puzzling factor in the appearance of Gustave's body was the burning of the mouth. Stas killed two dogs with nicotine, but neither showed the same burning. He then poured acetic acid into the mouth of one dog and this produced the same dark burns as had been seen on Gustave's body. Finally, after further searches of the chateau, de Bocarmé's chemistry apparatus was found concealed behind panelling. The case was complete.

As with so many careful plans, the perfect murder had unravelled as soon as it was put into operation. Gustave must have been thrown to the floor by the count, who tried to hold his mouth open while the countess poured extract of nicotine into her brother's throat, but the young invalid had been stronger than either had imagined, and put up a fight, biting his brother-in-law's finger. As the poison took effect, Gustave, in his convulsions, had gripped his tongue between his teeth. De Bocarmé's efforts at holding his victim's jaws open to receive the poison had caused the deep scratches on Gustave's cheeks and left trace evidence under the assailant's fingernails. In the violent struggle, drops of liquid nicotine had been scattered on to the floorboards.

When questioned, it was impossible for de Bocarmé to deny having made the nicotine, which he claimed he had done for an innocent purpose. He said he had been studying chemistry using, amongst others, the works of M Orfila. The couple sought to escape the consequences of their joint perfidy by turning on each other. The countess openly accused her husband of having committed the murder, saying that, if she had assisted him in any way, it was because he had forced her. De Bocarmé wrote to a friend from prison accusing his wife of committing the murder and then trying to blame him. At a practical level, however, it must have been apparent that immobilising the victim, holding his jaws open and then pouring poison into his mouth was a task that could only be accomplished by two people working together.

The case attracted international interest, not only because of the unique method of murder, but also the 'rank of the parties'.[8] The events came at a crucial time in the career of Professor Orfila. Following the establishment of the Republic in France in 1848, he had lost his political influence and his post as Dean of the Faculty of Medicine of Paris; nevertheless, he was still a respected figure. It has been suggested that Orfila – known to have expressed the view that it would always be impossible to extract alkaloids from human tissue – had been engaged by the defence to contradict Stas's findings.[9] Other sources suggest that Orfila took up the challenge on his own account.[10] Orfila later wrote a memorandum on his work with nicotine and coniine in which he stated that he and Stas had worked on the alkaloid problem quite independently of each other. Not only that, but he awarded the laurels to Stas, since he, Orfila, had been using animals he had poisoned with nicotine, and therefore knew what he was trying to identify, whereas Stas had been working with the unknown. Orfila had nothing but praise for Stas, for his learning, his expertise, and the care and precision of his work, which he believed would be the model for future investigations.

If Orfila had been employed as a consultant for the defence, he would, as his work progressed, have found this position impossible to maintain. The fact that his results agreed with those of Stas only added further proof of the correctness of the Belgian professor's conclusions. On 20 March, Orfila supplied a sealed report to the Academy of Medicine in Paris.

The trial opened on 27 May at the Court of Assize in Mons, where the crowds were able to note that the countess, who was dressed in black, was a handsome woman, her husband 'tall, good-looking, and of aristocratic bearing',[11] with blue eyes and flowing blond hair, although his voice 'resembled that of a boy of 15'.[12]

The countess's defence was that all blame should be laid at the door of her husband. She testified that the family's debts were due to his licentious habits and extravagance. He had made her write under an assumed name for poisonous plants and, when she had questioned him, he struck her. He had also made her help prepare the nicotine and, when she asked what it was for, he had replied, 'To do the business of Gustave'[13] – a reply that caused a stir in court, as did her contention that her husband had told her on 20 November that it was Gustave's 'last day'.[14] She claimed that, after dinner, her husband had rushed at Gustave and taken him by the shoulders, and she had been so frightened that she had run into an adjacent room, and was not a witness to subsequent events. Later, she said, her husband had confessed to the crime and begged her not to betray him. 'I am sure that nothing can happen to me,' he had added, 'as they can discover nothing, for nicotine leaves no trace behind it.'[15] She also revealed that the papers her husband had burned were his correspondence regarding nicotine and the purchase of apparatus and materials.

De Bocarmé, tearfully asserting that he had never struck his wife, changed his story yet again, and claimed that she had given the poison to Gustave by mistake, thinking it was wine. Lydie at once retorted that there was not a word of truth it this. It was pointed out to de Bocarmé that Stas had come to the conclusion that, when the nicotine was ingested by Gustave, the young man had been lying on his back on the floor.

Crucially, the nursemaid and the cook, who had both heard Gustave's cries, were adamant that the countess had been in the dining room at the time and had not left it until several minutes afterwards. The prosecution made the point that, as carried out, the crime needed the co-operative efforts of two people and that the countess, knowing her husband's intentions and having ample opportunity to warn her brother of his danger, had failed to do so.

Orfila's statement was read out in court at the session held on 4 June, before Stas testified. Orfila stated that he had initially hoped it would not be necessary to read the report for fear of influencing the trial that was about to open. His anxiety on that point had been relieved, however, because he had attended the first three days of the trial, and saw that his *mémoire* would not make the position of the accused any worse. Orfila reported that he had been able to reveal the presence of nicotine in the tissues of poisoned animals; describing the methods he had developed, he gave complete credit to Stas for his discovery, and confirmed that he had had no prior knowledge of Stas's work.[16] The timing of the reading of this paper, however, before Stas had even given evidence, must have made it look as if Orfila were trying to steal his former pupil's thunder. The Brussels newspapers later published articles accusing Orfila of attempting to deprive Stas of his priority, a claim that Orfila in his later *Mémoire sur La Nicotine et sur La Conicine* dismissed as a lie to which he would not lower himself to respond. Since de Bocarmé had already admitted that he had made nicotine and that Gustave had died from its effects, Orfila's results made no difference to the trial, and Stas's findings could not be contested.

There were signs that the public fear of poisoning was undergoing a shift away from anxiety about the uneducated using arsenic to the educated trying to fool doctors with rare poisons. Orfila had words both of warning and encouragement: '[T]he public need be under no apprehension. No doubt intelligent and clever criminals, with a view to thwart the surgeons, will sometimes have recourse to very active poisons little known by the mass and difficult of detection, but science is on the alert, and soon overcomes all difficulty.'[17]

On 9 June, Professor Stas gave evidence. Asked if someone would notice the smell of pure nicotine if it were offered in a glass, he produced a phial and suggested the jurors try it. There were visible signs of repugnance amongst the jurymen, only one of whom took up the challenge.

The de Bocarmé trial lasted seventeen days, at the end of which it took the jury an hour and a half to return a verdict of guilty against the count, and acquit his wife. Shortly afterwards, sentence of death was pronounced. On 19 June, Hippolyte Visart de Bocarmé was led to the place of execution in a public square at Mons. His main anxiety was that the blade of the guillotine should be well sharpened.

The importance of the case was not lost on the popular press. On 18 June, the *Manchester Guardian* told its readers: 'The progress which medical jurisconsults have made recently is so great, that poisoning by morphine, strychnine, prussic acid, and other vegetable substances, hitherto regarded as inaccessible to our means of investigation may now be detected and recognised in the most incontestible [*sic*] manner.'[18]

One feature of Mme de Bocarmé's defence – that the family debts were due solely to her husband's extravagance – can be viewed in the light of her later career. She inherited her brother's property, but, within a year, she had dissipated a third of her fortune and was planning to remarry. Her family, concerned for the welfare of the children, applied to a tribunal in an attempt to have her declared incapable of administering her own property.[19] Eventually, in the spring of 1853, she was married in London to a Belgian gentleman, Adolph Joseph Van Duerne.

Vegetable poisons continued to be a major concern for doctors. In October 1851, Orfila was in Prague when a Dr Jacques Ellenberger told him that he had discovered an antidote for morphine and its salts. Ellenberger proposed to swallow a dose of morphine, followed by his antidote, in Orfila's presence, and Orfila agreed to witness the demonstration. Ellenberger said that he had done this many times without ill effect and, in some cases, had substituted salts of strychnine and brucine. The powder Ellenberger proposed to take appeared to Orfila to have some of the properties of morphine, but he thought it had been mixed with something else.

Ellenberger took twenty-three gains of the powder, followed immediately by his antidote, and suffered no symptoms. Orfila, thinking of the application of the antidote to cases of poisoning, asked Ellenberger if he had ever left an interval between the dose and the antidote, and Ellenberger said he had and the results were the same. Orfila warned him to take care as he did not know of a case where someone had recovered after enough time had passed to absorb a fatal dose. In March the following year, Ellenberger repeated the demonstration but this time allowed a long interval to pass before taking the antidote. He died.[20] The powder he had taken contained morphine and the supposed antidote was identified as mainly magnesium carbonate and a little magnesium sulphate.[21]

Following a coup by Louis-Napoleon in December 1851, Orfila was back in favour in French society and the medical fraternity, and he was elected president of the Academy of Medicine. His health, however, was in decline, and, in March 1853, he died of consumption.

Stas's method, with later modifications by Friedrich Julius Otto, professor of chemistry at Brunswick, opened the door to a new era of toxicology. Chemists all over the world worked for many years devising reagents that would produce colour changes to identify specific alkaloids in extracts produced by the Stas–Otto procedure. It remained the standard means of extracting alkaloids for over a hundred years and its fundamental principles are still in use today. In 1851, however, Orfila had sounded an important note of caution. The nature of the poison used by de Bocarmé had already been proved by the prisoner's confession before Stas gave evidence. Science had yet to be tested in a case in which poisoning was in doubt.

CHAPTER TWELVE

Nil By Mouth

After the passing of the 1851 Arsenic Act, no further legislation for the control of poisons was considered for some time. While available statistics show an increase in the number of poison murders brought to trial in the 1840s and 1850s, poison as a method of murder still remained rare compared to the various forms of physical violence.[1] In 1855 and 1856, however, a number of murder trials, ranging from the mystifying to the sensational, attracted extensive newspaper coverage. This led to the perception that another epidemic of poisoning was sweeping the land (or a continuation of the old epidemic, unabated), which required the attention of parliament. Unlike the Essex cases, this time the accused were middle-class men, a new breed of sophisticated scientific poisoners – men who could read and learn from newspaper reports of trials, or even study texts on toxicology; and there were worrying signs that imaginative murderers hoping to evade detection were turning either to new poisons or to unusual methods of administering the old ones.

'It should be remembered,' wrote Robert Christison in 1829,

> that poisons may be administered in many other ways besides by mixing them with articles of food or drink, or substituting them for medicines. They may be introduced into the anus; they have been introduced into the vagina; they have also been introduced by inhalation in the form of vapour; and there is no difficulty in introducing them through wounds.[2]

He related a case that had occurred in Toulouse of a lady who had died suddenly after receiving treatment for a trifling ailment. Her servant, who had been unsuccessful in administering arsenic in the conventional fashion by dissolving it in soup, had added it to the enemas that formed part of the course of treatment. Arsenic applied by enema, said Christison, would have 'all the usual effects' and, unless the victim died soon afterwards, evacuation made it unlikely that it would be found in the body after death.[3]

Another case described by Christison occurred in France in 1799. A woman was taken very ill with vomiting, diarrhoea, swelling of the genitals and uterine discharge. Before she died, she told two neighbours that her husband had once tried to poison her by putting arsenic in her coffee, but more recently he had introduced a powder into her vagina 'while in the act of enjoying his nuptial rights'.[4] In a Finnish case, a farmer attracted suspicion when he married his maidservant six weeks after the sudden death of his wife. A few years later, he started an affair with a new maidservant and engaged her help in poisoning his second wife. When the usual methods failed, he succeeded one morning, after coition, in introducing a mixture of arsenic and flour into her vagina on the point of his finger. She was taken ill at midday and died next morning. The farmer was now free to marry once more and his co-conspirator became his third bride; but, some years later, he tired of this wife, too, and administered arsenic to her by the method he had used to murder the second. Realising that she was suffering in the same way as her predecessor, she demanded and got a confession from her husband. However, despite repeated applications of lotions to the vagina, it was too late to save her. Pain and vomiting were succeeded by delirium and she died in twenty-one hours. The ability of arsenic to cling to membranes was shown all too clearly, as, despite the constant washings, there were visible grains remaining in the vagina after death. 'The labia were swollen and red, the vagina gaping and flaccid, the os uteri gangrenous, the duodenum inflamed...'[5] Only the stomach was unaffected. Dr Manger, the medical inspector for Copenhagen, wishing to settle any doubts that arsenic could be fatal when applied to the vagina, conducted some experiments with mares, which produced 'violent local inflammation and fatal constitutional derangement'.[6]

Other cases of poisoning reported by Christison involving the application of arsenic to broken skin, ulcers, the lining of the nostrils or by inhalation of the fumes, were accidental, as were two deaths by sulphuric acid enema.

The respectable male middle-class poisoner posed a new problem. With poisoning having been represented as a crime of the uneducated, and especially females, there was an unwillingness to believe that a man with some position in the community could be a secret murderer. One had to be very sure of one's grounds to accuse such an individual of murder, since the repercussions on the accuser could be extremely damaging. The educated poisoner was not only in a position to carry out his plans in a subtle way but he could use his social status to influence the investigation, and was wealthy enough to take action for libel and slander.

Another alarming trend was the inclusion of poison not in food, but medicine. The secret poisoner, adopting the guise of carer and healer, was giving his poison to a vulnerable and trusting victim, who would not balk at being given something unpleasant to the taste. This perversion of the respected profession of medicine was especially uncomfortable to nineteenth-century sensibilities. When the *Provincial Medical and Surgical Journal* was launched in 1840 (it was renamed the *British Medical Journal* in 1857), it was dedicated not only to the dissemination of medical knowledge amongst practitioners, but also to 'the maintenance of the honour and respectability of the medical profession'.[7] Doctors were more than medical attendants: they were felt to hold a special rank in society due to 'their intellectual acquirements . . . general moral character . . . and the importance of the duties entrusted to them'.[8] Ann Daly's study of the *Dublin Medical Press* demonstrates how the period 1850–90, by establishing the foundations of the modern health care system, placed the doctor as a central and important figure to the public, and 'bolstered the perception of the medical practitioner as moral guardian of society in general'.[9]

Medical status was probably a factor in the acquittal of Donnall (Chapter 2) and – as it did in that case – it conferred upon the murderer the opportunity to interfere with the post-mortem examination of his own victim.

Medical attendants were to become the subject of harsh public criticism in one of the most controversial trials of 1855. The Great Burdon slow poisoning case, which has all the elements of the classic Victorian murder, was described by Christison as possessing 'a rare interest' and as 'one of the most instructive cases of poisoning with arsenic that has ever been published'.[10]

Joseph Snaith Wooler, son of a landed proprietor, was born in 1810 and, in 1836, he married Jane Brecknell, a surgeon's daughter five years his junior. The couple spent some time in India but, by 1855, they had been settled in the township of Haughton-le-Skerne, Great Burdon, near Darlington, for about seven years, and were well known to the local residents. There were no children of the marriage, but, to all appearances, they were a happy, affectionate couple, and had never been known to quarrel. Wooler was 'a gentleman of independent means',[11] living off farm rents, and his wife was entitled to a small annuity. The only other member of the household in 1855 was the twenty-seven-year-old servant, Ann Taylor. Ann had worked for the Woolers for six months in 1848 and, after leaving service for a time, had returned to work for them in 1854.

The Woolers had carried a great many medicines with them on their travels and, on their return, brought back a large selection, which they stored in two baskets.

From the beginning of May 1855, Mrs Wooler had complained of feeling weak and she had been vomiting. On 8 May, her husband called in Dr Thomas Hayes Jackson, who had practised in the district for twenty-one years. Jackson found the patient short of breath, with a dry throat and a cough, quick pulse and redness about the eyelids. Intestinal symptoms included griping, tenesmus[12] and a discharge of mucus from the bowels. He advised her to rest in bed, and, for the first week, treated her for influenza, but the symptoms only became worse. Vomiting and diarrhoea came on severely, and the mucus from the bowels was now streaked with blood. In addition, there were hiccups, dry throat and difficulty in swallowing. Jackson addressed the symptoms individually, giving sedatives to allay vomiting, and opiate injections to check diarrhoea, but was unable to determine their cause. Mrs Wooler was by now bedridden, and attendants were called in to care for her: schoolmistress Miss Elizabeth Lanchester, and a Miss Middleton, who alternated days and nights, while Ann Taylor also helped look after the patient.

During his attendance on Mrs Wooler, Dr Jackson needed some medication and Wooler brought him a round Indian basket containing a large number of medicine bottles. Jackson examined the bottles and found several poisonous items, including a one ounce bottle of liquor arsenicalis, or Fowler's solution. There was about a teaspoonful left in the bottle. Wooler explained that he had twice been by sea to India and on board a ship for six months together, and had therefore been in the habit of keeping some medicines by him. Jackson's impression was that Wooler had a better knowledge of medicines and their properties than most people.

Mrs Wooler was given the typical invalid diet of the day – milk, arrowroot, rice and barley, and sometimes beef tea and broth – and Dr Jackson also supplied a bottle of tonic, but the vomiting and purging increased. In view of the patient's difficulty in retaining nourishment, it was decided to feed her by enema (referred to as 'injections' in the contemporary accounts). Since the rectum has no digestive enzymes, the practice supplies very little that can be usefully absorbed, but this was not known at the time, and Mrs

Wooler's enema was composed of the tonic mixture, two eggs, a tablespoon of cod liver oil and a tablespoon of milk. Ann made up the liquid, to which Wooler added a dose of laudanum to help his wife rest. The enemas were administered by Wooler with Ann's assistance.

Mr George Harle Henzell, a twenty-three-year-old druggist who acted as Dr Jackson's assistant, also called on Mrs Wooler; his first visit taking place on 16 May. When he called again on 4 June, he found her much worse. It crossed his mind that her symptoms could be accounted for by an irritant poison, but, at the time, he didn't think there was anyone in the house acquainted with it. He did not mention this uncomfortable thought, but, two days later, Dr Jackson revealed that he had been thinking along the same lines. Jackson had decided to say nothing about his suspicions to the family:

> [I]f it had proved false, I was rendering myself liable to an action and damages. They had a lawyer in the family, and it would have suited them exactly.[13]

With his wife's condition deteriorating Wooler became dissatisfied with Jackson. A Dr Devey came to see her on 6 June, and said it would be 'a tedious case, but he saw no danger'.[14] On 8 June, however, Jackson told Wooler that his wife was consumptive and he did not think she would recover. Wooler was understandably annoyed that he had not been told this before. Jackson later claimed that the husband's reaction to the news was 'very indifferent',[15] but, according to Ann, her master often stood by his wife's bed weeping.

On 8 June, Dr William Haslewood, a practitioner of eighteen years' experience, was called in and told that Jackson suspected consumption. There was a brief professional spat when Haslewood sent for Mr Henzell to listen to the patient's chest, as he thought he had a better ear, and Jackson, insulted by his assistant and not himself being asked for, refused to send him.

Haslewood found Mrs Wooler suffering from irritation of the alimentary canal from the mouth downwards, as well as from hiccups and soreness of the lower bowels. Her skin, including the mucous membrane of the eyes and nostrils, was red and rough. Haslewood told Wooler he did not think the patient's lungs were affected, made some modifications of Dr Jackson's treatment and attended her every day. Apart from occasional remissions, she continued to decline. Small allowances of food and medicine were sometimes retained for two or three hours, but were usually vomited immediately.

Eventually, the medical attendants relied almost entirely on the enemas for the patient's sustenance, and Wooler persuaded his wife to take as little into her stomach as possible, the idea being to 'keep up life until her stomach got strengthened'.[16] The result was that the sick woman was, unknown to her doctors or family, receiving almost no nourishment. After a time, the egg oil and milk enema began to clog the injection apparatus, and Wooler called on a surgeon, J. R. Fothergill, explaining the problem and asking to borrow his apparatus. Fothergill lent him one, which was subsequently used successfully about a dozen times.

Henzell began to test Mrs Wooler's vomit and urine. He got no result from the vomit but, using Reinsch's test, obtained a metallic deposit from the urine, which led him to suspect the presence of arsenic. By 17 June, Haslewood too was suspecting poison and so prepared an antidote, which was included in the injection. Two days later, he discussed the case with Dr Jackson and found they were of the same opinion.

On 20 June, Mr Wooler was away visiting his father, and Mrs Wooler was attended by Dr Haslewood and Mr Henzell. They administered the antidote and she showed some improvement as she was able to take food without vomiting. From then on, the antidote was given daily but, after that brief rally, she again started to decline again.

Samples of the patient's urine were being stored in bottles in the Woolers' coach house and, on 22 June, Henzell, wanting some to test, went to the coach house but found none there. He asked for some, and Ann Taylor later gave a bottle of urine to Mr Wooler, who sent it to Henzell. On making the usual tests, however, Henzell found the character of the sample markedly different from before; so much so that it seemed to have come from another person. He mentioned this to Wooler, who asked Ann where she had got the bottle from and she said from the coach house as usual.

Haslewood was looking for other symptoms that would have suggested arsenical poisoning – in particular, spasms and a tingling in the extremities. On 23 June, Wooler mentioned tingling sensations in his wife's hands, which he said had lasted only a day. Mrs Wooler, however, said she had had the tingling for two or three days and she had asked her husband to tell the doctor but he had forgotten. There were also the first indications of the onset of spasms – a stiffness and numbness in the arms: '[W]e could not have a more complete chain of symptoms of arsenical poisoning.'[17] Dr Haslewood later commented, and Jackson also believed, that Mrs Wooler showed every symptom of slow poisoning with arsenic.

Still none of the medical men said a word about their suspicions to Mrs Wooler, her family, friends, attendants or the authorities, and it later became evident that they were guarding their own professional reputations more carefully than they were the life of their patient. When Wooler asked Haslewood to write to his nephew Mr Brecknell, who was a pupil of the eminent surgeon Sir John Fife, Haslewood did so, but told Wooler that he would not ask the nephew to put the case before Fife, since it would be 'unprofessional to get his opinion by a side wind'.[18] He could not, he said, prevent the nephew from doing as he liked with the letter, which described Mrs Wooler's symptoms. The letter did not, however, state Haslewood's opinion as to the cause of Mrs Wooler's condition, and he covered himself by deliberately omitting to mention the tingling of the hands and the start of spasms, which would have been tantamount to saying he suspected poison.

Finally, the medical men decided to seek expert advice and wrote to Robert Christison, professor of materia medica and clinical medicine at the University of Edinburgh. Expressing their concerns about poison, they said that they did not feel their suspicions were well grounded enough to make an accusation, especially as the husband was a potential suspect.

Christison had, for some years, given up dealing with medico-legal enquiries that were located at some distance from his home, as travelling to give evidence interfered with his professorial duties, but he was happy to give free advice on the basis that he would not be involved as a witness. Christison replied, advising that the symptoms might be those of either poisoning or natural disease and they should be cautious in divulging their suspicions until they had strong evidence against an individual, and positive proof of arsenic. He offered to assist by examining a specimen of urine prepared by concentrating it with hydrochloric acid.

Thirteen ounces of urine were concentrated for analysis and sent to Christison but, in view of the doubts about the unusual sample received by Henzell, a second one, which more closely resembled that of the patient, was sent on 26 June. The medical men, satisfied that they had done all they could or should do, waited to hear from Dr Christison.

It seems inconceivable that Jackson, Haslewood and Henzell all suspected that Mrs Wooler was being poisoned, yet did nothing to alert anyone other than Christison to what they believed was happening. The three of them regularly discussed whether their course of action was right and whether they should tell the authorities or the family of their suspicions. 'We came to the conclusion,' said Henzell, 'after long and anxious deliberation, that the best step we could take was that advised by Dr Christison, and that, unless we had the best oral evidence, or strong legal proof of arsenic being in vomited matter or evacuations, we must hold our tongues for the present.'[19]

By 26 June, the patient was sinking and, that night, she began to suffer almost continuous tetanic spasms, during which she retained both consciousness and her mental faculties. At 10.30, she died. Wooler showed little outward emotion. After talking with Henzell for a while, he invited him into the garden where some bees were hiving, and they spent some time watching the bees.

It wasn't until after Mrs Wooler's death that Christison started to analyse the second urine sample. He found arsenic.

On 29 June, Wooler informed the registrar of deaths that his wife had died from 'ulceration of the bowels and stomach'.[20] The post-mortem examination took place on the same day carried out by Drs Haslewood and Jackson and Mr Henzell. There was considerable ulceration of the bowels, the ulcers varying in size from that of a small pea to a crown piece,[21] some of them so deep that they extended through to the outer coat of the gut. The stomach was healthy but inflamed inside, especially towards the gullet, where it broke easily with the touch of a fingernail. Unknown to the Wooler family, a portion of the viscera including spleen and kidneys was removed, sealed in a bladder and oil silk, and taken home by Dr Jackson, who locked it up. The intestines, apart from a small piece of liver that would not fit, were later sealed in a jar.

The conclusion was communicated to Wooler not by personal visit, but in a letter from Dr Jackson. Immediately after reading it, Wooler summoned Ann Taylor, Miss Lanchester and his niece, Miss Brecknell, and told them that Dr Jackson thought his wife had died of poison. He wondered where the poison could be, and Ann said it wasn't in the food. Wooler then thought of the medicine bottles and asked where they were. Ann said she didn't know, but some of the medicines had been poured out and the vomits put in the bottles to be sent to the doctors. Wooler told her to find the remaining bottles and lock them up in her box, which she did.

Jane Wooler was buried on 30 June and, on the same day, Dr Jackson took the jar of intestines by train to analyst Thomas Richardson in Newcastle. Two days later, after applying the standard tests, Richardson was in no doubt that he had detected the presence of arsenic.

Stories were already circulating in the neighbourhood that Mrs Wooler had been poisoned. Wooler asked the chemist Mr Abbott if he had heard the scandalous reports about him in Darlington, and Abbott said that he had, and that Wooler should have stayed at home until they subsided. Wooler angrily accused Abbott of lacking sympathy for a broken-hearted husband, 'to which I replied', Abbott was later to comment, 'that I did not think, candidly speaking, that he was'.[22] Wooler returned the borrowed enema syringe to Mr Fothergill, saying it had worked well. Fothergill saw that it was clogged but did not trouble to clean it. He put it in a drawer.

The inquest on Jane Wooler was held on 14 July. Wooler gave evidence, and, even on the printed page, he sounded highly aggrieved. He said that Jackson had given his wife powders that had brought on 'the first bowel attack'[23] and those never ceased until she died: '[S]he vomited quarts of pure bile, and otherwise emitted it; I never saw anything like it except on board ship.'[24] Jackson, he said, had then told him he had made a mistake in treatment, suggesting that Mrs Wooler was consumptive, and 'from the first he had doomed her',[25] upon which Wooler had demanded to know why he had not been told this before. Jackson's refusal to allow Henzell to attend in his place as a matter of professional etiquette was the last straw. Wooler, angry at Jackson's 'gross and unfeeling' behaviour,[26] said that he 'had ample means of calling in the first advice in the kingdom . . . and had reason to feel hurt to find the symptoms so aggravated before I was made aware there was danger'.[27] Wooler, recollecting his last private conversations with his wife, broke down in sobs. He denied having any great knowledge of medicine. The Indian basket, he said, had never been opened until it was opened for Dr Jackson. He could not recall any arsenical preparation amongst the medicines, and, if he had had any, it had been in his possession since 1840. The medicines were kept under lock and key, and Mrs Wooler had the key. Wooler confirmed that he had no insurance on his wife's life, and her annuity had ceased with her death.

The Woolers' baskets of medicines were produced in court and the bottle of arsenical solution previously observed by Dr Jackson was missing.

The verdict was that Jane Wooler had died from an irritant mineral poison and the jury made a point of exonerating the medical attendants from all blame, but did not, as they were entitled to do, suggest who the murderer might be. The medical opinion was that the poison had been administered in small doses over a period of time.

The finger of suspicion was pointing only one way. Joseph Snaith Wooler had arsenic and other poisons in his possession and was 'acquainted with the skilful and scientific use of them'.[28] Jane Wooler, said the doctors, 'had been poisoned scientifically, and not by any person ignorant of the properties of poison'.[29] Of all the persons who had attended the sick woman, only one had been present throughout the illness and had the requisite knowledge – her husband.

Meanwhile, Dr Fothergill was having some thoughts about the enema apparatus returned to him by Mr Wooler. He took it out of the drawer, removed some of the matter clogging the tube, and tested it. He obtained a positive result for arsenic. He later wrote to the manufacturer asking if arsenic was used in the manufacture of the apparatus and they wrote back to say it was not. Mr Fothergill also went to Dr Haslewood's house; here, the piece of leftover liver was subjected to the Reinsch test, which demonstrated the presence of arsenic.

On 27 July, Jane's brother William Henry Brecknell made a deposition before Justices of the Peace, giving the reasons he believed that Wooler 'did feloniously, of malice aforethought, and with intent to kill, administer poison to the said Jane Wooler'.[30] On 29 July, Wooler was arrested on suspicion of murder.

Proceedings opened at Darlington Police Court, on 30 July, and, after the first day's evidence, an order was made to exhume the body. A portion of the liver still remained in the body, and Mr Henry Marshall for the defence wanted it to be sent to Professor Taylor or Mr Herapath or some other eminent man. Wooler is reported to have said, 'It is stated that the body exhibits traces of poison; if so, it has been contained in the medicines administered by the medical men.'[31]

Haslewood told the court that the poison was administered 'by someone well acquainted with the nature of arsenic and of its effects in small doses – an amount of knowledge very rare except with a few educated men who have made scientific study their pleasure or business'.[32]

On 4 August, Mrs Wooler's body was exhumed, and samples of liver, intestines, lungs, heart, and liquid from the abdomen and rectum were sent to Professor Taylor. He examined all the remains with a magnifying glass but did not detect any arsenic in powdered form. By analysis, however, Taylor was able to produce metallic arsenic from all the organs and viscera and the bloody fluid. In the liver, heart and lungs, the arsenic was incorporated into the structure of the organs, which showed that it had been absorbed during life. The greatest proportion of arsenic to weight of organic matter was in the rectum. He concluded that the arsenic had been taken in small doses during life, and the quantity found was consistent with death by arsenic. He also said that it must have been taken in a soluble state, though he was unable to say for certain how the arsenic had been administered. Asked about the possibility of normally occurring arsenic, he assured the court that this was an error, although 'many barristers and lawyers still believe in it'.[33]

Taylor gave his evidence on the fourth day of the hearing, when he described Fowler's preparation as a solution of arsenic with a little potash, adding that he had not found any trace of the other ingredient of Fowler's. He was succeeded by Mr Fothergill, who was a little embarrassed. Since making his tests on the matter found in the enema tube, Fothergill had tested the acid he had used and found a minute trace of arsenic in it. He repeated his tests and felt sure that they confirmed his original opinion that there was arsenic in the tube. Dr Haslewood now listed the items found in the round Indian basket: there were thirty-four, including some that were unlabelled and some empty bottles. Another oblong basket contained nine items. There were two more baskets, which contained items obtained more recently from local chemists. None of the bottles contained arsenic.

On 25 August, Joseph Snaith Wooler was formally charged with the murder of his wife and committed for trial. He declared his innocence trusting that Almighty God 'will bring to light the atrocious criminal who has perpetrated this foul deed'.[34] The medicine bottles removed from Wooler's home and the two enema syringes were sent to Professor Taylor, who found no arsenic in any of the bottles, and traces of it on the syringes.

The trial of Joseph Snaith Wooler for the murder of his wife opened at Durham on 7 December, before Mr Baron Martin. Christison later observed: 'The proof of poisoning by arsenic was so perfect, in very nice and difficult circumstances, that even the prisoner's counsel evidently surrendered that point, without attempt at dispute, from the very beginning. How different was the case only five-and-twenty years ago, when the main efforts of counsel were invariably directed to deny and disprove the poisoning?'[35] Christison disposed of any suggestion that the poisoning could have been accidental, as the medicines and the components of the injections were found to be uncontaminated. The theory that Mrs Wooler had committed suicide by slow poison was so improbable as to be dismissed.

There had been an anonymous suggestion in the press that Mrs Wooler had been 'in the secret habit of taking arsenic'[36] and had suffered after giving it up, which Christison described as 'a mess of absurdity'.[37]

This theory was based on a revelation made in *The Chemist* in 1854.[38] The journal had published letters by naturalist Jakob von Tschudi, who, while travelling in the region of Austria known as Styria, had discovered the practice of arsenic-eating amongst the peasantry. Tschudi claimed that the Styrians, who believed that arsenic improved the complexion and assisted respiration while out walking in the mountains, gradually habituated themselves to it until they were able to consume up to four grains daily, and actually became ill with symptoms such as pain and vomiting if they stopped taking it. The report was treated with derision by the medical establishment, and there was considerable concern about a scientifically unproven legend muddying the waters of a murder trial. The legend endured into the twentieth century when investigation suggested that the Styrians had consumed arsenic in an impure form, and would not have absorbed the amounts suggested by Tschudi.[39] In 1855, however, it was clear that there was no evidence whatsoever to support the idea that Mrs Wooler was a secret arsenic eater.

The judge, in his closing remarks, said that he found the conduct of the medical men 'reprehensible',[40] adding that it was for the jury to decide if they could safely conclude that the prisoner had given his wife arsenic. 'I am unable.' According to *The Times,* he then said, 'I may observe that, if I were to make a surmise, there is a person upon whom my fancy would rest rather than the prisoner.' Christison, despite the 'dark tales and guesses of the neighbourhood',[41] claimed to have no idea to whom Baron Martin referred. It took the jury ten minutes to bring in a verdict of not guilty, after which Martin intimated that he would have stopped the trial sooner, but thought it better to have the case fully heard.

The three medical men were heavily criticised by Wooler's counsel, censured by the judge and condemned by the press for not revealing their suspicions while their patient was alive. Baron Martin suggested that they should have gone to her husband if they did not suspect him or to a magistrate if they did, 'and not have gone on, from the 8th of June to the 27th, seeing the woman murdered before their eyes'.[42] The medical press hastened to the defence of the practitioners. 'We are, alas!, too accustomed to see the medical man treated like pantaloon, the moment an Old Bailey lawyer can put him into the witness box,' lamented the *Association Medical Journal.*[43] 'Once there, he is considered a fair butt for forensic impertinence.' Christison's opinion was that Mrs Wooler's doctors had acted properly in seeking conclusive tests before they did anything, since country districts did not have the resources at hand to meet the analytical requirements of such a case. Science had kept ahead of practice, but the science was known only to a few. He ascribed the inconclusive results of Henzell's test to his preference for the Reinsch test, Marsh being better for delicate analysis.

Christison advised that the best course of action in such cases was to mention any suspicions to the patient, who might very well have had suspicions of their own and be able to point to a culprit. Another course, suggested by a Dr Gairdner, if the doctor wished to act on suspicions without divulging them, was to place a confidential person in the house who would supervise everything. It might raise the alarm to the poisoner, but the main duty of the doctor was to protect the patient.[44]

The uncomfortable result, however, was that no one was ever brought to account for the murder of Mrs Wooler, which remains unsolved to this day. Under the circumstances, Christison must have felt it necessary to utter a word of warning to the would-be poisoner. The case showed, 'in these days of growing refinement in crime, that the secret poisoner is not to expect to produce, by slow poisoning with arsenic, the obscure pining, the imperceptible progress, and nameless death, which, in the age of Secret Poisoning [*sic*], arsenic was supposed to occasion in the hands of the skilful; but that on the contrary, he will in all probability excite the most characteristic and aggravated symptoms'.[45]

Baron Martin must have had his own ideas as to the identity of the guilty party, although he later wrote to *The Times* disputing the newspaper's account of his words, saying he had pointed out that there was no place in law for 'fancy or surmise or suspicion, but facts that amounted to proof . . . It appeared to me that there was no proof against any one; but that if I were to indulge in mere surmise and fancy, not the prisoner but some other person would first occur to my mind.'[46] What he claimed to have meant was that, by using fancy, anyone else could be a suspect, and he was not, therefore, pointing at any individual. Nevertheless, only two people were constantly present at all stages of Mrs Wooler's illness and had the opportunity of administering poison, and they were Mr Wooler and Ann Taylor.

During the course of the investigation, the assumption was made, and firmly adhered to throughout, that Mrs Wooler could only have been poisoned by someone skilled in the use of medicines. The image conjured up was that of an educated man who had made poison his special study. Ann, as a female servant with probably no more than a basic education, was, by this mindset, not a plausible suspect, whereas Mr Wooler precisely fitted the picture. Leaving aside any question of whether Ann had any special knowledge of medicines (something that never seems to have been explored), it does appear that the prosecution had become so wedded to the idea of the expert scientific poisoner that they never considered that the answer could have been a great deal simpler than supposed.

Medical men certainly liked to protect their professional status by suggesting that the handling of poisons was a matter for experts, but the long history of poisoning has shown that anyone can employ it successfully. Only four years previously, Sarah Chesham, a labourer's wife who was unable to write her own name and signed her statements with a cross, had killed her husband by slow poisoning with white arsenic. Where the non-medically trained poisoner often errs is in underestimating the power of the dose, something Sarah had mastered after many uses, and in overestimating the poison's solubility in fluids; however, in the Wooler case, the probable source of the arsenic was not pure powdered arsenic, lethal at only three grains, but the Fowler's preparation, arsenic in solution, probably about 1 per cent. One tablespoonful of such a solution (half a fluid ounce) would provide about two grains of arsenic, enough to make an adult very ill, and enough, eventually, to kill a woman weakened by lack of nourishment.

On the day that Mrs Wooler was first taken ill, she had dined off pig's cheek and soup. Mr Wooler never had soup, as Ann must have known. He later said that it was only his wife and the servant who had it, but, as servants did not, as a rule, eat with the family, he might not have seen Ann take the soup. Mrs Wooler's initial illness could well have arisen from a dash of arsenical solution in the soup. It would have been hard for Wooler to pour it into his wife's plate at the table, but easy for Ann to dose the pot during the soup's preparation and then have none herself. If Wooler had slipped into the kitchen to dose the soup pot without Ann's knowledge, she would have eaten some and been taken ill, too.

Ann was responsible to a great extent for caring for Mrs Wooler and mixing the nutritional enemas. It did not require the expert scientific hand of an educated man to add a spoonful of arsenic solution to an enema. The very fact that a dilute solution was the vehicle meant that the illness was a slow one, and not the acute poisoning episode that would have aroused immediate suspicion. During Wooler's brief absence from home, Mrs Wooler was closely supervised by medical men and given an antidote, and her condition improved. It was assumed that Wooler's absence was a key factor here, but the close supervision of the patient might have discouraged Ann from administering poison during this time.

Regarding the urine sample sent to Henzell, which was clearly not that of the patient, it was suggested at the trial that Wooler had had the opportunity to make a substitution. Another explanation offered was that the urine was that of Miss Lanchester, who slept in the same room as the patient, and it was simply sent in error. Ann, however, had lied about where the sample came from, perhaps not knowing that Henzell had already searched the coach house, and had ample opportunity to supply some of her own.

The most probable source of arsenic was a bottle of solution already in the house. The ounce bottle seen by Dr Jackson, even if it had been full before Mrs Wooler was ill, would not have contained enough arsenic for a long campaign of slow dosing, but there might have been others. If, as Wooler suggested, the medicines had been in his possession since 1840, it would have been an old or even overseas formulation, which could account for the absence of potash. There was no evidence of any recent purchase of arsenic and there was none in the prescription medicines.

Did Ann have a motive to murder Mrs Wooler? Had she been prosecuted, it might have been suggested that she wanted to remove Mrs Wooler in order to supplant her, or was taking revenge for a scolding. Wooler, if we assume he was innocent, does not seem to have suspected Ann, but blamed the doctors. There has never been a suggestion of a relationship between Wooler and Ann. During his testimony before the inquest, Wooler referred to her impersonally as 'the servant' or 'the girl' and never by name.

Ann Taylor, in all probability the murderer of Jane Wooler, remained with the widower as his housekeeper until his death in 1871. In his final will, made in January that year, his estate was valued at £600.[47] Ann was obviously a valued servant, since she was bequeathed two bonds worth £50 each and a lifetime annuity of £50. In 1881, she was still in Haughton le Skerne, living off her annuity and taking in lodgers.[48] In 1891, she was a 'retired housekeeper' in Stockton-on-Tees.[49]

The Wooler case, one of the more fascinating yet curiously forgotten poisonings of the 1850s, was soon to be eclipsed by another, which rapidly became a public sensation.

CHAPTER THIRTEEN

The Palmer Act

In the same month as the trial of Joseph Snaith Wooler, an inquest opened at Rugeley in Staffordshire on the body of twenty-eight-year-old John Parsons Cook, who had died under dramatic circumstances the previous November.

Cook had once been articled to a solicitor, but, on coming into an inheritance of some £12,000, he lost any inclination to work for a living, and plunged into the world of the turf, purchasing racehorses and betting heavily. He had made the acquaintance of William Palmer, a thirty-one-year-old surgeon, who was so devoted to the turf that he had largely abandoned his practice, the work of which had been transferred to his partner.

Palmer had been in grave financial straits since 1853, and, the following year, he raised money on loans, one of which, in the sum of £2,000, was guaranteed by his mother's signature. Mrs Sarah Palmer was a woman of property, and there was no difficulty in Palmer obtaining the money. His financial affairs were handled by a London solicitor, Mr Pratt. It was not to emerge for some time that Mrs Palmer knew nothing of the loan and that her signature had been forged by her son.

Palmer's debts continued to mount, but when his wife Ann died after a short illness he received life insurance of £13,000, which enabled him to pay his most pressing creditors. In March 1855, he bought two racehorses called Nettle and Chicken. As the year progressed, he took out a series of further loans, which were used to pay for the renewal of earlier loans, all of which bore the forged signature of his mother. It was a precarious situation, with Palmer constantly hovering on the brink of bankruptcy.

In May, Palmer was being pressed for a payment of £500 but had only £310 lodged with Pratt, who refused to loan him the difference without security. Palmer offered the guarantee of his good friend Mr Cook, and duly obtained £200. When the time came to repay the loan, however, Palmer was unable to do so and Cook was obliged to stump up the money.

In August 1855, Palmer asked Pratt for a loan of £500, which he said was required by Mr Cook, and this was provided using two of Cook's race-horses, Polestar and Sirius, as security. Whether or not Cook required the money or even knew about the transaction is not known; the only certain thing is that Palmer forged Cook's endorsement on the cheque he received from Pratt and paid it into his own account.

The following month, Palmer tried to induce George Bate, a stable-hand, to insure his life for £25,000, describing him in the proposal as a gentleman of independent means. Encouraged by Cook and a promise of payment, Bate signed the application. When enquiries were made about Bate by the insurance offices and it was found that he was not a man of means, the insurance was declined.

By November 1855, Palmer owed £11,500, all of which fell due during the following two months and which he was unable to meet. The loan he had obtained in Cook's name would soon be due, and, unless it was paid, it would be apparent that Palmer had forged his friend's signature. No more money was forthcoming from Mr Pratt who, on 6 November, issued two writs, one against Palmer and one against Palmer's mother, which Palmer was able to intercept. Palmer was now facing not only financial ruin, but also exposure as a forger and a thief, and the prospect of a long prison term.

On 13 November, Cook's horse Polestar, on which he had placed substantial bets, won at Shrewsbury. He then had between £700 and £800 in his pocket and was entitled to receive over £1,000 the following Monday from his account with Tattersalls. A week later, Cook was dead.

Cook had celebrated his win with champagne at the Raven Hotel, Shrewsbury, and had been in good health when, the following night, he sat in his room with friends including Palmer, who encouraged him to drink down a tumbler of brandy and water. Shortly before, Palmer had been observed in the lobby looking at a tumbler of drink by the light of the gas lamp. He had then gone into his own room, returned with the glass in his hand, and entered Cook's room.

Cook swallowed the tumblerful at a gulp and then suddenly complained that it had burned his throat. Palmer picked up the glass, drank down the teaspoonful of liquid that remained, and declared that there was nothing wrong with it. He offered the glass to others to taste, but there was none left. A few minutes later, Cook was seized with violent vomiting, which continued for several hours. By the following morning, Cook had recovered and was well enough to go out to the races again. Palmer's horse Chicken was running that day, and the cash-strapped owner, who had been obliged to borrow money to enable him to travel to Shrewsbury, backed his horse heavily. It lost.

Palmer and Cook travelled to Rugeley, arriving on the night of 15 November. Palmer went to his home on Rugeley High Street, and Cook took a room at the Talbot Arms opposite. Cook and Palmer were out together on the 16th and, early the next morning, Palmer arrived at the inn to see his friend and ordered coffee. Cook drank the coffee and, soon afterwards, was seized with the same violent vomiting he had suffered at Shrewsbury.

From then onwards, Palmer was in constant attendance on the invalid. Foods such as toast and water or broth were either prepared in Palmer's house or brought by him. Cook remained unable to keep down any food, although Palmer insisted he try to eat, though when the chambermaid sampled some broth sent for Cook, she vomited so violently all afternoon that she was obliged to retire to bed.

Cook's medical attendant was Mr William Bamford, a gentleman of about eighty. Palmer told Bamford that his friend had had a bilious attack after drinking too much champagne after his racing win, but Cook protested that he had only had two glasses. Palmer continued to minister to Cook but whatever he provided was vomited up.

On 19 November, Palmer, after bringing Cook a cup of coffee that was immediately vomited, went to London. It was settling day at Tattersalls, and Palmer met with a friend of Cook's called Mr Herring and was able to obtain the payment due to Cook, which he used to satisfy his most pressing creditors. Palmer returned to Rugeley the same night and went to surgeon's assistant Charles Newton, from whom he obtained three grains of strychnine.

During Palmer's absence, Cook's condition improved. Palmer went to see Cook, who had been left some sedative pills[1] by Mr Bamford. Palmer left Cook at about 11 p.m., and Cook was alone in his room when, at about midnight, the servants heard loud screams and rushed to see what the matter was. Cook was rolling about in intense pain, his whole body jerking convulsively, his eyes bulging out of his head, and he was beating at the bed with outstretched arms. He gasped for breath, hardly able to speak, and, when he did, he cried murder and called on Christ to save his soul. Cook begged the maid to bring Palmer, and one of the servants ran to fetch him. The screams continued for several minutes, then Cook's arms and legs became rigid and he asked the maid to rub his hands, which were stiff as if paralysed. Palmer arrived, sent for some medicine, and gave Cook an opiate mixture and some pills. Cook vomited the medicine. The maid tried to spoon-feed him toast and water and he snapped at the spoon, trapping it between his teeth. Gradually, the terrifying symptoms abated, and the patient calmed and was able to sleep.

Next morning, Palmer was at a druggist's buying prussic acid when who should enter the shop but Charles Newton. Palmer told Newton he wished to speak to him and drew him outside the shop for a conversation. Another acquaintance passed by and stopped to talk, and, while Newton was so engaged, Palmer quickly slipped back into the shop and purchased six grains of strychnine and Battley's liquor of opium, a common sedative. When Newton returned to the shop, Palmer had gone, and Newton, out of curiosity, asked what Palmer had purchased.

That same morning, Palmer induced an old friend, former schoolfellow Samuel Cheshire, then postmaster of Rugeley, to write out a cheque purporting to be from Cook in his, Palmer's, favour. He said that Cook was too ill to do it himself, but he would get Cook to sign it. Palmer provided more coffee and broth for Cook, who vomited all afternoon. Earlier, Palmer had written to a medical friend, a Mr Jones, asking him to attend Cook, saying that the patient was suffering from a bilious complaint. Jones arrived that afternoon, and, on examining Cook, was dubious about the diagnosis. Palmer, Jones and Bamford were about to confer when Cook said he didn't want to take any more pills as he thought they had caused his illness the night before. Palmer insisted he should take them and accompanied Bamford when he went to make up a further supply. At Palmer's request, Bamford wrote on the box that they were to be taken at bedtime. When Palmer returned to the inn with the pills, he showed the box to Mr Jones, making a point of remarking on how clear Bamford's handwriting was. Cook at first objected to taking the pills, but, at Palmer's insistence, he did. It was now about half past ten. Cook vomited again but did not bring up the pills. There was a second bed in the room, and it was agreed that Jones would sleep there that night. Palmer went home, and, after supper, Jones went to bed at about midnight. Not long afterwards, Cook suddenly started up with a loud scream that he was going to be ill as he had been the night before and asking for Palmer to be sent for. Palmer was summoned at once, and, despite the lateness of the hour, he was fully dressed when he arrived two minutes later.

Cook was screaming and gasping for breath, all his muscles painfully convulsed. Palmer gave Cook some more pills, which he immediately threw up. Cook's body stiffened with cramps, and he asked to be raised up to help him breathe, but it was impossible, since his body was rigid, bent back like a bow, resting on the back of the head and the heels. He begged to be turned over and they complied, placing him on to his right side, but, a few minutes later, his heart stopped. From the first paroxysm to death was hardly ten minutes.

Jones sent for women to lay the body out, and Palmer lost no time in searching Cook's coat and under the pillow and bolster. Jones took charge of Cook's watch, some coins, and family letters, but found no banknotes or a betting book. Once the body was laid out, the room was locked.

On 22 November, Palmer managed to gain admission to Cook's room and spent some time rummaging amongst his things. Prior to the Shrewsbury races, Palmer had had no money, and Cook was carrying over £700 in cash. After Cook's death, Palmer is known to have banked or paid out at least £400. He also had a paper that he said bore Cook's signature, which he wanted to be attested, showing that he was owed £4,000. He asked Cheshire to sign it, but Cheshire was dubious and refused.

Cook's stepfather, retired merchant Mr William Vernon Stevens, arrived that day and, when he viewed the corpse, he saw that the muscles of the face were still locked in spasm. Palmer claimed that he was owed £4,000 by Cook's estate, which Stevens doubted would be forthcoming. Palmer was clearly in a hurry to see the body buried, and offered to arrange the funeral, but Stevens told him the body was to be buried in London. Palmer then went out and, when he returned half an hour later, Stevens asked him for the name of a local undertaker and was astonished when Palmer said he had already been to one and ordered a coffin. Stevens was further surprised and concerned to be told that his stepson's betting book was missing, a vital document, which Palmer airily dismissed as unimportant.

Stevens returned to London, where he went to see a solicitor. Although he did not as yet suspect murder, he had decided that, for the sake of Cook's sister and half-brother, he should arrange for a post-mortem to determine the cause of death. On Sunday the 25th, Palmer persuaded Mr Bamford to sign a death certificate for Cook, giving the cause as apoplexy. On the same day, he got into conversation with Newton about using strychnine to kill a dog and, on asking what appearances he might expect to see in the stomach after death, was told it left none. 'It is all right,' said Palmer, as if to himself, with a snap of his fingers.[2]

Next morning, John Thomas Harland, a physician of Stafford, travelled to Rugeley to conduct the post-mortem. Harland had already heard that there was a suspicion of poisoning but, on arrival, was met by Palmer, who told him that Cook had had an epileptic fit and there was old disease in the heart and the head. Palmer said that 'a queer old man',[3] by whom he must have meant Mr Stevens, suspected him of having the missing betting book. Palmer later invited Newton to drink brandy with him and confided that Cook was diseased in the throat and had syphilis.

There were several men at the examination, including Harland, Palmer and Bamford. A young medical man, Charles John Devonshire, carried out the operation, assisted by Newton. The body was rigid, the head bent backwards and the feet so distorted they gave Cook the appearance of being club-footed. The tightly closed fists could only be opened by force, whereupon they at once closed again. When the body was opened, the medical men could find no disease to account for the death.

It would have been better had the stomach been removed, tied at each end, and sent for analysis intact, but, instead, Devonshire was asked to open it over a jar. As he scissored it open, Palmer, who was standing near Newton, gave him a push, which caused him to knock against Devonshire and some of the stomach contents spilled on to a chair. Harland said, 'Do not do that.'[4] The stomach and intestines showed no signs of inflammation, and Palmer smiled and said to Bamford, 'Doctor, they won't hang us yet.'[5] The stomach, which, according to Devonshire, had been suddenly turned inside out by Newton,[6] was dropped into the jar together with its contents. Harland covered the jar with two bladders, tied and sealed it and put it to one side, but, a few moments later, found it had disappeared. When he called out for it, everyone looked around and saw Palmer holding the jar, standing near a door at the other end of the room. He was asked to bring it back, and Harland saw that two slits had been cut in the bladders with some sharp implement, although nothing had been removed. He demanded to know who had done this but everyone in the room denied it. Harland replaced the cut bladders with new ones and re-sealed the jar. Palmer was very anxious to know what was being done with the jar and seemed unhappy about it going to London for analysis. The jar was taken to the house of a Rugeley surgeon, Mr Frere.[7]

Later on, finding that Mr Stevens was to be driven up by fly to take the train to London with the jar, Palmer attempted unsuccessfully to bribe the driver to overturn the fly. At this point, it might be hard to imagine anything Palmer had left undone that might have further invited suspicion, but his ingenuity in that respect had still not been fully realised.

The jar containing the stomach and intestines of John Parsons Cook arrived safely at Guy's Hospital where it was delivered to Professor Taylor, who was assisted in his examination by Dr George Owen Rees, lecturer in materia medica. Taylor later commented that the condition of the specimens was 'the most unfavourable that could possibly be'[8] for the detection of strychnine. 'The stomach had been completely cut from end to end; all the contents were gone; and the fine mucous surface on which any poison, if present, would be found, was lying in contact with the outside of the intestines, all thrown together.'[9] The contents of the jar had also been much shaken about during the journey to London. The remains did not exude any of the characteristic odours that would have identified poisons such as opium or prussic acid. The surface of the stomach was examined with a magnifying glass and no signs were found of ulceration, perforation or disease. No suspicious substances were observed. The yellowish mucus found in the intestines and the 'bilious-looking liquid of the surface of the stomach'[10] were tested for all the poisons that might cause sudden death, including strychnine, but the only poison identified was antimony, and there were only small traces of this. The failure to detect strychnine Taylor later attributed solely to the condition in which the mishandled remains had been received.[11]

The result was sent to Mr Gardner, an attorney employed by Mr Stevens. It was a private letter, and yet Palmer was able to discover what it said and wrote to the coroner Mr Ward telling him that Taylor had written from London to say that he and Rees had completed their analysis and found no traces of strychnine, prussic acid or opium. His letter ended, 'I hope the verdict to-morrow will be that he died of natural causes, and thus end it.'[12] To ensure a favourable verdict, Palmer purchased a turkey, a brace of pheasants, a codfish and a barrel of oysters, which were packed into a hamper and sent to Mr Ward.

The inquest opened at the Talbot Arms on Wednesday 12 December, and continued on Friday and Saturday. Palmer declined to appear. His debts had finally caught up with him, and, following a civil action, he had pleaded illness and taken to his bed. His optimism had been misplaced, his attempt at bribery only making him appear guiltier. Where science had failed, experience triumphed. Based on the detailed description of Cook's illness, Taylor attributed the death to tetanus, caused by medicine given shortly before death. Although he had not found any strychnine, he believed that only this could account for the symptoms, the poison being 'so speedily absorbed in the blood that in the course of an hour after administration no chymical [*sic*] test at present known could detect it'.[13]

It took the jury barely seven minutes to find that Cook died of poison administered to him by Palmer, who was committed for trial at the Spring Assizes. The police searched Palmer's house and found a medical book with a memorandum in Palmer's writing. It read: 'Strychnia kills by causing tetanic fixing of the respiratory muscles.'[14]

The inquest verdict, according to Palmer's counsel, 'flew upon the wings of the press into every house in the United Kingdom'.[15] The case was a national scandal before it came to trial. It involved the arrest of a medical man, the very kind of individual who should have inspired trust and respect, the curious mystery of how Palmer had learned the contents of Taylor's private letter, large sums of money, and rumours of numerous other suspicious deaths, including those of several of Palmer's associates and children. It was also the first prosecution for murder in the UK in a case of poisoning involving strychnine. Palmer had overcome the difficulty imposed by its bitter taste by incorporating it in pills.

If there is anything the press likes better than a poisoner it is a mass poisoner. Newspapers filled not merely columns but pages with the story, wallowing in thrills of horror as new details emerged of the many loathesome transactions of William Palmer. 'Disclosures of a startling character, and of a nature almost unsurpassed in the history of crime, are expected to be made before the termination of the enquiry into this extraordinary case,'[16] wrote the *Times* journalist, in an uncharacteristic appetite-whetting piece. 'I hear on every side of disclosures, that will startle and affright the public,'[17] revealed the *Manchester Advertiser*, although the *Staffordshire Advertiser* refused to publish 'rumours of the most exaggerated description ... derived entirely from the gossip of the neighbourhood'.[18] The *Daily News* went further. 'Every sudden death that has happened for years back seems likely to be raked up with a view to try whether some connexion cannot be established between Palmer and the deceased. The finger of suspicion is pointed at imaginary accomplices. Sweeping insinuations are thrown out against medical practitioners and insurance offices.'[19] The worry was that the deluge of indiscriminate gossip would make it impossible for Palmer to receive a fair trial in Staffordshire.

Further enquiries solved two questions. Taylor's letter had been intercepted and opened on Palmer's behalf by his friend Samuel Cheshire, an offence for which the postmaster was later dismissed and imprisoned. Meanwhile, medical men, hoping that Palmer might turn out to be an unqualified quack and therefore not 'one of ourselves',[20] were disappointed to discover that he had been awarded the diploma of the College of Surgeons in 1846.

The poisoning of John Parsons Cook suggested that a closer examination should be made of the deaths of both Palmer's wife Ann and brother Walter. Ann Palmer had died on 29 September 1854, and, during the nine months prior to this, Palmer had made proposals for insuring her life with eight different companies for a total of £33,000, only £13,000 of which was accepted. It was not then the practice of insurance offices to check with other insurers to see if multiple proposals had been made. By the time of his wife's death, Palmer had paid premiums of £338, which he had had to borrow on the security of the policies. There was some suspicion that Ann had died from poison, but no inquest was held and the body was hastily buried: '[T]he seeming respectability of *Palmer* [*sic*], his social and professional position, together with the two medical certificates of the cause of death of the wife, checked any intention which might have existed on the part of the offices to resist the payment of the policies.'[21]

Palmer next set about insuring the life of his brother Walter, an habitual drunkard, with proposals totalling £82,000. One office agreed to insure Walter's life for the sum of £13,000, and Palmer instructed the man who had charge of his brother to allow him as much alcohol as he wanted. This may have been too slow a method and, with money troubles looming, Palmer was thought to have hurried Walter to his grave with poison, probably a narcotic. Walter died suddenly on 16 August 1855. The insurance office, finding that Walter's state of health had been misrepresented, refused to pay.

On 21 December, the coffins of Ann and Walter Palmer were exhumed and carried to the Talbot Inn for the commencement of the inquests. The body of Mrs Palmer was relatively dry and its odour 'endurable',[22] but the stench as Walter's coffin was opened provoked vomiting and fainting amongst the jurors, and the revolting smell spread throughout the inn, which had to be fumigated to make it habitable. The spectacle of the corpse was, of course, 'absolutely frightful'.[23] A week later, the room where the coffins had been brought still smelt so strongly it had to be redecorated and part of the floor relaid.

Ann's remains, buried some fifteen months previously, were in an excellent state of preservation. Taylor found the body 'saturated with antimony, which I never found before in my examination of 300 dead bodies'.[24] The antimony had even been absorbed into her uterus and ovaries. Her death certificate had given the cause as 'bilious cholera' but, on exhumation, there was no trace of this or any other disease. Her death was entirely consistent with poisoning by antimony, which had not been prescribed to her. Taylor was unable to find evidence of either disease or poison in Walter's remains, and gave probable cause of death as 'apoplexy resulting from excessive drinking'.[25] It would have been impossible to find either morphine or prussic acid five months after death, and the symptoms caused by drink would have been the same as those caused by a narcotic poison.

In January, Professor Taylor was interviewed by the press, something he was later to bitterly regret. While newspapers regularly reported statements made by chemists and toxicologists during inquests and trials, paying particular attention to the words of celebrities like Taylor, it was unusual to seek out a great man for private interview, and Taylor was to demonstrate his inexperience of the press. Journalist Henry Mayhew,[26] who was conducting an enquiry into suspicious deaths associated with life insurance offices, called on him on behalf of the *Illustrated Times* with a letter of introduction from Professor Faraday. An article based on their subsequent conversation was published in the issue of 2 February,[27] which also included Mayhew's study.

> [Taylor] was decidedly of the opinion that secret poisoning was on the increase; but in order to strike the country with no unnecessary alarm he wished us to state explicitly, that the progress of chemical science had done more than keep pace with the progress of the hateful crime, and that if there was more poisoning now than in former years, the poisoner had more to fear from the skill of the analyst.[28]

Taylor had asked for all references to the Rugeley case to be struck out of the article before publication, but, on seeing it in print, he found that some remained. He was, in any case, naïve to imagine that even if they had been removed, his comments would not have been interpreted as referring to the case that was dominating the press.

Taylor also found his statements to the coroner widely misrepresented in other newspapers, which suggested, wrongly, that he had said that strychnine could not be detected in a dead body and that it was destroyed by putrefaction, when what he had actually said was that it could not be separated after being absorbed into the blood. Anxious that this misinformation would lead to a spate of murders, Taylor wrote a letter to *The Lancet* and unwisely commented on Palmer's case.

The *Illustrated Times* declared insurance poisoning to be 'The Crime of the Age'. The often-told historic poison dramas of Italy had been prompted by high passions, but now murder was 'done simply to turn a penny'. This new brand of murderer did not hate his victim, he might even like him, 'but a commercial calculation makes him pick him out as the best fellow for his purpose, and he kills a human being as coolly as a rat-catcher kills a rat'.[29] Chilling words indeed.

Money has long been one of the prime motives for murder: for insurance, inheritance, to conceal other financial crimes such as fraud, or simply to save on living expenses. The *Illustrated Times* had, however, identified a sea change that was taking place in poison murder and was to continue throughout the latter half of the nineteenth century – the gradual disappearance of the archetypical female poisoner who operated in an impoverished domestic setting, and the emergence of the educated middle-class male. This trend was also apparent in the fiction of the period, both novels and plays, which responded to and was often directly inspired by specific crimes. The murderer was a gentleman; cultured, unemotional, subtle, secretive and audacious. The reality was far less glamorous.

* * *

While press and public awaited the trial of William Palmer, two more unusual deaths attracted attention to the fact that even someone without medical qualifications could easily obtain powerful poisons. On 17 February, the body of disgraced MP, banker and fraudster John Sadleir was found on Hampstead Heath. He had committed suicide with prussic acid. The financial scandal that followed brought ruin and catastrophe to enormous numbers of small investors.[30]

In the same month, twenty-eight-year-old Harriet Dove of Leeds began to suffer repeated attacks of violent spasms, and died on 1 March. Her medical attendant Mr Morley learned that, on 10 February, one of his assistants, Elletson, had been approached by Harriet's husband William, who discussed the Palmer case and said (perhaps basing his opinion on the very reports to which Taylor had objected) that he believed strychnine could not be detected in the body after death. He asked for a supply to kill stray cats and was given ten grains. On 15 February, a cat was found poisoned and, a few days later, William obtained a further supply of strychnine. A post-mortem examination found strychnine in Harriet's stomach and intestines, and, on 17 March, an inquest committed William for trial on a charge of murdering his wife.

The Rugeley deaths aroused such violent public prejudice against Palmer that it was felt he could not have a fair trial in Staffordshire, and an Act of Parliament,[31] colloquially known thereafter as either the Palmer Act or Palmer's Act, was hurriedly introduced to enable the hearing to be transferred to the Central Criminal Court. The trial opened on 14 May 1856.

The Attorney General, Sir Alexander Cockburn, remarked in his opening statement on the national interest in the case. 'The peculiar circumstances of this case have given it a profound and painful interest throughout the whole country: there is perhaps scarcely a man who has not come to some conclusion upon the issue which you are now to decide. The details have been seized upon with eager avidity – there is scarcely a society in which the merits of it have not been discussed.'[32]

The prosecution's case was that Cook had first been weakened by antimony and then given the strychnine that killed him. Cockburn was careful to explain to the jury the particular symptoms, the convulsions and spasms, occasioned by strychnine, which have some similarity to the effects of tetanus, but pointed out that traumatic tetanus arises from a wound, and there was no sign of a wound on Cook's body. Idiopathic tetanus (i.e tetanus not arising from a wound), which arises from a chill and also produces spasms, was, he told them, rare and hardly ever fatal in England. Surgeon Sir Benjamin Brodie, described by Cockburn as 'probably the most distinguished medical man of the present age',[33] testified that the progress of Cook's condition was 'entirely different'[34] to that of tetanus. Cook's ability to swallow was a sure sign that he was not suffering from tetanus, where lockjaw is an early symptom.

The defence did its best to maintain that Cook had died of natural causes. One difficulty for both sides was that death from strychnine poisoning was rare. A great deal was made of the tragic demise of Agnes Senet, who had died in Glasgow Infirmary in 1845 after swallowing pills intended for another patient. The description of her symptoms – the paroxysms, the clenched hands, the rigidity of the limbs and the fact that she remained conscious throughout her ordeal – were similar to the death of Cook. In October 1848, Mrs Georgiana Sergison Smith died screaming with violent spasms after taking a mixture in which strychnine had been used in error for salicine, the two being kept side by side in the dispensary. The chemist concerned was tried for manslaughter at Winchester Assizes the following March and acquitted.

With William Dove incarcerated in York Castle awaiting trial, his case was referred to frequently, since it was a suspicious death in which strychnine had been unequivocally found in the remains. It also gave the defence counsel the opportunity to ask Elizabeth Mills, the chambermaid of the Talbot Arms who had described Cook's horrifying paroxysms, if she had read about the Dove case and was thereby familiar with the symptoms of strychnine poisoning, the insinuation being that she had simply repeated what she had read. She said that she was not familiar with the symptoms, and her testimony was clearly given and convincing.

The important question to be determined was whether after such a poisoning, strychnine would be found in the stomach. George Morley, a surgeon, had attended Mrs Smith, but had also conducted experiments giving strychnine to animals, chiefly frogs and guinea pigs. He had managed to find strychnine in cases where death had taken place as much as two months previously.

The fifth day of the trial opened with the evidence of Professor Alfred Swaine Taylor. Taylor, then in his fiftieth year, was a towering figure in the science of toxicology. Frequently called upon to advise in criminal proceedings, for the preceding ten or twelve years he had been engaged in all the most important cases of poisoning. Tall, dignified and assured, he cut an imposing figure in court, where he was a compelling expert witness. The Palmer trial, however, was to be a less comfortable experience.

Taylor testified that he had carried out experiments using strychnine with rabbits, giving it in both a solid and a liquid state, and saw that it acted much faster when given as a liquid. The poison operated by being absorbed into the blood and circulated around the body, and especially acted on the spinal cord.

His description of the suffering rabbits – the paroxysms, the head thrown back, the screams, the stiff limbs, the protruding eyes, the spasmodic closing of the jaws, the quiet death after a series of such fits, and the rigidity of the corpse continuing for several days after death – matched exactly the circumstances of the last illness of John Parsons Cook, although Taylor agreed that the symptoms of tetanus were very similar. He had, however, never seen the effect of strychnine on a living human subject.

Taylor told the court that vegetable poisons were far more difficult to detect than mineral. He had carried out tests on animals he had destroyed with strychnine and had not always been able to detect it. The antimony had certainly been given to Cook less than three weeks before death.

Taylor suddenly found himself under personal attack by Palmer's defence, since he had written the letter to *The Lancet* at a time when he was legally bound not to discuss the case in public. Taylor protested that he had not expressed an opinion on Palmer's guilt, and stressed that his real concerns were for public safety:

> I would observe that during a quarter of a century which I have now specially devoted to toxicological inquiries, I have never met with any cases like those suspected cases of poisoning at Rugeley. The mode in which they will affect the person accused is of minor importance compared with their probable influence on society. I have no hesitation in saying that the future security of life in this country will mainly depend on the judge, the jury, and the counsel who may have to dispose of the charges of murder which have arisen out of these investigations.[35]

He was especially annoyed by incorrect accounts of his evidence at the inquest: '[I]f those statements which I have seen circulated in medical and other periodicals were to have their weight, there is not a life in this country that would be safe. . .'[36]

The defence then introduced the article in the *Illustrated Times*, which, it was suggested, might have unduly influenced the jury. Taylor, indignant to see a caricature of himself and Dr Rees in the paper, claimed that Mayhew had misrepresented the purpose of his visit and that this falsehood was 'the greatest deception ever practised on a scientific man'.[37]

Henry Mayhew later wrote to *The Times*[38] denying the allegation that he had lied about the object of his visit, and provided the text of Taylor's letter to him dated 30 January, which granted him permission to publish the article, the proofs of which Taylor had seen and corrected. It was an extremely unedifying squabble in which the only conclusion was that one of the two respected gentlemen had not been telling the truth.

The defence brought a series of medical witnesses, who had varying amounts of knowledge and experience of convulsive disorders, in an attempt to persuade the jury that Cook had died a natural death. Several claimed, without fear that they would be asked to make a demonstration, that had strychnine been in the remains, they would have been able to find it.

The trial was a particular disaster for the Bristol chemist Mr William Herapath, whose fame, derived from the 1835 Mary Ann Burdock case, had since been overshadowed by Taylor. Although appearing for the defence, Herapath was obliged to admit that he had often expressed the opinion that the case was one of poisoning by strychnine, which Dr Taylor had simply been unable to detect. The Attorney General was scathing. He had every respect for men of science who came forward to state what they honestly believed was the truth, whichever side they took, but said, 'I have seen him mixing himself up as a thoroughgoing partisan in this case, advising my learned friend, suggesting question upon question, and that on behalf of a man whom he has again and again asserted he believed to be a poisoner by strychnine ... I abhor the traffic in testimony to which I regret to say men of science sometime permit themselves to condescend.'[39]

On 27 May, the jury found Palmer guilty of murder, and Lord Campbell directed that the prisoner should be executed in the county of Staffordshire. There were a few who spoke in Palmer's favour, one of whom even attempted to draw a parallel between Palmer and that proven innocent, Eliza Fenning.[40] As the day of execution approached, Palmer was repeatedly exhorted to make a confession, but he declared his innocence to the end. Perhaps the doubts and confusion he left behind were his final revenge. He was hanged in front of Stafford Gaol on 14 June. On 19 July 1856, William Dove was found guilty of the murder of his wife. He was hanged on 9 August.

The Palmer trial was to have lasting reverberations. The failure of science to find strychnine in Cook's body 'tended to shake the confidence of the public in chemical analysis', and the public disagreements at the trial 'conferred no honour upon the professors of chemistry'.[41] Taylor, who had cultivated a reputation as a dispassionate man of science, had revealed that he was not untouched by personal vanity. His report written the following October, in which he strongly criticised Herapath, quoted one commentator who said that the contradictions of the medical witnesses were enough to make 'wise men tremble, good men sad, and bad men bold'.[42] The problem, Taylor believed, was that the public, encouraged by the press, expected that if it were alleged that death was due to poison, then the experts should always be able to demonstrate its presence. It also left the unfortunate impression that '*with a little search medical men might be got to prove anything*'.[43] Taylor, emphasising the importance of the observation of symptoms in establishing the presence of poison, asserted that 'an absolute and blind trust in chemistry, as all-sufficient to settle a disputed case of death from poison ... would lead to the most serious consequences'.[44]

It was *The Lancet* that summed up the situation most comprehensively: 'Let a man be ever so honest, what surer method of biasing him can be devised, than to pay him a heavy fee, pit him against a scientific opponent holding diametrically opposite opinions, pique his self-love by setting him to prove himself right and everybody else wrong, and lastly, to stir up his bile with the vexatious cross-questioning and superficial impertinence of men whose peculiarity it is to be vexatious and impertinent.'[45]

The trial did, however, stimulate further research on the detection of strychnine, and, at the same time, poison murder was once again elevated to a matter of national concern. In 1859, a royal commission recommended that coroners should receive a salary and thus be independent of the financial control of Justices of the Peace. This passed into law in 1860 as the County Coroners Act. The result was a welcome increase in the number of inquests. In the next three decades, further reforms gave coroners greater powers and required that inquests must be held in all cases of sudden or suspicious death.

Meanwhile, in the wake of the Palmer trial, William Herapath must have been stinging with annoyance at his failure to outshine Taylor. It would take him three years to get his revenge.

CHAPTER FOURTEEN

Expert Witnesses

During the nineteenth century, there was a steady and substantial rise in standards of adult literacy. The provision of affordable newspapers for the masses was initially restricted by taxation, but, when stamp duty on newspapers was abolished in 1855 and excise duty on paper in 1861, penny newspapers became available, many of them illustrated with bold and eye-catching engravings, and these reached a wide readership. Dramatic press accounts of poison murders, which were often happy to repeat rumour as fact, aimed for the sensational at the expense of truthfulness, and encouraged the popular belief that homicidal poisoning was on the increase and that most poison deaths were the result of murder.

When poisoning featured as a murder method in Victorian fiction, there was rarely any great effort by the writer to depict its action with scientific accuracy or even to employ a known named poison. It is easy to ascribe this to laziness, or the writer not wanting hard facts to get in the way of the plot, but there may also have been an element of caution – concern that if details of an easy murder method were supplied, the reader might be encouraged to copy it. The emphasis in these works is not so much on the poison as on the character of the poisoner, who is usually depicted as clever, scheming and heartless, therefore eliciting terror in the reader.

Following the high-profile cases of Wooler, Sadleir, Palmer and Dove, legislation to control the sale of poisons, demand for which had languished after the 1851 Arsenic Act, became a hot topic in all sectors of the press, and an issue for debate in parliament.

On 10 July 1856, Lord Campbell, in the House of Lords, responding to the flood of publicity, 'was shocked to say that for several years past the crime of poisoning had become remarkably common, and in his opinion some new law was imperatively required for the regulation of the sale of poisons'.[1] Although arsenic was 'out of fashion . . . nux vomica had taken its place'.[2]

The Pharmaceutical Society, while still concerned with overall poison control, was pursuing its own professional demands, and opposed a succession of bills designed to restrict the sale of poisons. The main objection of the society was that legislators were seeking to make substantial impositions on its members without any corresponding benefit to their status. Another important question being debated at this time was the adulteration of food and drugs, a problem that ultimately led both to the appointment of public analysts, from whose ranks many future expert toxicologists were to emerge, and to enhanced status and powers for the Pharmaceutical Society.

Doctors remained anxious not to restrict the availability of cheap medicine for the poor. Dr G. Goddard Rogers, medical registrar of St George's Hospital, referred to the sale of poisons as 'a necessary evil in the druggists' trade' and thought it 'not very desirable' that non-professional persons should be prevented from obtaining laudanum. It would be hard to prohibit chemists from supplying sugar of lead to 'the poor man, who purchases a penny or twopenny worth to dissolve in water, and regards it as a panacea for almost every kind of superficial injury'. The sale of essence of bitter almonds, he thought, might be forbidden, although its culinary use was both 'harmless and agreeable'.[3]

An unusual case of poisoning sent a ripple of alarm through Bolton in 1856. Thirty-eight-year-old Betsey McMullan[4] was tried for the murder of her forty-four-year-old husband, Daniel, by giving him small regular doses of antimony. The couple had lived in extreme disharmony, and there were violent quarrels during which knives and rolling pins became airborne. From time to time, Daniel would go on a drinking spree lasting several days. There was, however, a remedy readily to hand for Bolton wives with drunken husbands. For a penny, they could buy five doses of a powder called 'quietners', sometimes rendered 'quietness' in the press, which consisted of tartar emetic mixed with cream of tartar. It was a remedy that was only ever asked for by women.

In the nine months prior to his death, Daniel had been seriously drunk on three occasions. These alcoholic bouts would begin on the Saturday and he would stay pretty well soused until the following Wednesday. The last such bender was three weeks before he died. Betsey had been giving her husband the powder on a regular basis for several months, producing sickness and burning pains in his stomach. Because of the small doses, many of the usual symptoms of poisoning, such as diarrhoea and cramps, were absent. Daniel had first consulted a surgeon, James Dorian, about his sickness in April, and, towards the end of June, he was again very ill. Dorian came to see him, and called in a Dr Chadwick, who asked both Betsey and her husband if the patient had had any medicine apart from what had been prescribed. Both said that he had not. 'You have not been giving him any powder at any time?' Chadwick asked Betsey, adding meaningfully, 'You know you ladies do practise a little on your husbands.'[5]

The McMullans' servant had suspected poisoning for some time. After seeing Betsey put powder into the bottle of mixture prescribed by Dorian, and finding a white deposit at the bottom of the medicine glass, she took it to be analysed. It was tartar emetic. Shortly before Daniel died, Betsey admitted giving a powder to her husband, but said that she did not know what was in it. Later analyses of Daniel's urine, liver and kidneys left the cause of death in no doubt.

The main defence at Betsey's trial for murder at the Liverpool Assizes in August was that she had not known that quietners were poisonous, and had no idea what antimony was. She had only done what other women in Bolton did: she gave her husband powders 'sold indiscriminately . . . to ignorant women and by stupid and ignorant shop boys'.[6] The jury was kind, although it was apparent that Betsey had been giving her husband the powders over a period of several months even when he was not drunk. She was found guilty of manslaughter and transported for life.

Mr Justice Willes, finding the case 'repugnant', stated that, if druggists sold poisons to married women 'without the knowledge of their husbands, well knowing that they are intended to be administered to their husbands, no matter for what purpose, and death ensues, [they are] equally guilty of manslaughter'.[7]

The case excited considerable comment in the newspapers, which deplored the reckless sale of poisons to the uneducated. The sale of quietners was all the worse because the poison was, unlike arsenic, sold not as a vermin killer or for use in a trade, but for the express purpose of being administered surreptitiously to the purchaser's husband in order to make him ill, an action that was itself a crime.

In May 1857, a bill was brought before parliament to regulate the sale of poisons in much the same way as the 1851 act had applied to the sale of arsenic. It was proposed to repeal the 1851 Arsenic Act and create a new act to include other poisons. One of the bill's recommendations was that an order should be obtained from a clergyman or Justice of the Peace for the supplying of certain substances. From June to August of 1857, a select committee of the House of Lords met fourteen times to consider the Sale of Poisons Bill and hear the evidence of witnesses from the Pharmaceutical Society and the medical profession.

The Pharmaceutical Society, whose deputation of witnesses was headed by one of the society's founders, Jacob Bell, raised a series of objections to the measures in the bill, many of which related to its overall practicability, and stated that 'security of the public would be better effected by an attention to the intelligence and qualification of the vendor'[8] than by the kinds of regulations currently before the Lords. The society believed that the Arsenic Act had 'done some good'[9] and that poison murders had diminished as a result, but, if other substances were added and certificates required, it would be too complicated and unworkable, and would amount to a prohibition. It was the chemist who was the greatest safeguard. 'It would be a great hardship upon the public to be deprived of the use of antimony and certain preparations of opium...'[10]

The society was mainly concerned with the measures to prevent accidental poisoning and, when the committee asked for any suggestion 'for the purpose of preventing intentional poisoning by evil designing persons',[11] the reply was, 'I do not know of any positive method of preventing it. If you prevent the sale of poison, you might as well prevent persons selling cutlery and ropes.'[12]

Many witnesses felt that the Arsenic Act had not been effective. A deputation from the Liverpool Chemists Association believed that it was 'frequently ignored',[13] and William Herapath said that the act, while reducing poisoning by arsenic, had simply 'driven the secret poisoner into other poisons which are more difficult to detect'.[14] He believed a limited schedule under a new act might be practicable, and listed the substances most often used for secret poisoning as arsenic, Fowler's solution of arsenic, prussic acid, strychnine and nux vomica.

Mr Frederick Crace Calvert of the Sanitary Association of Manchester dismayed the committee by revealing that arsenic was commonly sold uncoloured: '[T]hat is the great fault of many of these measures; the laws are very good, but there is nobody to see that they are carried out.'[15] Laudanum, he added, was frequently given to children. Asked if this was with the intention of killing them or 'merely as a quieting draught', he replied, 'For both purposes.'[16] Unlike the pharmacists, he saw no difficulty in a system of certification, such a system having already been in operation in France, where, he believed, judging from the French journals, the incidence of both accidental and homicidal poisoning was lower.

The most important witness was Alfred Swaine Taylor, who had made a detailed study of death statistics. He estimated that the average annual poison deaths for the previous six years was 536, a substantial increase on his 1837/38 figures, although greater accuracy of reporting might have been a factor. There was, unfortunately, no means of determining which of these deaths were due to murder, and the returns did not analyse the kind of poison used. From the admissions to Guy's Hospital, he believed that the incidence of poisoning with arsenic had declined since the 1851 act; however, he still believed that 'there is too great facility for procuring poison for criminal purposes, even now, although the Arsenic Act has done some good in that respect; I think it has reduced the number of deaths'.[17] He did not know personally of any instances where penalties had been imposed under the act, which he described as 'much violated'.[18]

Although it was generally agreed by all those with an interest in the subject that reform was required, year after year bills were introduced, discussed, opposed and dropped.

During the hearings of the 1857 select committee, a dramatic trial was taking place in Edinburgh. Madeleine Smith, daughter of a wealthy Glasgow architect, was accused of murdering her secret lover Pierre Emile L'Angelier, who had died from arsenical poisoning. Madeleine had purchased arsenic, and she certainly had motive. L'Angelier was ardent but poor and she had received a more promising offer of marriage. When she tried to end the romance, the spurned paramour threatened to show her father some highly compromising letters. The shock that the case excited was less about the poisoning than the scandal of Madeleine's passionate love-notes in which she referred to L'Angelier as her husband, and which were read out in court.

On Madeleine's behalf, it was claimed that she had bought the arsenic as a cosmetic, and it was possible that the disappointed lover had committed suicide. There was also confusion about the dates of the incriminating letters. Importantly, no evidence was presented that the two had met at the time the arsenic had been taken. The defence raised an old ghost, that of Eliza Fenning. It had been suggested by the prosecution that the courage shown by Madeleine when charged with the crime was not incompatible with guilt, since a woman with the nerve to commit such a murder would also have the nerve to meet the accusation calmly. During his defence, John

Inglis, the Dean of Faculty of Advocates, put forward Eliza as an example of a young girl who was as brave as she was innocent, who died on the scaffold 'as serene as an angel . . . The true perpetrator of the murder confessed too late to avoid the recalling of that bloody action. Gentlemen, that case is a matter of history.'[19] Unfortunately, Inglis did not enlighten the court as to which one of the several rumoured confessions he was referring. His closing speech for the defence was held by onlookers to be masterful in its emotional eloquence. 'Perhaps the most telling part of the whole was his introduction of the story of Eliza Fenning, and the warning which he drew from it.'[20] Reasonable doubt enabled the jury to return a verdict of 'not proven' and Madeleine was released.

Press reports of the Dean's moving speech prompted a letter to *The Times* from J. H. Gurney, Rector of St Mary's Marylebone, whose late uncle William Brodie Gurney had been a shorthand writer to parliament. W. B. Gurney had kept a little notebook in which he jotted interesting anecdotes, one of which concerned Eliza. Shortly after her execution, he was told that the Revd James Upton, a Baptist minister, had visited her while she was under sentence of death, and, when he entered her cell, she at once greeted him with great earnestness and tears, assuring him that she was innocent. He had spoken to her gently, explaining that he had not come to her about the charge, but as a minister, and that, unless she repented of her sins, she could have no good hope of eternity. 'Before I had done she was quite melted down, and then it all came out.'[21] Eliza confessed everything to Upton, but, to subsequent visitors who doubted her guilt and suggested that she might still hope for a reprieve, she continued to assert her innocence. Not everyone was convinced by Gurney's revelation. The author of a piece about Eliza in Charles Dickens's *All the Year Round*, published in 1865, declared 'a deep conviction that she was entirely innocent' and refused to accept the statement of Revd Upton. 'Weak ministers, unaccustomed to giving spiritual advice to persons condemned to death, get flurried and confused in a prison cell,' commented the article, suggesting that, out of previous prejudice, Upton 'may have mistaken or exaggerated' what Eliza said.[22] On what evidence the author had decided that Upton, who was in his mid-fifties when he saw Eliza, was weak, inexperienced and easily confused was not revealed.

By the 1850s, expert witnesses and their detailed and complex testimonies had become a staple feature of the high-profile poison trial. Scientists have always disagreed, but they had previously done so in private, or in the pages of scientific journals, with the occasional sarcastic letter to *The Times*.

In the adversarial criminal trial in the UK and US, the prosecution and the defence appointed and paid their own teams of doctors and chemists, who were, therefore, obliged to disagree with each other on a public platform, an event widely reported and commented upon in the local and national press. (By contrast, in France, experts were appointed by the court from a list of approved specialists. A similar system was used in Germany.)

The confident certainty of a witness could evaporate in court where his expertise was being challenged in a way to which he was unaccustomed and he had no time to make a considered response. Men of science, who thought of themselves as trusted and respected members of society, were most reluctant to make a public admission of error or of acting in any but the most correct professional manner. Time and again, juries were subjected to the sight of an unfortunate doctor squirming in the witness box and saying everything he could think of to preserve his reputation. A high-profile trial could not only make or break an individual's career, but also influence public perception of the reliability of medicine and chemistry. This concern was increasingly being addressed by lectures, articles and books on medical jurisprudence, which gave advice to faint-hearted men of science on how to give evidence in court. The ability to give convincing evidence is, however, as much a factor of personality as expertise, and assurance of manner is no guarantee that the speaker is not in error.

In mid-nineteenth-century England, three men stood above the rest: William Herapath, Henry Letheby and Alfred Swaine Taylor had testified at many trials;[23] their names were known to the public; and they were listened to with particular attention. A man of Taylor's stature especially, the author of a standard text on medical jurisprudence, enjoyed an authority greater than his contemporaries. The opinion of a distinguished man whose work was at the forefront of medical knowledge gave the jurors a sense of safety as they explored the mass of evidence before them; he provided a path they felt they could follow with confidence. This image had been dented by the Palmer trial and, together with the entire science of toxicology, was to be severely tarnished just three years later.

* * *

The main thing that everyone was agreed upon regarding Dr Thomas Smethurst was that he was a scoundrel. In 1859, however, a jury was asked to determine whether or not he was a murderer. In 1828, at the age of twenty-three, he had married Mary Durham, who was some twenty years his senior. He became a licentiate of the Society of Apothecaries in 1830. Later, he obtained a medical degree from Erlangen,[24] probably, as was widely believed, by purchase, and practised for several years before setting up a successful water cure establishment. By 1853, he had retired from medicine, with income from property of £240 a year.[25] In 1858, he and his wife were living in comfort at a boarding house in Bayswater. Another resident of the house was Isabella Bankes, an attractive forty-two-year-old spinster of good family and in possession of private means.[26] Her personal fortune amounted to £1,740 and she also had a life interest in £5,000, which would, on her death, pass to members of her family.

It soon became apparent to the landlady that the friendship between Dr Smethurst and Miss Bankes had passed well beyond the bounds of propriety, and Miss Bankes was asked to leave the house. She did so, and, soon afterwards, Smethurst joined her. The pair were bigamously married on 9 December and took up residence together in Richmond.

Smethurst continued to support his wife, writing her letters and paying occasional visits. Miss Bankes was in generally good health, although occasionally subject to bilious attacks. Towards the end of March 1859, however, she was taken ill, and, on 3 April, Smethurst called in a Dr Julius, saying that his 'wife' was suffering from violent diarrhoea and vomiting. Julius was shown some vomit of a peculiar grassy green colour, but did not then suspect anything out of the ordinary, and prescribed a mixture. Two days later, the symptoms had become more distressing, and the patient was complaining of a sore mouth and a burning sensation in the throat and intestines. Julius prescribed further medicines, including bismuth and grey powders composed of mercury and chalk. He began to suspect poisoning, and asked his partner Dr Bird to see the patient, which he did on 18 April. No remedy alleviated her condition and she became steadily weaker. Julius and Bird discussed the case, and agreed that the symptoms could be attributable to irritant poison, possibly arsenic or antimony given in small regular doses.

Although Smethurst constantly attended on Miss Bankes, giving her medicines and food and taking away any eliminations, he sometimes travelled to London, and the doctors called to see her while he was away. The symptoms continued undiminished in his absence. On 18 April, Smethurst wrote to Miss Bankes's younger sister Louisa asking if she would visit. Louisa found Isabella very ill and much changed since they had last met. Isabella complained that some tapioca in the room had a nasty taste, but Louisa's offers to make up more tapioca or blancmange were rejected by Dr Smethurst. He then gave Isabella a saline draught, which immediately made her very sick. Louisa suggested that a medical relative call to see

Isabella, but Smethurst objected. Isabella was very anxious to see her sister again, but, after Louisa's return home, Smethurst wrote several letters to her saying that Isabella had been made ill from the excitement, and the doctor had forbidden any more visitors.

On 28 April, a consultant, Dr Todd, was sent for. Noticing Isabella's 'peculiar expression . . . a terrified look, such as I have never before observed in a patient',[27] Todd also came to the conclusion that Miss Bankes was suffering from the effects of a mineral poison. She failed to improve under his treatment.

On 30 April, Smethurst went to a solicitor with a draft will asking him to come next day and execute it. The will, signed on 1 May in her maiden name by Miss Bankes, who described herself as a spinster, left all her possessions, except for a brooch, to 'my sincere and loved friend, Thomas Smethurst'.[28]

Smethurst wrote to Louisa again, this time asking her to come down to Richmond and take rooms nearby. She arrived the following day to find her sister very weak. Louisa brought some jellied soup, to which Dr Smethurst added hot water, saying he would take it out of the room to cool. She heard him stirring it, and he returned a minute later and gave some of the soup to Isabella, who immediately vomited it up again. Some arrowroot he brought in from outside the room had the same effect. Smethurst told Louisa that she had better leave, adding that pills prescribed by Dr Todd had given Isabella great pain and caused burning sensations all over her body.

The doctors agreed that Isabella's stools should be analysed. Dr Julius told the Richmond magistrates of his suspicions and they supplied an order for the analysis. Samples were collected, bottled and sent to Professor Taylor, who performed a test on 1 May using Reinsch's process. He found arsenic. On 2 May, Louisa again called to see her sister, and Smethurst sent her out to get some medicine. When she returned, Inspector M'Intyre of the Richmond police was there to take the doctor into custody.

Smethurst was brought before the Richmond magistrates, protesting that any poison was in the medicines sent by Dr Julius. A letter was found on him addressed to his wife saying that he would see her soon, having been delayed in town on a medical case. This was later interpreted to mean that he planned to go back to live with his wife, but it could equally have shown that he simply divided his time between his wife and his mistress.

Smethurst pleaded that he needed to be at home as the lady might die in his absence, and he was released on his own recognisances and returned to Isabella's side. The police, however, came with him and took possession of a number of bottles, which were sent to Taylor. Smethurst explained to Louisa that Dr Julius had accused him of poisoning Isabella when really it was Julius and the other doctors who were killing her. Louisa had, meanwhile, prepared some invalid food for Isabella, who was able to eat it without being sick, and she stayed with her sister overnight. On 3 May, however, Miss Bankes died, and, when M'Intyre learned of this, he arrested Smethurst for wilful murder.

At the post-mortem examination carried out on 4 May, Miss Bankes was found to be between five and seven weeks pregnant. The viscera were sent to Taylor, who found a large patch of effused blood in the stomach, but no ulceration or inflammation. The large bowel, however, showed very extensive ulceration. The mucous membrane of the caecum was almost destroyed by inflammation and ulceration, the effect diminishing towards the colon and rectum. All the medical men who saw the remains were of the opinion that Miss Bankes's death was probably the result of an irritant. The pregnancy alone could not account for the symptoms, and the only natural disease that might have produced them was acute dysentery.

No arsenic was found in the body, but tests on the caecum, the small intestine, one of the kidneys and some blood taken from the heart revealed small quantities of antimony. Nothing untoward was found in any of the medicine bottles except one. This bottle, which was to become the infamous 'bottle 21', contained a solution of potassium chlorate, a common diuretic sometimes prescribed for bilious complaints. Although the bottle had come from Dr Julius's surgery, it had originally contained a quinine mixture. Taylor, who had long preferred the Reinsch test to that of Marsh, experienced an unusual difficulty with the contents of bottle 21, as the liquid dissolved the copper gauze he used in his tests. He continued to add more and more copper gauze until it could no longer be dissolved, and, on the next one, there finally appeared a deposit of arsenic. The original copper, he decided, had dissolved because of the presence of the potassium chlorate, which, he deduced, had been used by Smethurst to facilitate the absorption of small regular doses of arsenic and speed up its elimination from the body. This explained the failure to find arsenic in the remains. It was another case of the slow scientific poisoner.

At both the inquest and the committal proceedings before the Richmond magistrates, Professor Taylor gave evidence in his usual assured and authoritative style, confirming that he had discovered arsenic in a solution of one grain to the ounce, in bottle 21, and that, before conducting the test, he had first tested all his reagents and found them to be pure. 'I can only account for death by supposing that it had been the result of antimony and arsenic administered in small doses and at intervals.'[29] Dr Smethurst, who was now being openly described as a 'second Palmer' in the press,[30] was committed to be tried for murder. The obvious motive assigned to Smethurst was financial. During the month of April 1859, he had about £100 or £150 in his bank account. On 16 April, £71 5s., a dividend due to Miss Bankes, was paid into his account.

The trial opened at the Central Criminal Court on 9 July before Lord Chief Baron Pollock, but, in view of the illness of a juror, it was postponed to 15 August.

As was usual in poison trials, the first task of the prosecution was to show that the deceased had been poisoned, and then to show that the prisoner had administered the poison. Since dysentery was the only natural disease that had been mentioned as a possible cause of Miss Bankes's symptoms, a number of doctors who had practised abroad and were familiar with it were brought to give evidence and agreed that, on the basis of the symptoms described, Miss Bankes had not suffered from dysentery. The details of the post-mortem examination were so unpleasant that the foreman of the jury fainted and had to be taken out into the fresh air to recover. The Lord Chief Baron made the extraordinary observation that it was 'quite unnecessary to go into those matters with such minuteness particularly as the jury probably would understand very little of such a subject'.[31]

The dispenser to Drs Julius and Bird testified that there was no arsenic or antimony in any of the medicines sent to Miss Bankes, and that all poisons were kept in a locked cupboard.

When Taylor was called to give evidence, he made an astonishing and professionally embarrassing revelation. There had, after all, been no arsenic in bottle 21. After the inquest, he had re-tested samples of chlorate of potash and copper in case they contained impurities, and, by 7 June, he had established that the copper gauze he had used in his analysis contained arsenic. He asked William Brande, veteran professor of chemistry at the Royal Institution, to carry out further tests on the contents of the bottle. The conclusion was inescapable: all the arsenic found by the Reinsch test had been introduced by the gauze. Taylor at once informed the crown prosecutor, but consoled himself with the assertion that, although his conclusions had been disputed before, he had never previously been proved wrong, and that the ability of copper to contaminate the Reinsch test was previously unknown. He remained of the opinion that Miss Bankes's death was due to an irritant poison.

The defence brought a number of respected medical and scientific witnesses, who testified that death was due to natural causes and, had Miss Bankes been poisoned, then poison would have been found in her remains. A Dr Thudicum, lecturer in chemistry, said that he had analysed grey powder and bismuth and found that they contained impurities of arsenic and antimony, which he believed could account for the traces discovered by Dr Taylor. Mr Serjeant Ballantine, for the prosecution, while stating that Miss Bankes had died from poison, was unable to say whether that poison was arsenic or antimony, and when or where it had been administered.

Mr Serjeant Parry, conducting the defence, pointed out that Miss Bankes had not been under the exclusive control of the prisoner. No poison had been traced to Smethurst, and no arsenic had been found in the body. The same test that had mistakenly suggested that there was arsenic in bottle 21 had also shown arsenic in the stool, and that might also have been mistaken.

Despite this apparently compelling defence, which raised substantial reasonable doubt, on 19 August, Dr Smethurst was found guilty of the murder of Isabella Bankes and sentenced to death. The sentence was to be carried out on the Tuesday fortnight. A barrister who had been present for most of the trial later described the proceedings as a 'sham' and a 'mockery' in which spectators, witnesses, counsel, judge and jury 'seemed weighed down, absolutely unable to escape from some mysterious weight hanging over their imaginations, which impelled them to a belief in the prisoner's guilt'.[32] Physician Leonard A. Parry, in his introduction to *Trial of Dr Smethurst,* believed that the atmosphere was created by 'the terrible preliminary mistake made by Dr Taylor, a mistake which grossly prejudiced the position of the prisoner'.[33]

There was an immediate outcry in the press, not so much because Smethurst was believed to be innocent, since many commentators agreed with the jury, but because it was felt that the evidence was flawed, and the trial had not succeeded in proving him guilty. Although the Baron had not dissented from the verdict, he had, when passing sentence, omitted to tell the prisoner, as judges often did, that he should have no hope but should prepare himself to leave this world, a statement usually taken to indicate a judge's intention to advise against any commutation.

The case was vigorously argued in *The Lancet*, the *British Medical Journal,* the *Medical Times and Gazette* and *The Times*. No one approved of Smethurst; the *BMJ* echoed public opinion by referring to him as 'a man with whom no honest mind can feel the least sympathy – a liar, a cheat, a scoundrel of the blackest dye in every relation of life',[34] and suggested that Miss Bankes's pregnancy, the start of which coincided with the commencement of her illness, accounted for all the symptoms. The *BMJ* also pointed out that, in almost every recent case of poisoning, unmistakable quantities of poison had either been found in the body or been traced to the possession of the murderer. In dealing with small traces, one had to be very sure that there was no error. 'We are very far from having reduced chemistry to an exact science. This trial, indeed, proves how far.' If Smethurst had been tried immediately after his committal, 'his life would have fallen through the illusive gauze wire, and Dr. Taylor might possibly, when too late, have discovered his lamentable error'.[35]

Dr Letheby, while making no comments on the verdict, believed that the errors exposed by the trial would 'tend to encourage crime by leading the evil-minded to believe that the researches of chemical science are so beset with difficulties as to leave its conclusions open to doubt',[36] offering both a defence for the guilty and the means of bringing an unjust charge against the innocent.

Even Taylor's claim that the error was due to a fact previously unknown to chemistry was repudiated. J. E. D. Rodgers, former lecturer in chemistry at St George's School of Anatomy and Medicine, had given evidence for the defence at Smethurst's trial. He did not believe that potassium chlorate would speed absorption in a case of slow poisoning, as Taylor had suggested, and now stated that the mixture of potassium chlorate and the hydrochloric acid used as a solvent in the Reinsch test 'has long been known as one of the most powerful solvents, actually used to dissolve and separate copper from its ores'.[37]

The outspoken *Dublin Medical Press*, which concerned itself not only with medical matters, but also with wider issues such as public morality, tore into Dr Taylor without mercy: '[T]he man without whose assistance no criminal suspected of poisoning could be found guilty in England; the man whose opinion was quoted as the highest of all authorities at every trial where analysis is required, is the same who has now admitted the use of impure copper in an arsenic test where a life hung upon his evidence, the same who has brought an amount of disrepute upon his branch of the profession that years will not remove . . . We must now look upon Professor Taylor as having ended his career, and hope he will immediately withdraw into the obscurity of private life . . . He can never again be listened to in a court of justice. . .'[38] Not only was Taylor's twenty-nine-year career (for almost twenty of which he had used the Reinsch test) coming under serious scrutiny, but his sniping against rivals such as Letheby and Herapath was being openly criticised. Taylor's failure was also toxicology's failure, when, commented *The Times*, in cases of secret poisoning, we are to 'pin our faith upon the conclusions of chymists, and hang a fellow-creature because a small crystal, so minute it can only be recognised by the microscope, is exhibited on a scrap of copper wire'.[39] Toxicology, which had laboured for so long to make the invisible visible, had, over the years, become so refined and sensitive that the clues it relied upon had once again become invisible to the non-scientific observer.

The *BMJ* was particularly outraged that professional jealousy should intrude into a matter of law:

> The farce must no longer be exhibited to the world of the three most celebrated toxicologists of the country contradicting each other in matters where there should be no possibility of doubt. Dr Taylor, throughout his ample volume on poisons, never allows an opportunity to escape of sneering at the scientific attainments of Mr Herapath; and the latter gentleman, we must confess, is not behindhand in returning the compliment. Dr Letheby is equally complimentary to Dr Taylor . . . The public should no longer have exhibited to them the spectacle of professional men stabbing each other's reputation over the bodies of malefactors.[40]

The *BMJ* also raised a point that was to cause increasing concern as the science of toxicology advanced. Was a jury of non-scientists able 'to thread the delicate paths of such a case as Smethurst's'?[41]

Herapath, dining on a nicely chilled dish of revenge, wrote to *The Times* saying that the amount of arsenic that Taylor had extracted from bottle 21 was stated to be more than could have been contained even in the copper gauze, adding triumphantly, 'the fact is, the whole set of operations were a bungle'.[42]

There was no court of criminal appeal, and no legal error had been committed that might have given grounds for a retrial. The only possible route to upset the verdict was through the Home Secretary Sir George Cornewall Lewis, who was bombarded with correspondence on the case, much of it from doctors who believed that Miss Bankes had died of natural causes. Thirty medical men sent a petition asking him to exercise the royal prerogative of mercy, and twenty-nine barristers petitioned for the sentence not to be carried out. With only days to go before the execution, Smethurst was respited.

The documents were passed to the Lord Chief Baron, who advised that further enquiries ought to be made and, on 15 November, the Home Secretary took an unprecedented course of action. All the medical evidence was turned over to Sir Benjamin Brodie. Following receipt of Sir Benjamin's report, a copy of Lewis's reply to the Baron on 15 November was forwarded to *The Times*.

The Home Secretary had decided, in view of the doubts about Smethurst's guilt, to grant him a free pardon. The necessity had arisen not from the conduct of the trial, but 'from the imperfection of medical science, and from the fallibility of judgement, in an obscure malady, even of skilful and experienced medical practitioners'.[43]

Was Dr Smethurst guilty? Reading the evidence today and comparing it with other poisoning cases, the most damning testimony is that of Isabella's sister Louisa, who described the excuses and evasions of Dr Smethurst to keep her away from the invalid, the removal of items of food briefly from the room on some pretext and his return with them, and feeding of them to the patient, who instantly vomited. These are some of the classic behaviour patterns of the secret poisoner. On the other hand, if he had been trying to acquire Isabella's money, he left it very late in the day to arrange for a will. On any analysis, however, the prosecution had failed to prove its case beyond reasonable doubt, and Smethurst might well have been acquitted had Taylor's initial error not prejudiced opinion from the start.

In a conversation between the Home Secretary and the Lord Chief Baron, the judge expressed the opinion that Smethurst had administered drugs to his mistress not to kill her, but to obtain an abortion. Smethurst objected strongly to this suggestion, making the valid point that Isabella had only been five to seven weeks pregnant when she died, the illness beginning on 27 March and ending with her death on 3 May, a period of five weeks and two days. No one could have known at the end of March that Isabella was pregnant. Smethurst also challenged the idea that a medical man would have administered mineral poisons to obtain an abortion, adding, in a comment that even today would be considered monumentally tasteless, 'it is generally known in the profession that a stiletto would have done the business in the course of an hour or so'.[44] Although criminal history is rife with cases of men murdering pregnant wives and mistresses, the same objection arises – it is probable, especially with Isabella's illness masking any natural symptoms, that it was not known that she was pregnant until the post-mortem.

Perhaps, however, the pregnancy was a vital factor in Isabella's death, since her illness commenced very soon after she conceived. Isabella was known to have a bilious complaint before her cohabitation with Dr Smethurst. Her Bayswater landlady commented that she had sometimes had to retire from the dinner table because of this. If Isabella had been prone to what is now known as ulcerative colitis, the onset of pregnancy might have created a flare-up of this condition that Victorian doctors were unable to treat, and, over the five weeks of her illness, she would have wasted away from malnutrition and dehydration.

On 1 December 1859, Smethurst was tried at the Old Bailey for bigamy and sentenced to a year's imprisonment with hard labour. After his release, he returned to live with his wife. On 25 April 1862, he applied for proof of the will of Miss Bankes, the deceased's family claiming that her illness rendered her not competent to make the will and that she had been acting under the influence of Dr Smethurst. Much of the trial evidence was repeated, with Dr Taylor admitting to a mistake in one analysis, adding, in stubborn denial of the point made by J. E. D. Rodgers, that it was 'in consequence of a fact then new to chemists'.[45]

Although Dr Smethurst was deemed 'a man against whom human nature must rise up',[46] the will was held to be valid. On 11 November, he obtained grant of probate and took possession of Isabella's legacy.

Smethurst later petitioned the Home Office for compensation for the six months of imprisonment between his arrest and reprieve, saying that he had been financially ruined by the case and was unable to resume his profession. The application was refused with a comment on the file: '[M]ost people think that he had a very lucky escape.'[47]

CHAPTER FIFTEEN

Deadly Doctors

Public concern about the prevalence of poison and the dangers of homicide, accidental poisoning and adulteration of foodstuffs led to new demands for legislation, either to control the sale of poisons or to make chemists legally responsible for harm caused by negligence, or both. The mass poisoning that occurred in Bradford in 1858 from the accidental inclusion of arsenic in peppermint lozenges was still very fresh in memory.

Despite his error in the Smethurst case, Alfred Swaine Taylor was still considered to be the leading UK expert on poisons, although his reputation was undeniably dented. He never again achieved the status of an implicitly trusted authority, and he was not awarded the honours in recognition of his long and distinguished career that he might otherwise have expected.

He continued to be called as a trial witness and, in 1864, he was commissioned to advise the Privy Council about the dangers of poison. 'So long as a person of any age has the command of threepence he can procure a sufficient quantity of one of the most deadly poisons to destroy the lives of two adults,' he reported. If refused by a druggist, he could go to a grocer, and if refused by a grocer, he could go to 'a village general shop, where poisons are retailed by girls and boys, and no questions are asked'.[1] Despite the 1851 act, arsenic, even pure white arsenic, was, he said, still freely available at any general shop at a penny or twopence an ounce. The other poisons were sold without restrictions and were 'readily obtainable by the poorest or the most casual applicant'.[2] The annual report of the Privy Council's medical officer concluded unhappily that the extent to which the administration of poison caused disease or death was 'not even approximately known'.[3]

Deploring this 'disagreeable intimation', *The Times* commented: 'If we could be guided by the returns of the Registrar General, we should conclude that from 400 to 500 persons a year fall victims to either the accidental or the wilful administration of poison.'[4] The number of poisonings that remained undiscovered could hardly be guessed at. 'Of course the murderous poisoner plans to avoid detection, and too often, as familiar cases have proved, with considerable success.'[5]

In 1865, Taylor reported to the select committee on the Chemists and Druggists Bills, urging the necessity of legislation to further restrict the sale of poisons. No legislation, however, could prevent murder by someone who was qualified to obtain poison for legitimate purposes, and had both the knowledge to use it and the confidence afforded by his class and education to enable him to outwit authority. Sometimes, however, a doctor who thought he was clever enough to get away with murder, especially when using a poison he felt sure was undetectable, met a forensic expert who was more than a match.

Auguste Ambroise Tardieu, born in Paris in 1818, was a distinguished and respected figure in nineteenth-century forensic medicine and toxicology. Known for the clarity and precision of his publications, he also had the ability to look beyond outward appearances and use his knowledge and ingenuity to outwit criminals who thought they had covered their tracks. He first achieved public fame in 1847, when he examined the crime scene following the brutal murder of the Duchesse de Praslin. Her husband claimed that she had been murdered by burglars, but a reconstruction of the crime exposed the Duc as the killer. Praslin poisoned himself rather than face shame, ruin and the guillotine. In 1863, it would take all of Tardieu's skills to trap a far more cunning murderer.

Edmond-Désiré Couty de la Pommerais, born in Neuville-aux-Bois around 1830,[6] was a homeopathic doctor practising in Paris. Claiming to be a count, he found that neither this subterfuge nor his practice brought him the riches he needed to pay his debts and support his mistress Séraphine de Pauw. Her husband, a painter, who had been a patient of Pommerais, had died in 1858, leaving his thirty-six-year-old widow struggling to support three children. In 1861, Pommerais abandoned Séraphine and her children to marry the wealthy Mlle Clotilde Dubizy. For two years, he did not see Mme de Pauw, and even refused to call on her children when they were ill. Unfortunately for his ambitions, his mother-in-law, certain that Pommerais had married solely for money and that his claims to be a man of property were spurious, kept a firm hold on her daughter's fortune, and neither his pleading nor his threats would make her change her mind. Mme Dubizy went so far as to inform M Antoine Claude, head of the

Paris Sûreté, that her son-in-law wished to murder her. Claude had encountered Pommerais previously, when he had asked to be appointed as a prison doctor, and was well aware that the young homeopath's claims to a title were fraudulent. Many years later, Claude described Pommerais in his memoirs as a wretch who had descended upon the Dubizy family like a mad wolf on a flock of sheep,[7] but, since the angry mother-in-law could not prove her allegations, there was, at the time, nothing he could do.

Not long after this visit, and just two months after her daughter's wedding, Mme Dubizy fell ill with sudden violent vomiting after dining with her son-in-law, and died the same night. The cause was assumed to be cholera, but Claude was sure she had been poisoned. Pommerais was now able to live in comfort, but he wanted more – he wanted to be wealthy. He was looking for a way to make another fortune, and, having got away with murder once, thought he could do so again.

In June 1863, Séraphine de Pauw was living in poverty with her children, the eldest of whom was eight, when her former lover unexpectedly returned, bringing her a small gift of money, seductive ways and a scheme to provide for her children.

The first part of the plan was for Mme de Pauw to take out several policies insuring her life. She was then to assign the policies to Pommerais, informing the insurers that she did so because he had loaned her large sums of money and wished to provide for her children in the event of her death. The next step, he explained, was for her to feign illness, and then, when she was apparently dying, arrange for the policies to be exchanged for an annuity of 6,000 fr. (£240), something to which the insurance company would no doubt agree. Finally, she would make a miraculous recovery and the two would live together on the proceeds of their fraud.

The lady was medically examined and certified to be in good health, and, in July, eight policies of insurance were accepted for a period of three years, which would pay out 550,000 fr. (£22,000) in the event of her death. The annual premium payable, in two instalments, was 18,840 fr. (about £750). Pommerais had to stretch all his resources to pay the first instalment, and, in August, the policies were assigned to him. He had no intention of making the second payment, which would become due in January and was, in any case, beyond his means. To ensure that there would be no claims from Mme de Pauw's family, he asked her to make a will in his favour.

Mme de Pauw duly played her part. In September, she summoned several doctors to examine her, claiming that she was suffering from stomach pains after a heavy fall down some stairs. They found nothing seriously the matter. On 12 November, in preparation for the final part of the scheme she declared that she felt very ill and took to her room. Four days later, however, she was in the best of spirits and, after lavishing careful attention on her appearance, spent the evening with Pommerais. Next morning, a woman who arrived to deliver bread found her in great pain, the bed and floor stained with vomit. Although Mme de Pauw dismissed her illness as indigestion, she died the following day. Pommerais obtained a certificate to say that the cause of death was gastritis and perforation of the stomach, probably as a result of internal injury from the fall. He then arranged his mistress's burial and tried to collect on the insurance.

Unfortunately for the schemer, Mme de Pauw had previously confided in her sister, Mme Gouchon, that she and her lover were planning to defraud the insurance company and that the fall was a fake. The police were informed, and Pommerais was arrested. The body of Mme de Pauw was exhumed thirteen days after her death, and the remains examined by Tardieu, who had been appointed professor at the University of Paris in 1861, and pharmacist François-Zacharie Roussin. Nothing in the appearance of the corpse suggested that Mme de Pauw had been suffering from a chronic illness, and there were no external injuries. The viscera did not present any unusual appearances, and neither the standard chemical tests of the day nor microscopic examination revealed any traces of poison. She was about eight weeks pregnant.

Tardieu and Roussin felt certain that the victim had died from a poison of vegetable origin, which they were unable to positively identify. Pommerais had, however, recently made large purchases of digitalin, far beyond the requirements of his medical practice, only a small amount of which remained. Digitalin, the active constituent of the foxglove, is a powerful poison, causing, amongst other symptoms, nausea and a lowering of the pulse. The analysts' attempt to separate poison from organic matter by dialysis was unsatisfactory, so they were obliged to fall back on an old technique, the use of animals, albeit with some modern twists. They prepared alcoholic and aqueous extracts of the contents of Mme de Pauw's stomach and intestines, and administered some to a pigeon, which died very quickly.

A frog was then operated on to expose its heart, and it was demonstrated that the frog was able, under these conditions, to live on for some time. When another frog was similarly operated upon and a drop of extract was placed on the heart, the beat slowed and the animal soon died. A drop of digitalin had the same effect. The mixed extracts were then introduced into the thigh of a dog. The animal vomited twice, and its pulse slowed and became irregular and intermittent, its respiration both deep and painful. By the following day, the dog had made a complete recovery. When the extracts were administered to a rabbit through a funnel, the animal died in a few minutes.

Since Mme de Pauw had been violently sick during her short illness, it was thought that the bulk of the poison would be in the vomit. The wooden floor on which she had vomited was scraped, and residues of vomit were also found lodged between the planks. Tardieu and Roussin tested these samples and found no mineral poison. Their prepared extract, which was brown, with a rancid oily odour and bitter taste, was introduced into the thigh of a dog and resulted in vomiting and depression of the action of the heart. The animal was dead in twenty-two hours. A rabbit fed the extract by funnel died in less than three hours. Scrapings were also taken of the floor in an area free of vomit, and an extract of this material caused no harm to an animal.

Although it was never positively identified from either the contents of the victim's viscera or her vomit, Tardieu and Roussin inferred from their experiments that Mme de Pauw had been poisoned with digtalin. They also suspected that Pommerais's mother-in-law had died by the same method.

Pommerais undoubtedly had means, motive and opportunity for murder, and letters from Mme de Pauw in his possession showed that he had anticipated his mistress's death and the questions that he might be asked.[8] Mme de Pauw had written that the insurances were her idea, as a way of repaying the debt she owed Pommerais, and also described her illness following the supposed fall. One letter stood out, since it included the comment that she had taken digitalin as a stimulant. It was thought that she had written the letters at the request of Pommerais, who had convinced her that they were necessary to support the insurance fraud, but he must have had it in mind that they would also answer the charge of murder if it arose.

While in prison awaiting trial, Pommerais unsuccessfully attempted suicide three times. He was tried for the murder of Mme de Pauw and his mother-in-law in Paris in May 1864. The medical evidence was attacked robustly by the defence, since the claims that Mme de Pauw had been poisoned were unsubstantiated by any physical proof of poison. Counsel for the defence, Maître Lachaud, proposed a bold alternative explanation, the idea that deadly poisons could arise spontaneously in the body from the decay of proteins, a fantastical concept that was not given credence. Lachaud, as was later to be proved, was well ahead of his time. The possibility of the formation of cadaveric alkaloids would not be seriously addressed in a murder trial for another eighteen years.

Pommerais was acquitted of the murder of Mme Dubizy but found guilty of murdering Mme de Pauw and sentenced to death. He awaited his fate in Paris's La Roquette prison, hoping that Emperor Napoleon III would grant a reprieve. At half past five on the morning of 9 June, he was awoken and told that there was to be no reprieve. Half an hour later, he was led out and executed by guillotine in front of the prison gates.[9]

* * *

'When a doctor does go wrong he is the first of criminals. He has nerve and he has knowledge,'[10] wrote Arthur Conan Doyle, doctor of medicine and creator of Sherlock Holmes, in 'The Adventure of the Speckled Band'. The two medical murderers said by Holmes to be 'at the head of their profession' were Drs Palmer and Pritchard. Published in 1892, the story shows how these two cases from 1856 and 1865, respectively, resonated for a generation in public memory as the prime examples of doctors gone to the dark side. It was even believed that Pritchard had studied Palmer and thought he could improve on his methods. In doing so, he brought a little-used poison into new prominence.

Pritchard, like Palmer, relied on his position in society, which he hoped would disarm suspicion, as well as on the close personal supervision of his victims. 'The secret poisoner is the most dangerous of malefactors;' stated William Roughhead in his introduction to the trial of Pritchard, 'and he is specially to be dreaded when, as here, he prosecutes his subtle design in the two-fold disguise of loving relative and assiduous physician.'[11]

Edward William Pritchard was born in 1825 and was apprenticed in a surgical practice at the age of fifteen. He later served as an assistant surgeon in the Royal Navy. In 1850, he married Mary Jane Taylor, the daughter of an Edinburgh silk merchant. The following year, he obtained a medical practice in Yorkshire, but achieved 'a very indifferent reputation. He was fluent, plausible, amorous, politely impudent, and singularly untruthful.'[12] In 1857, he purchased the diploma of Doctor of Medicine from the University of Erlangen, and, a year later, became a Licentiate of the Society of Apothecaries. After spending a year abroad, he returned to start a practice in Glasgow. Despite his good manners, there was something untrustworthy about him that made him instantly disliked, and the stories he told of his previous career and adventures marked him out as a boastful liar and a fantasist. He was also said to be 'grossly ignorant of his profession, while daring and reckless in its practice'.[13] Despite strenuous efforts to achieve popularity, he was unable to gain admission to any medical societies because he could not find a proposer. A tall, well-built man, he was, by the 1860s, completely bald, something he attempted to conceal with an artful comb-over, while compensating with a long, luxuriant beard of which he was especially proud. Vain, with ingratiating manners, he managed initially to attract and later repel. It became a matter of public knowledge in Glasgow that he took advantage of his position to make advances on female patients, both married and single.

In May 1863, a fire broke out at the Pritchard home in Berkeley Terrace and a servant girl was found dead in her attic bedroom. The positioning of the body led to suspicions that the girl had been either dead or drugged when the fire broke out, and rumours spread that she had been pregnant by Dr Pritchard. No action was taken, however, and the Pritchards engaged a new maid, fourteen-year-old Mary M'Leod.[14]

By 1864, the family had moved to Sauchiehall Street. There were five children, the eldest of whom lived with her mother's parents in Edinburgh, so the household consisted of Dr and Mrs Pritchard, the cook, the maid, four children, and two medical students who lodged there. It was in October that year that Mrs Pritchard first became ill with vomiting and, the following month, she went to stay with her family in Edinburgh, where she made a good recovery.

The two poisons that Pritchard was known to have had in his possession at that time were tartarised antimony (the popular tartar emetic) and tincture of aconite, a highly poisonous preparation of the monkshood root. The active ingredient was the alkaloid aconitine, and it was used in doses of just a few cautious drops for pain relief and reduction of fever in many common complaints. Two teaspoonfuls could be fatal to an adult. Another preparation, Fleming's tincture, was some six times as concentrated,[15] and was mainly used in liniments.

The bitter, intensely burning taste of aconite made it an unlikely choice for the secret poisoner, and deaths from its use were usually due to error. The resemblance of monkshood root to that of horseradish sometimes led to fatal results following a roast beef dinner, and there were instances of accidental consumption of the toxic tincture.

Although the medical profession suspected that homicide with aconite did occur, it was very hard to prove. The earliest known case that came to trial was in 1841.[16] Twenty-five-year-old Mary Anne McConkey of Clones, County Monaghan, had grown tired of her husband Richard, a forty-year-old weaver, and was having an affair with another man. On 31 July 1840, she served her husband a dish of greens. Richard complained about the sharp taste of his greens, which Mary had seasoned separately in a small pot after taking a portion from the main one, but everyone else at the table – all of whom had been served from the main pot – said they were good. Soon after the meal, Richard was taken ill with a burning sensation on his tongue and tingling flesh. An hour later, he was found vomiting a greenish matter, purging, frothing at the mouth and nose, with a locked jaw and clenching of the hands. When able to speak, he said that his 'heart was on fire'.[17] Two hours later, he was dead. A visitor who had sampled a spoonful of greens from the same dish as Richard had also become very ill with similar symptoms, and took five weeks to make a full recovery.

Richard McConkey's body was examined the day after death. The stomach was empty, its interior the colour of mahogany with blackish-brown patches, and smeared with a yellowish-grey mucus. The small intestine contained a yellowish-brown mucus. The viscera and the contents were subjected to chemical examination but no trace of poison could be found.

The symptoms of both the dead man and his neighbour strongly suggested poison, and rumours pointed to monkshood, a plant known locally as 'blue rocket', which grew near the accused's residence and was generally understood to be poisonous. It was used by jockeys, who mixed it with ginger and placed it in a horse's anus 'in order to render him lively'.[18] Mary Anne was found guilty of her husband's murder, and sentenced to death. She confessed that Richard had been poisoned by the root of blue rocket, and, in a last-minute (and unsuccessful) attempt to escape the noose, blamed it on an acquaintance who, she said, had mixed the root with pepper and used it to dress the greens. Murder in which the alkaloid extracted from the root was cited was unknown, however, until the Pritchard case.

Mrs Pritchard came home for Christmas 1864, but, a fortnight later, her sickness returned and it was worse than before. She was usually taken ill after meals, and was soon obliged to stay in her room, where her food was brought to her by her husband. In February, she began to suffer violent and painful cramps. Doctors who were summoned were mystified by the illness, and it was suggested that Mrs Pritchard's mother, Mrs Jane Taylor, a healthy and active lady in her early seventies, should go to nurse her daughter. Mrs Taylor arrived on 10 February and proceeded to care for her daughter night and day, attending to her meals and sleeping in the same bedroom, Dr Pritchard being relegated to the spare bedroom. Pritchard did sometimes dine with his wife in her room, and occasionally poured her tea or buttered her bread.

Some tapioca was bought from the grocer's to prepare a dish for the invalid, and a portion was cooked in water and taken to the dining room, where it stood for a while before Mrs Taylor brought it to her daughter. Mrs Pritchard did not eat any, so Mrs Taylor ate some and, about an hour and a half afterwards, was seized with vomiting, which she thought was very like the attacks her daughter had been suffering. The bag of tapioca was put away and was not used again. Despite Mrs Taylor's attentions, Mrs Pritchard continued to be very ill and was sick almost every day, complaining of thirst, heat in her head and pains in her stomach.

Mrs Taylor, though otherwise temperate in her habits, was a regular user of Battley's sedative liquor, a preparation of opium, which she took for neuralgic headaches. She carried a bottle with her and regularly had it refilled at the chemist's. On 24 February, she was taken very suddenly and violently ill. Dr James Paterson, a practitioner with thirty years' experience, who lived in the same street, was summoned and told by Pritchard that his wife and mother-in-law had both become ill after drinking beer and that Mrs Taylor, who he claimed 'was in the habit of taking a drop',[19] had had a fit. When Paterson examined Mrs Taylor, who was lying fully clothed on her daughter's bed, it was immediately obvious to him that she was not intoxicated, but dying from the effects of a powerful narcotic. Pritchard said that she was in the habit of using Mr Battley's preparation, and 'might have taken a good swig at it'.[20] Paterson, however, seeing that Mrs Taylor was a stout and robust lady, felt sure that she could not, as Pritchard was suggesting, be addicted to opium.

While attending Mrs Taylor, Paterson was particularly struck by the exhausted appearance of Mrs Pritchard. 'Her features were sharp and thin, with a high, hectic flush on her cheeks, and her voice was very weak and peculiar.'[21] At first he attributed her condition to gastric fever and grief over her mother's illness, but was unable to banish from his mind a conviction that she was under the influence of antimony. Dr Paterson suggested some palliative treatments for Mrs Taylor, and left. He was roused at 1 a.m. and asked to come and see Mrs Taylor, but declined to do so, as he was tired and thought he would be of no use. The following morning, Mrs Taylor died. Paterson refused to sign a death certificate since he had only seen the deceased briefly; moreover, he wrote to the registrar to say that he regarded her death as 'sudden, unexpected and, to me, mysterious'.[22] Paterson later claimed that he had hoped and assumed that this letter would be acted upon and an inquiry would be launched that would save Mrs Pritchard's life, but, in the event, his comments did not arouse suspicion. The letter was destroyed, but Paterson, who had worded it with great care, was able to remember its contents. Mrs Taylor's death certificate was eventually signed by Pritchard, and gave as the cause paralysis and apoplexy.

On 1 March, Dr Paterson met Pritchard in the street by chance, and was asked to call on Mrs Pritchard, which he did the following day, when he suggested that she should be given some soothing powders and easily digested invalid food.

On the evening of 13 March, Pritchard sent up a piece of cheese for his wife's supper. Mrs Pritchard did not eat it, but asked Mary M'Leod to taste it. Mary sampled a tiny piece. She found it 'hot, like pepper',[23] and it caused a burning sensation in her throat. The cheese was removed and the cook later tried some, also finding it burning and bitter. She suffered stomach pains and vomited several times. Two days later, Dr Pritchard asked the cook to make up some egg flip and himself added two lumps of sugar to the tumbler. The cook, who had recovered from the sample of cheese, tasted a spoonful and remarked on its horrible taste. Despite this, it was taken up to Mrs Pritchard, who drank a wineglassful and was immediately sick. The cook later suffered the same symptoms she had experienced after trying the cheese.

Two days later, the condition of the patient worsened. Dr Paterson was called in and was shocked by her altered appearance since his last visit. There was a wild expression on her face, her eyes were red and sunken, her tongue dark brown and foul. Mrs Pritchard said that she had been vomiting but her husband denied this, saying that his wife was 'only raving'.[24] She also declared that her mother, while she took Battley's for headaches, was not an habitual user. Paterson suggested a sedative prescription that the doctor might easily make up himself, only to be told by Pritchard that he kept no medicines in the house, except for Battley's and chloroform, which struck Paterson as extremely strange.

The following morning, Mrs Pritchard was dead. Her murderer, who certified her death as due to gastric fever, made extravagant displays of grief over the corpse, imploring his beloved wife to return to him. On 20 March, he travelled with the body to his father-in-law's house in Edinburgh where the coffin was opened for viewing. With a great exhibition of emotion the widower kissed his victim on the lips. After arranging the funeral, Pritchard returned to Glasgow, and, as he stepped from the train, was arrested for the murder of his wife.

During Pritchard's absence from Glasgow, an anonymous letter had been received by the Procurator Fiscal. Written on 18 March, it pointed out that Dr Pritchard's mother-in-law had died under 'circumstances at least very suspicious',[25] and his wife had died that day under 'equally suspicious circumstances',[26] the letter-writer adding: '[W]e think it right to draw your attention to the above as the proper person to take action in the matter and see justice done.'[27] Dr Paterson, thought of as the obvious person to have sent this missive, always denied having done so.

A search was made of the Pritchard residence, and numerous medicines, including Mrs Taylor's personal supply of Battley's, later referred to in court as bottle 85, were recovered. Mrs Pritchard's clothes and bed linen were taken possession of by the police, as was the packet of tapioca. The case was a newspaper sensation, reported upon daily in great detail. Pritchard, confident and unruffled, made a highly plausible declaration of innocence and both his family and his wife's were firm in his support. Popular feeling was running in his favour when, on 28 March, it was revealed that antimony had been found in Mrs Pritchard's remains. Two days later, Mrs Taylor's body was exhumed, and tests again revealed the presence of antimony. Pritchard was charged with her murder. Awaiting trial in Glasgow's North Gaol, Duke Street, Pritchard remained confident of acquittal, and carried with him a photograph of the family group including his wife and mother-in-law, which he proudly showed to the warders.

The trial opened at the Edinburgh High Court on 3 July. The prisoner in the dock looked pale and worn but remained calm and composed throughout, except when, as one newspaper reported, 'as a fond husband, it was proper that he should be moved – [he] wept, or did something dextrous with his pocket-handkerchief which might very well pass for weeping'.[28]

At first, the outward appearance and demeanour of the man – intelligent, mild, sad and thoughtful – seemed to be at odds with the crimes of 'refined and consummate villainy and diabolic cruelty'[29] laid at his door, but, as the murderous plot and his many lies were revealed, a change came over the mood of the onlookers. The crucial evidence was that of Mary M'Leod, and reporters saw 'a certain vulpine look' stealing over the prisoner's face as he gazed keenly at the girl who, under searching questioning, 'rent aside the curtain which had hitherto veiled the inner life of that apparently happy home'.[30] Mary admitted to the court that the doctor had seduced her in the summer of 1864. Pritchard had given Mary gifts of jewellery and portraits of himself, and promised to marry her if his wife died. There was an occasion when they had been in one of the bedrooms, kissing, and his wife had walked in and seen them. Mary had gone to Mrs Pritchard saying she wanted to go away, but Mrs Pritchard said her husband was 'a nasty, dirty man',[31] and would not allow the maid to leave. Mary also very hesitantly confessed to the court that she had become pregnant and, when she had told Pritchard, he said 'he would put it all right'.[32] She miscarried that autumn. The intimacy continued and went on during the time Mrs Pritchard was away in Edinburgh, although it did not resume after her return.

Dr Paterson admitted in court that, despite his impression that Mrs Pritchard was being poisoned with antimony, he had felt that, as she was not his patient, he was under no obligation to do anything about it. He had not mentioned his conclusions to either Mrs Pritchard or any member of her family and, since her husband was a doctor, had assumed that it was he who was treating her. His second visit, he said, had been a social call and not a medical consultation. To go back and check that the advice he had given on that occasion had been carried out would have been a breach of professional etiquette.

'Do you mean', asked Lord Justice Clerk, one of the presiding judges,[33] 'that you believed that some person was engaged in administering antimony to her for the purpose of procuring her death?'

'Yes,' replied Paterson, 'that was my meaning.'[34]

As the trial progressed, all doubts about Pritchard's guilt were swept away. A Sauchiehall Street apothecary listed the medicines purchased by Pritchard in person between September 1864 and February 1865, which included two ounces of tartar emetic, three of tincture of aconite and one of Fleming's tincture. A chemist testified to selling Pritchard one and a half ounces of Fleming's tincture. The medical report on Mrs Pritchard revealed that the stomach and intestines contained little apart from mucus. The application of both the Reinsch and Marsh tests to the contents produced unequivocal evidence of antimony, as did tests on the urine, bile, blood, liver and the solid part of the organs, as well as stains on the deceased's chemise and bedsheets. The Stas process applied to the half-ounce of matter in the stomach failed to reveal any vegetable poisons. Substantial evidence of the administration of antimony, but not of any vegetable poison, was also found in the body fluids and organs of Mrs Taylor. Antimony was detected in the tapioca and the Battley's solution, but establishing the presence of aconite was more complicated. Dr Frederick Penny, professor of chemistry at the Andersonian University of Glasgow evaporated a portion of the Battley's to dryness and carefully tasted it, which produced the tingling sensation followed by numbing that was characteristic of aconite. He then compared the effect of purchased Battley's with Battley's to which he had added Fleming's tincture, and with the Battley's found in bottle 85. Standard Battley's injected into rabbits produced a stupor from which they recovered. Battley's to which aconite had been added produced a flow of saliva followed by 'piteous and peculiar choking cries . . . and the breathing is painfully laborious. Convulsions now set in.'[35] The animals invariably died. The effects produced by injection with some of the contents of bottle 85 were almost identical to those seen after the aconite-spiked Battley's.

The conclusion of the medical witnesses was that Mrs Pritchard had died from the effects of taking antimony over a period of time in repeated small doses. Mrs Taylor had also taken antimony in a succession of doses, and her final illness could be attributed to the effects of a combination of opium, antimony and aconite.[36]

The only mystery was Pritchard's motive for murdering his wife, since the financial rewards, two-thirds of a life rent interest in £2,500, or the desire to marry the servant girl appeared inadequate. Pritchard was, however, deeply in debt, all his bank accounts being overdrawn. He had borrowed on his insurance policies, and there was an outstanding loan payable on his house. Had his finances collapsed into bankruptcy, the effect on his public standing would have been disastrous. Mrs Taylor had undoubtedly been removed because she was an obstacle to the scheme to murder her daughter. The defence was able to provide no witnesses to the prisoner's character and could only point to a lack of sufficient motive and the unlikelihood of such a crime being committed by a man of his position and education.

Pritchard was found guilty of both charges. As he awaited execution, he made three separate confessions. In the first one, he told a visiting clergyman that he had murdered his wife with chloroform and that Mary M'Leod had been present and knew that her mistress was receiving poisoned food. On 11 July, he admitted he had been intimate with Mary since the summer of 1863 and that both his wife and Mrs Taylor suspected the intimacy. He declared that Mrs Taylor died from an overdose of Battley's and he had put the aconite in afterwards to prove death by misadventure. He had given his wife chloroform to help her sleep and then succumbed to the temptation to give enough to cause death. Eight days later, he finally confessed to both murders, which he put down to 'terrible madness' and 'ardent sprits', exonerating Mary of any involvement.[37]

No one had suggested that chloroform had been used to kill Mrs Pritchard, and none had been looked for, although if inhaled it will leave traces in brain and lung tissue. Chloroform was found in the house, and it is possible that, at the last, unable to witness his wife's prolonged suffering, he finally put her out of her misery.

Ten thousand people came to watch the execution of Dr Pritchard, which took place at Glasgow Green on 28 July. In his last words, he acknowledged the justice of the sentence. It was the last public execution to take place in that city.

CHAPTER SIXTEEN

Matrimonial Causes

During the 1860s, the Pharmaceutical Society and the United Society of Chemists and Druggists continued to call for restrictions on the sale of poisons, not only to protect the public, but also to bring the practice of selling and dispensing medicines under the exclusive control of qualified professionals. New parliamentary subcommittees were formed to receive the evidence of experts, and Professor Taylor was once again called upon to reiterate that he was often consulted in cases where drugs were 'sold by persons who are incompetent to know the nature of them',[1] and death or injury had been the result. Two bills placed before parliament in 1865 were considered unsatisfactory, since it was believed that, as worded, the proposed legislation would tempt pharmaceutical chemists to undertake medical practice for which they were not qualified. It took three more years to resolve the issues.

The Pharmacy and Sale of Poisons Act finally became law on 31 July 1868, and came into force in January 1869. No person was permitted to sell, compound or dispense poisons except a pharmaceutical chemist, a chemist and druggist, or an assistant with at least two years' experience who was registered under the act. The legislation did not affect the business of qualified medical practitioners or veterinary surgeons, or manufacturers and dealers in poisons for industrial use. All poisons sold had to be labelled with the word 'POISON', the name of the poison and the name and address of the seller. Poisons were not to be sold to any person who was unknown to the seller, unless introduced by someone who was known, and all sales were to be recorded in a book, together with the name and address of the purchaser.

The initial proposals included the following schedule of the substances deemed to be poisons:

Arsenic and its preparations
Oxalic acid
Prussic acid
Chloroform
Cyanides of potassium and mercury
Strychnine, and all poisonous vegetable alkaloids and their salts
Aconite and its preparations
Emetic tartar
Corrosive sublimate
Belladonna and its preparations
Essential oil of almonds, unless deprived of its prussic acid
Cantharides
Savin and its oil[2]

Opium and its preparations are noticeably absent from this list. Regarded as a special case because of the frequent sales of small quantities of laudanum to the poorer members of society, it was feared that a restriction of its sale would be impossible to enforce and lead to an illegal market. A later amendment to the proposed legislation included opium and its preparations in a second schedule, which made them subject to labelling restrictions, but patent medicines were not included. Following this enactment, there was a substantial decline in the number of infant deaths from overdoses of opium from 20.5 per million of the population in the years 1863–67 to 12.7 per million in 1871 and about six or seven per million in the 1880s.[3] An initial fall in the adult death rate was not maintained, and it had actually increased by the end of the century. The reasons for this are unclear, but, overall, the majority of deaths from opium were accidental, while it remained the leading cause of death in poison suicides until overtaken by carbolic acid in the 1890s.

The 1868 act was a considerable advance both in controlling the sale of poisons and in the status of the profession of pharmacists and chemists, who became a defined body of qualified persons listed in a register. There were, of course, ways of getting around the restrictions. A would-be murderer could still buy lethal quantities of poison as long as he or she exhibited good manners, appeared respectable and had a cunning plan.

The deficiencies of the legislation were exposed in 1871 by architect's daughter Christiana Edmunds. Christiana, born in 1828, was a spinster who moved in respectable circles in Brighton society, where she lodged with her widowed mother in Gloucester Place. Their family doctor was

Charles Izard Beard, born in 1827, who, with his wife Emily, family and servants, occupied a substantial town house at 64 Grand Parade. Christiana was a skilled draughtswoman and Beard engaged her to make drawings of anatomical subjects to be displayed in the library at Sussex Hospital. She became a family friend and a guest in his home. Unfortunately, Miss Edmunds, who concealed a passionate and savagely ruthless nature beneath a calm and ladylike exterior, became obsessed with Dr Beard and jealous of his wife, whom she would undoubtedly have liked to displace. In 1869, the amorous spinster began to write affectionate letters to the doctor. Whether he returned her feelings, or there was any actual intimacy, as has sometimes been alleged, has never been proved. He was later to claim that he did not reply to the letters.

In September 1870, Miss Edmunds paid a social call on Mrs Beard, bringing a gift of chocolates, one of which she playfully popped into Mrs Beard's mouth. The chocolate had a peculiar metallic flavour and Mrs Beard spat it out, but, during the night, she suffered from copious salivation and an attack of diarrhoea. The nature of the poison in the chocolate was never determined, but the symptoms are suggestive of arsenic. When Emily mentioned the chocolate incident to her husband, he realised that his admirer had gone too far, and he went to see Christiana to discuss his concerns. He did not, at this stage, accuse her of deliberately trying to poison his wife, since he had no proof that she was responsible, but commented in what he hoped was a light-hearted manner that it was now possible to detect poison in animal tissues. His suspicions must have been very plain. Christiana denied that she had had anything to do with the contamination of the chocolate, saying that it was the fault of Maynard's, the West Street shop from which she had purchased it. The matter was dropped, but Dr Beard nevertheless made it clear that Christiana was no longer welcome at his home.

Dr Beard was away from Brighton for a while, but, when he returned at the start of 1871 and found Christiana anxious to renew the friendship, he refused, saying that he could not help thinking she had tried to poison his wife, something she indignantly denied. She continued to send him letters couched in such inappropriately affectionate terms that he was obliged to ask her to stop.

The spurned admirer was distraught, and her mother found her pacing the floor in agitation, claiming that she would go mad. She had certainly lost touch with reason. Christiana decided that the only way to restore the friendship was to convince Dr Beard that she had not tried to poison his wife, by proving that the source of the poison was Maynard's. She may not have been aware at the time that Mr Maynard did not manufacture the chocolates sold in his shop, but, even after she realised this, her single-minded campaign against him never faltered.

In March, Christiana called on Mr Maynard to complain about his chocolate creams, saying she had bought some last September and again that month and they had tasted nasty and made her and a friend ill. He brought some chocolates from his stock and they sampled them but found nothing wrong. Christiana took the chocolates she was complaining about to a chemist, Julius Schweitzer, to be analysed, and his report, dated 23 March, concluded that they contained a metallic irritant poison, which he thought was zinc. He theorised that it was a contamination from the vessel in which the cream filling had been prepared.

Christiana did not send this report to Mr Maynard. Perhaps it was a weapon to be held in reserve as she stepped up her campaign. In doing so, she didn't care who she distressed, injured or even killed to achieve her aim. This time, her poison of choice would be strychnine.

Christiana had been an occasional customer of Isaac Garrett's chemist shop at 10 Queens Road for some four years, and was known to him by sight but not name. She was always veiled, and said that she suffered from neuralgia. On 28 March 1871, after making a few small purchases, she asked him for some strychnine as she said her garden was infested with cats. Garrett, with due attention to the provisions of the 1868 Sale of Poisons Act, declined to supply it without a witness who knew the customer and whom he knew. Christiana suggested Mrs Stone, a milliner whose shop was only three doors away, and Garrett said that would be acceptable.

Christiana entered Mrs Stone's shop, purchased a second veil, and asked the milliner if she would witness her purchase of strychnine as she and her husband wanted it to stuff birds. She was a stranger to Mrs Stone, who asked for her name and was told it was Mrs Wood. Despite this short acquaintance, Mrs Stone obligingly accompanied Christiana to Mr Garrett's and signed the poison book. Ten grains of strychnine were supplied to the customer, who gave her name as Mrs Wood of Hillside, Kingston.

Christiana's tactic was to buy chocolates from Maynard's, where they were displayed in a glass case and sold loose by weight in paper bags that had Maynard's name printed on them. She would then insert poison into the sweets and leave the bags in various locations, hoping that someone would be tempted to eat them. She first left a bag of chocolate creams in a stationer's shop called Halliwells. She was known there, and it was assumed that she had left them by mistake, so they were put aside for her, but when she called again, she denied that she had left them and told the Halliwells' thirteen-year-old son that he might as well eat them. He did and was ill for several days. She left another bag there on a later occasion, but this time they were thrown away. She left chocolates at the greengrocer's shop of a Mr Cole, and his wife Harriet and a friend of hers ate some and became ill. Christiana also approached children in the street with gifts of sweets, which made them ill.

On 15 April, 'Mrs Wood' once again asked Mr Garrett for ten grains of strychnine to poison cats and the transaction was again witnessed by Mrs Stone. On 11 May, however, 'Mrs Wood' was now sufficiently well known to Mr Garrett for him to sell her ten grains of strychnine without requiring a witness. Perhaps concerned that her repeated purchases were looking suspicious, Christiana decided to retire her Mrs Wood persona by telling Garrett that she was moving to Devon.

The mysterious poisonings had so far received no public attention, and perhaps Christiana decided to test the efficacy of her supplies of strychnine. On 27 May, a dog belonging to a lady who lodged at the Gloucester Place address died suddenly in great agony shortly after Christiana was seen playing with it. The body was taken to a taxidermist who, on opening it, saw a 'peculiar rigidity and inward bending of the backbone' and 'a saliva about the throat and mouth of an offensive character';[4] he was convinced that the dog had been poisoned.

On 8 June, a messenger boy brought Mr Garrett a note purporting to come from local chemist's Glaisyer & Kemp asking for a quarter of an ounce of strychnine. Garrett sent the boy back with a note saying they should send him an order, but that he could only supply them with a drachm. The boy returned with half a crown and another letter saying that Glaisyer & Kemp would be satisfied with a drachm, adding that their signature had always been sufficient before. Garrett, via the messenger boy, sent a bottle containing a drachm of strychnine at a cost of 1s. 3d.

Despite all her efforts, Christiana had still failed to convince anyone, including Mr Maynard, let alone the unattainable Dr Beard, that Maynard's chocolates were at fault. The analysis of Mr Schweitzer was all very well, but there was no proof that the chocolates he had examined came from Maynard's, and the same applied to those in the bags she had distributed. Realising that she needed to place the poison directly in the shop, she continued to make purchases of sweets from Maynard's via messenger boys, then sent them back to be exchanged for others on the pretext that they were the wrong ones. The returned chocolates did not appear to have been tampered with and were put back into stock.

On 12 June 1871, a four-year-old child, Sidney Albert Barker, began to cry with pain soon after eating some chocolate creams purchased from Maynard's. The shivering child seemed to be having a fit, and he was wrapped in a blanket, and his legs placed in a bath of warm water with mustard. Surgeon Richard Rugg was sent for. With minutes, the boy was unresponsive, with clenched teeth, rolling eyes and stiffened limbs. His uncle Charles Miller, who had eaten several of the creams, collapsed with a spasmodic attack. Charles's brother Ernest had tried two, but, finding they had a metallic taste, spat them out. Rugg arrived very quickly and sent out for an emetic, but it came too late. Some twenty minutes after eating the chocolates, the boy died. Charles eventually recovered.

An analysis of the child's stomach contents by Professor Letheby showed that the cause of death was strychnine poisoning, and the remains of the original purchase of chocolate creams also contained strychnine. A second batch of the same creams bought from Maynard's shop was tested and no poison was found.

Many murderers just cannot leave their crimes alone. They feel a need to become involved in the investigation and think they can divert suspicion from their guilt by establishing their status as an innocent victim. The police are wise to such tactics, so the result is that the murderers only draw attention to themselves as potential suspects and ensure that they are carefully watched. So it was with Christiana, who claimed that she too had been taken ill after eating chocolates bought from Maynard's shop and provided Mr Rugg with Julius Schweitzer's analysis. She was called upon to give evidence at the inquest on little Sidney and did so with verve. The chocolates, she said, tasted 'frightful – she never tasted anything like it. It was a metallic taste. She became livid and trembled. Her throat burned frightfully, and she felt strange all over.'[5] She said she had not sent the analysis to Mr Maynard because she did not think he would believe her. The inquest verdict was accidental death, it being theorised that, since strychnine was a component of rat poison used to prevent infestation in warehouses, the chocolates had somehow become contaminated during the manufacture or storage of the raw materials. The court was told that Mr Maynard had agreed to destroy his entire stock of chocolate creams, an announcement that was met with restrained applause.

Christiana, believing she was fully vindicated, now felt justified in writing a gushing letter to Dr Beard in which she tried to engage his sympathies and renew their friendship.[6] It began '*Caro Mio*' (My Dear) when surely 'Dear Dr Beard' would have been more appropriate, and went on to address him as 'my dear boy' and 'darling'. Christiana pleaded that she was miserable as she was unable to speak to him, and demonstrated her pitiable feminine weakness by saying how afraid she had been when giving evidence at the inquest. 'That clever Dr Letheby, looking so ugly and terrific, frightened me more than anyone ... His physique is large and grand, like his mind.'[7] Convinced that her evidence had been carefully watched for errors, she felt that she had come through the ordeal well. She made sure to place the blame for the poisoning squarely on Mr Maynard. 'That man's chocolates have been the cause of great suffering to me.'[8] The letter, which ended with 'a long, long *bacio*' (kiss),[9] failed in its object. Christiana's campaign continued.

Between 27 June and 1 July, little Sidney Barker's grieving father received three letters of condolence, posted from different locations in Brighton, urging him to take legal action against Maynard's. Later that month, chocolate creams were found in a Maynard's bag at Mrs Cole's shop after a visit from Christiana. They were given to a boy, who took them home to his mother, who became ill after tasting one. The rest were handed to the police and, on analysis, the white centre was found to contain a lethal quantity of strychnine.

Christiana must have been worried that the death of little Sidney would arouse Mr Garrett's suspicions of the veiled cat-poisoner Mrs Wood and that the purchase might be traced to her. On 14 July, Mr Garrett received another note carried by a boy purportedly from the borough coroner, asking for the loan of his poison book to assist in an investigation. Trustingly, he sent the book and it was returned to him half an hour later.

Christiana also decided to change poisons. On 19 July, Garrett received another order supposedly from Glaisyer & Kemp, again by letter, enclosing a shilling and asking for two ounces of arsenic, or three if he could spare it. Suspicious at last, Garrett checked with Glaisyer & Kemp and found that they had not sent the order. He went to the police. When he looked at his poison book, he found that some pages relating to transactions between July 1870 and July 1871 had been torn out, although the pages relating to Mrs Wood's purchases remained. It was later confirmed that the coroner had not written to ask for the book.

On 21 July, a chemist named Bradbury received a letter carried by a boy, apparently from Glaisyer & Kemp, asking for three ounces of arsenic. This he supplied, and, since it was a wholesale transaction, it was not entered in his poison book.

When Dr Beard next saw Christiana – where, he did not reveal, but as they lived nearby it might have been a chance encounter in the street – he told her that her letters must stop. She was shocked to discover that he had shown her passionate and what she had thought to be private letters to his wife. Shortly afterwards, the poison campaign took a new and more organised turn, when gifts of sweets, cakes and both fresh and preserved fruit were received by several prominent Brighton residents anonymously in parcels either sent by post or, following a rail trip made by Christiana to London, conveyed in deal boxes by train from Victoria station. Mrs Beard, her near neighbour Mrs Elizabeth Boys and chemist Mr Isaac Garrett all received these items, which, when eaten by servants and Mrs Boys's children, made them ill with symptoms that suggested they had been poisoned. The remains of the cakes and fruit and two jars of a servant's vomit were sent for analysis and were found to contain common white arsenic. All the preserved fruits had arsenic in and on them, and one 'was literally stuffed with it',[10] containing enough to cause death. Mr Garrett had been the recipient of two peaches, the surface of which had been powdered with a lethal quantity of strychnine.

A police notice offering £20 reward for information leading to the discovery of the sender of the poisoned treats sent Brighton into a panic and there was a ferment of rumour and speculation: '[D]istrust generally prevailed, and it was thought that the time of the "Borgia's" [*sic*] had returned.'[11] Christiana, not to be left out, said she had also received a box that contained suspect fruit. It was handed to the police and, when Inspector Gibbs called on her, she was found reclining weakly on a couch saying that she had been poisoned again. When Mrs Beard was interviewed by the police, she suddenly recalled the incident of the dubious chocolate the previous September and suspicion hardened.

Inspector Gibbs wrote a letter to Miss Edmunds, ostensibly to request more details of her purchase of chocolate creams from Maynard's. She replied that she had bought them on 16 March, and had had them analysed the same day. The police now had a sample of her handwriting.

Dr Beard had thus far been silent on the question of Miss Edmunds's attachment to him, a serious matter for a man in medical practice, but he decided at last to overcome his embarrassment. He contacted the police and handed over some letters he had received from Christiana. On 17 August, she was arrested, and police enquiries succeeded in tracing many of the boys who had carried messages or purchased chocolates on her behalf. The anonymous letters sent to the murdered boy's father and the chemists, as well as the notes accompanying the parcels of poisoned food-stuffs, were examined by a handwriting expert, who concluded that they were written by Christiana.

Christiana Edmunds was charged with the murder of Sidney Albert Barker, and, after taking advantage of the Palmer Act, her trial took place at the Central Criminal Court, in January 1872. The main defence, the only one possible under the circumstances, was that she was insane. Those physicians who had examined her since her arrest had found her strangely indifferent to her position, and lacking in feeling and moral sense, but not actually mad, so the strongest evidence offered to the court in her defence was a family history of insanity. Medical witnesses testified that Christiana's late father had been subject to delusions and outbreaks of violence, and was taken to an asylum in a straitjacket in 1843. He had died there, aged forty-seven, just four years later. Christiana's brother Arthur Burns Edmunds had been epileptic since childhood, and died in an asylum in 1866 at the age of twenty-four. Mrs Edmunds went further, attributing insanity and imbecility to several other family members. Christiana, she said, had suffered from paralysis and fits of hysteria.

Mr Baron Martin, summing up, was clearly impressed with the defendant's carefully planned campaign, and observed that 'a poor person . . . was seldom afflicted with insanity, and it was common to raise a defence of that kind when people of means were charged with the commission of crime'.[12]

Christiana was found guilty and condemned to death. She attempted to delay her fate by claiming that she was pregnant, but a dozen respectable matrons were soon assembled, who examined her and declared that she was not. Following petitions to the Home Secretary pleading for mercy on the grounds that Christiana was insane, or at the very least of doubtful mental stability,[13] she was reprieved and confined to Broadmoor Asylum, where she died in 1907. The question of her sanity, taking into account her behaviour, family history, callous attitude to murder and detailed planning, has been debated ever since. Her crimes were brought home to her not because of the poison legislation, but by her tendency to draw attention to herself and her failure to disguise her handwriting. She was not insane in the legal sense, in that she knew full well that what she was doing was against the law, but her heartless disregard for the fate of others suggests a psychopathic personality disorder.

The aspect of the case that most exercised *The Lancet,* however, was how Christiana had managed to obtain four separate packets of strychnine and three ounces of arsenic from unsuspecting chemists in the space of four months without either chemist contravening the provisions of the 1868 act. *The Lancet* pointed out that the act was intended primarily to prevent crime, 'and the idea of a person intending to commit murder calling in a witness to his preparation for doing it being so entirely beyond all human probability, the necessity for his doing so seemed the greatest safeguard which could be enforced'.[14] The second object of the act, should the first safeguard fail, was to supply evidence to connect the perpetrator with any crime. Christiana's cunning had found a way around both of these apparently insuperable difficulties. *The Lancet* urged that chemists should be convinced that 'so dangerous an article as strychnia should never be entrusted to unskilled hands, even to rid houses of domestic nuisances'.[15]

Despite the legislation, many deadly poisons remained on open sale as ingredients of other items, presumably on the assumption that it was impossible for an unqualified buyer to extract the poison. Mary Ann Cotton, who was hanged at Durham Gaol in 1873, was one of the United Kingdom's most prolific serial arsenic murderers, with possibly as many as twenty-one victims, many of whom she had insured. It was thought that she had extracted pure arsenic from a mixture of arsenic and soft soap she had purchased, saying she wanted it to rub on bedsteads to kill bugs. Tartar emetic, commonly used to treat horse ailments, and strychnine, as an ingredient of vermin killers, also remained easily available.

Marriage – either the desire for it, as in Miss Edmunds's case, or the wish to escape it or its consequences – is a common motive for murder, and the incidence and nature of domestic crimes can be influenced by changes in the laws of property and divorce. The 1870s ushered in an era of gradual reforms in the rights of women. The first Married Women's Property Act in 1870 allowed married women to retain money they earned or inherited, which would previously have become the property of the husband, but a single woman's property still passed to her husband on marriage. Divorce in England and Wales had been easier to obtain since the Matrimonial Causes Act of 1857 had transferred jurisdiction from ecclesiastical to civil courts; however, a man could obtain a divorce on the grounds of adultery alone, whereas a woman had to prove additional offences such as cruelty. In Scotland, however, a woman had only to prove adultery. (This particular inequality was not rectified until 1923.) A woman who was financially dependent on a brutal husband and unable to prove his adultery or afford a divorce had very few options.

The death of Charles Bravo in 1876, poisoned with tartar emetic, is officially, on the basis of the inquest verdict, an unsolved murder. Over the years, many suspects have been suggested, and alternative solutions, including suicide and accident, offered, but a good case has been made for Bravo's unhappy wife, exhausted by repeated miscarriages and controlled by a cruel and domineering husband, being the culprit.

Even when a woman had grounds for divorce against an abusive, unfaithful husband, many shrank from the public exposure of such a relationship, the social disgrace of being a divorcée, and the effect it would have upon the children. In the case of Mme Elizabeth Chantrelle, her reluctance was to prove fatal.

Eugène Marie Chantrelle was born in 1834 in Nantes, where he was enrolled for an education in medicine. He was later obliged, for financial reasons, to abandon his formal studies. After a few years in America, he sailed for England, where he started a new career as a language teacher.[16] He maintained an interest in medical matters, and, over the years, was to purchase large quantities of medicines, attend families in the capacity of a medical advisor and even prescribe for them. By 1866, he was in Scotland, where his linguistic abilities secured him engagements at some of the most highly regarded schools in Edinburgh. One of his pupils was fifteen-year-old Elizabeth Cullen Dyer,[17] with whom he commenced an intimate relationship, and, by the summer of 1868, it was apparent that she was pregnant.

Reluctantly, for she must have realised that her lover had no affection for her, she married him on 11 August. A son was born two months later. Three more sons were born of this increasingly miserable marriage, but, by October 1877, only three were living, the eldest, Eugène John, seven-year-old Louis, and the baby of the family, less than a year old.

Chantrelle, for all his education and abilities, was essentially a selfish individual who lacked both decency and self-control. He drank a bottle of whisky a day, was known to frequent a brothel, and attempted to take liberties with his fourteen-year-old servant girl, Isabella Ness, who left the house before the end of 1877. The unfortunate Elizabeth, subjected to both verbal and physical abuse, was terrified by her husband's threats to murder her. He once pointed a loaded pistol at her head, and several times said that he would poison her, boasting that his medical knowledge would enable him to give her a fatal dose that could not be detected. These outbursts were witnessed by the children and the servants, all of whom agreed that Elizabeth never used hard words or violence towards her husband. The only good thing that could be said for Chantrelle is that it was never suggested that he was anything other than kind to his children.

Elizabeth sometimes fled her drunken, abusive and unfaithful husband to take refuge with her mother, but every time she returned to the family home because of her children. In May 1876, however, she called the police, and Chantrelle was placed in custody for assault and threatening behaviour. Elizabeth approached police surgeon Henry D. Littlejohn,[18] asking him to go and see her husband, saying that she doubted his sanity and was in terror of her life because he went about with pistols. 'She complained of his conduct as unnatural and outrageous, and such as she could not explain,'[19] Littlejohn later said. He spoke to Chantrelle but returned to tell Elizabeth that 'her husband was perfectly sane, and that, according to his account, there were faults on both sides'.[20]

Chantrelle was convicted at the police court and bound over to keep the peace, and Elizabeth consulted a solicitor about obtaining a separation on the grounds of adultery. Concerned that a scandal would disgrace both her and the children, she asked if, when seeking a separation, which would involve supplying evidence of her husband's visits to brothels, there would be any public exposure of the circumstances. She was told that this would probably be the case, and, for the sake of the family reputation, she decided not to proceed.

Chantrelle's once successful career was drowning in alcohol, and he found himself in financial difficulties. In October 1877, wholly against his wife's wishes, he took out a policy on her life that would pay out £1,000 if she died an accidental death. Before taking out the policy, he had been careful to check on the precise definition of accidental death under its terms. When Elizabeth discovered that her life had been insured, she told her mother that she would soon die.

Even in his drunken desperation Chantrelle was sensible enough to know that the sudden death of a previously healthy young woman might arouse suspicion. He could not rely on an assumption of natural causes. Even if he used a poison that left no trace, his victim's symptoms during life and appearances after death would point to murder as they had done in other cases. Chantrelle's refinement of the art of murder was to lay a false trail.

On New Year's Day 1878, Elizabeth was feeling unwell and went to bed early. She occupied the back bedroom, and, until recently, all three children had slept in the same room as her; however, Chantrelle, who slept in the smaller nursery bedroom, had recently asked for the two older boys to sleep in his room. At ten o'clock, Elizabeth asked the servant, Mary Byrne, for a glass of lemonade and a piece of orange. These were brought, and Mary left. The gaslight in the room was then burning.

Next morning, when Mary rose early as usual, she heard a moaning noise from her mistress's room. Going to investigate, she found the door open, the gas out and the baby gone. Elizabeth lay in bed, pale and unconscious, with greenish-brown vomit staining the pillow, sheets and bedgown and on the ends of her hair. Mary hurried to wake Chantrelle, who was sharing his bed with the three children, then returned to the back bedroom to try to rouse Elizabeth. Chantrelle arrived at his wife's bedside and sent Mary to tend to the baby as he said he could hear it crying. Mary obeyed, but found all three children still asleep, and returned to her mistress's room to see that Chantrelle had just opened the window. He asked Mary if she could smell gas. At first she could not, although soon afterwards she began to notice a smell, which became stronger, and she was obliged to shut off the gas supply at the meter. Chantrelle went to fetch a doctor.

Dr James Carmichael arrived within the hour, and was at once aware of a strong smell of gas in the bedroom. The patient was pallid and her lips were blue, her breathing shallow and intermittent. Vomited matter was oozing from her mouth on to her chin. Realising that the patient was probably dying, Carmichael sent for Dr Littlejohn saying that it was a case of coal-gas poisoning. While he waited, he had Elizabeth removed to another room, called for brandy and, since she was unable to swallow, had it administered to her by enema. Left within reach of M Chantrelle, the brandy supply dwindled rapidly. Littlejohn arrived to find the patient receiving artificial respiration and was told by Chantrelle that there had been an escape of gas in the bedroom. Elizabeth, though profoundly unconscious, was still breathing, and a faint odour of gas could be detected on her breath. Littlejohn said that Mme Chantrelle's mother should be sent for, and was both astonished and impatient when Chantrelle claimed that he did not know his mother-in-law's address. One of the children (presumably the eldest, Eugène) said he knew the address, and went to fetch her. At the order of Dr Littlejohn, Mme Chantrelle was removed to the Royal Infirmary, and he also wrote to the gas company asking for an inspection.

At the infirmary, Dr Douglas MacLagan, professor of medical jurisprudence at Edinburgh University, was asked to examine a case of coal-gas poisoning. Elizabeth was scarcely breathing, had almost no pulse and a barely perceptible heartbeat. MacLagan tried to detect the telltale smell of gas by sniffing the patient's breath and applying his nose to the skin of her chest. He smelt nothing and said that he did not believe it was gas poisoning at all, but more likely a narcotic such as opium. Despite the artificial respiration and brandy enemas, Elizabeth died at four o'clock.

The post-mortem was carried out by MacLagan and Littlejohn, who soon ruled out coal gas as the cause of death, since there was no trace of its scent in the blood, brain or lungs; but no other poison could be discovered. The symptoms strongly suggested opium. The absence of any residue in the body was not proof that she had not taken it. 'It is a fact well understood', Littlejohn was to explain, 'that opium is exceedingly apt to escape detection if the person survive long enough to allow diffusion.'[21] Elizabeth had been found unconscious at 7 a.m. and had been kept alive until 4 p.m., more than ample time for all trace of the poison to disappear. Analysis of the vomit stains on the sheets and bedgown did, however, reveal the presence of opium. On 5 January, Eugène Chantrelle was arrested for the murder of his wife. His defence was that she had died from the accidental inhalation of coal gas.

The gas pipe in Mme Chantrelle's bedroom was found to have been recently and deliberately broken by being wrenched back and forth, something that could have been done very quickly. Both dry and fluid extracts of opium were found in the house, and Chantrelle had purchased three drachms (three teaspoonfuls, amounting to ninety poisonous doses) of extract of opium as recently as 25 November. Concentrated extract of opium was never prescribed in that form; it was used as an ingredient when preparing pills or mixtures. Just one spot found on Elizabeth's sheet contained an amount that was very nearly a fatal dose. Dr Littlejohn thought it would be easy to administer the opium without its being detected either by inserting it into a piece of orange or in lemonade, the last two things Mme Chantrelle was known to have consumed.

On 10 May, an Edinburgh jury found Eugène Chantrelle guilty of the murder of his wife. He appeared surprised by the verdict, and, after sentence was passed, asked to be allowed to address the court. Having maintained throughout that his wife had died from gas inhalation, and having heard his defence counsel cast doubts on opium as the cause of death, Chantrelle astonished the court in his final rambling remarks. With much excited gesticulation, he admitted that the stains on the sheet and nightgown were evidence that opium was there, 'I go further,' he went on, his voice rising in pitch, 'I say opium *was* there . . . I am satisfied further, gentlemen, that I did not put it there; that it did not proceed from Madame Chantrelle's stomach; that it was rubbed in by some person for a purpose which I do not know.'[22] He continued to babble that opium had obviously been given in a solid form and could not have come there accidentally, until the judge stepped in and advised him to say no more.

The Times compared Chantrelle's case with a trial that had recently been concluded in France. A Parisian chemist, thirty-three-year-old Louis Danval, had been charged with the murder of his wife by slow poison with arsenic, a crime that, the editor felt, was far worse than Chantrelle's. 'Slow poisoning by arsenic is a prolonged torture, and involves a degree of cruelty as well as malice for which no punishment can be held excessive.'[23] Danval's wife had brought him a dowry, but he had been unable to obtain further funds from her father as he had hoped. The court was told that the wife had suffered cruelty at the hands of her husband and, as in the case of Mme Chantrelle, she had been obliged to seek refuge with her parents and ask for the intervention of the police. Mme Danval fell ill and slowly wasted away. After her death, only a small amount of arsenic (about 2mg or ⅓ grain) was found in her body; not enough to account for death. Her symptoms, however – burning pains in her stomach, and almost paralysed legs – were such as would be produced by small repeated doses. Danval had assiduously attended his wife during her illness, which he at first ascribed to typhoid and, after her death, suicide. The French jury, though finding Danval guilty, had decided that there were extenuating circumstances, which 'probably represents either a sentimental horror of the punishment of death or vague doubts whether the prisoner was really guilty at all',[24] suggested *The Times*, with the latter explanation considered more likely. 'There were some differences of opinion, it appears, among the medical witnesses at Danval's trial, and he may have had the benefit of it in a side-way . . . The doubt', added the editor with a touch of grim humour, 'has taken off the edge of the sentence.'[25] Danval was not guillotined but sentenced to hard labour for life and sent to a penal colony in New Caledonia. There he led an exemplary life, keeping up with medical advances and hoping one day to hear of some development that would exonerate him.

There were no such doubts about Chantrelle. Although the evidence against him was not of the strongest, his guilt, especially given his final outburst, was never contested, and there was little sympathy for such an abusive husband. *The Times* observed that, while circumstantial evidence was 'a kind of proof on which juries are not apt to look favourably', it was often the only kind available in such cases. 'It is enough if the whole chain is complete, and if each separate link of it holds well together with the rest,'[26] and those conditions had been satisfied.

Appeals for a reprieve mostly came from opponents of capital punishment. Chantrelle met his death stoically on 31 May 1878 without making a confession. The names of his three sons were changed to protect them. Eugène, who had had to go through the harrowing experience of giving evidence at his father's murder trial when only nine years old, later went into the navy, but committed suicide shortly before his twenty-second birthday.

It was not until the Amended Matrimonial Causes Act was passed in 1878 that women whose husbands had been convicted of violence against them could obtain a legal separation on those grounds alone; for many, this was a more acceptable solution than divorce. For the poorer sectors of society, divorce remained out of reach. It was not until the passing of the Married Women's Property Act of 1882, which became law on 1 January 1883, that a married woman had a legal identity distinct from that of her husband and was permitted to own property in her own right.

CHAPTER SEVENTEEN

Sleight of Hand

It is easy enough for a murderer to slip poison into food or medicine unobserved when under the same roof as the victim, but to poison at a distance needs imagination. Lethally laced food and drink can be sent by post, but it might arouse suspicion and be thrown away, be consumed in error by someone other than the intended victim, or worse still, from the criminal's point of view, be analysed and traced back to the sender. To kill one's chosen victim by poison from a distance requires careful planning and trickery.

In December 1872, Mrs Elizabeth Jenkins of Myrtle Cottage, Stapleton, near Bristol, received an unexpected visitor: a lady about thirty years of age, who said her name was Anne, but was strangely reticent about her surname. She asked if Mrs Jenkins had a daughter who was a dressmaker, as her sister would like to have a dress made. Mrs Jenkins's daughter, seventeen-year-old Susan, was in service. She had given birth to a child two months previously, a healthy girl called Sarah. Seeing the baby on its grandmother's lap, Anne wept, exclaiming that it reminded her of a child she had lost. Anne's visits to Myrtle Cottage continued, although she never pursued the dressmaking commission. She took tea with the family, brought gifts for the baby, and gave conflicting accounts of her address and her husband's occupation.

Shortly after Christmas, Susan visited her mother and met Anne for the first time. Susan had some good news: her employer was going to speak to a charitable lady at Cotham in Bristol on her behalf. Anne suggested that this lady might be a member of the Dorcas Society, a Christian women's organisation that provided clothing to the poor. Time passed, but nothing was heard either from the lady at Cotham or from the Dorcas Society.

Meanwhile, Anne doted on the baby, fussed over her, helped to nurse her and offered to take her out. This affection must have seemed excessive, since she was not allowed to take Sarah from the cottage, possibly for fear that she might not bring her back. Anne even offered to adopt the child, but Susan said that her mother was able to care for her.

Susan revealed to Anne that the father of baby Sarah was thirty-one-year-old Edwin Bailey, who kept a boot and shoe shop at Boyce's Avenue, Clifton, Bristol. Susan's employer had sent her to the shop to get some boots repaired and he had dragged her into an inner room and raped her, cramming a handkerchief into her mouth to stifle her cries. Despite the violence of this event, Susan, who might well have been unwilling to reveal to her employer what had happened, had returned to the shop on another errand, where a second assault had taken place, resulting in her pregnancy. Bailey had refused to admit that he was the father of her child or pay maintenance, and Susan had been obliged to obtain an affiliation order that required him to pay the cost of the confinement and five shillings a week until Sarah was sixteen. The payments were made via a policeman, Constable Critchley. Anne said that she did not know Bailey, but commented that he must be a great blackguard.

In the summer of 1873, baby Sarah was teething and Anne asked what medicines she was receiving. Mrs Jenkins said she gave magnesia, a common antacid and mild laxative. Anne suggested she try Steedman's soothing powders, one of a variety of preparations intended for children, especially to reduce the pain of teething, the active ingredient being opium. Mrs Jenkins had never seen or used one and said they were too expensive. Despite this, Anne persisted in recommending them.

On 13 August 1873, with Bailey's payments falling into arrears, Anne was confidently predicting that Susan would soon hear from the Dorcas Society, and, the same day, a letter arrived containing three Steedman's soothing powders and a shilling's worth of stamps. The sender was purportedly Jane Isabella Smith of Hope Cottage, Cotham, who described herself as 'a lady visitor of Cotham Dorcas Society'.[1]

When Anne called again the following day, she asked how the child was doing, and was told she was teething. Susan said that the expected letter had been received, and Anne enquired as to whether the child had been given one of the powders. She was told no, as the child was currently well and not suffering any discomfort. Anne nevertheless encouraged Susan to give her baby one of the powders, saying that it 'would do the child so much good in cutting its teeth'.[2] Anne, who appeared to be in unusually low spirits that day, then explained, with tears in her eyes, that she would not be able to call again as her husband was removing to another part of Bristol.

On 17 August, baby Sarah was fretful and her gums were red and swollen. Her mother and grandmother decided to try one of the powders, and stirred it into some soaked bread and sugar. The resulting mixture was light blue, but, never having used a Steedman's powder before, they saw nothing unusual in this. Neither of them thought to taste it. The mixture was fed to the child, who took it reluctantly, and, as soon as it was eaten, began to cry. Susan took her child out into the garden to soothe her, but she only cried harder. Soon, the cries became screeches of agony, which brought Mrs Jenkins rushing out into the garden to see what the matter was. The child's little body was rigid and arched, her face dark, tiny fists clenched. At first, they thought a pin had run into her, but no pin could be found, and, as the baby's condition rapidly worsened, the distraught women realised they needed a doctor. Susan ran to two local doctors, neither of whom was willing to come. She eventually found Dr Parsons, who lived a mile away. By the time he arrived, Sarah was dead.

The remaining two powders were shown to Dr Parsons. Cautiously, he touched one with his tongue, and noticed a metallic taste and a sickly sensation. Parsons's post-mortem on little Sarah confirmed that she had been poisoned. He placed the stomach and intestines in a jar, and this, together with the powders, was passed to the county analyst John Horsley. He saw at once that whatever was in the Steedman's wrappers was not the original soothing powders, which he thought had been removed and something else substituted. His analysis revealed that the powders were a product called 'vermin killers', each one containing a fifth of a grain of strychnine. One sixteenth was enough to kill a small child. This was not a case of accidental contamination. Baby Sarah had been murdered, deliberately targeted by someone who had cruelly duped her loving family into innocently administering poison.

Christiana Edmunds had had to use some ingenuity in order to obtain pure strychnine, but anyone could buy it openly as an ingredient of vermin killers, of which there were many formulations. Mixed with powdered rice or wheat starch, and averaging about 10 per cent strychnine, they were sold in quantities varying from about five to sixty grains and were usually coloured blue either with Prussian blue or with ultramarine, although there were some varieties that used carmine or soot. There were many proprietary brands, but most pharmacists liked to make up their own. The powders were usually sprinkled on bread and butter and left where rats and mice would consume them. The bright blue colour was intended as a warning that the material was not suitable for human consumption, and the wrappers clearly stated the nature of the contents, but tests carried out in 1889 revealed, to the concern of analyst Alfred H. Allen, that some vermin killers were barely tinted, and one variety not at all.[3]

The letter sent to the Jenkins family, supposedly from the Dorcas Society benefactress, and the envelope in which it and the powders had arrived were handed over to Constable Critchley. He soon confirmed that there was no Jane Isabella Smith at Cotham, but, importantly, he recognised the paper and envelope as the same as those used by Edwin Bailey, and the letter bore a Clifton postmark. Bailey gave evidence at the inquest on baby Sarah, but, after being bound over to attend the adjourned hearing, he disappeared.

Bailey, it transpired, had a long history of sexual assaults on women. When working in Gloucester, he had paid his addresses to a young lady, but her cautious father had made enquiries and discovered that the suitor already had a wife who was said to be an invalid and lived in London.[4] Bailey had made a hurried departure from Gloucester and opened his new business in Clifton. It was reputed that a young woman could not enter his shop without danger of molestation, and many servant girls, having once been sent there on an errand, refused ever to go again. At the time of Sarah's birth, he was already paying maintenance on another paternity suit.

A warrant was issued for Bailey's arrest, and attention was drawn to his charwoman, Anne Barry, who fitted the description of the woman who had visited Mrs Jenkins. Anne was arrested and admitted that she had gone on errands for Bailey. According to Anne, he had denied paternity of Susan's child and her purpose in visiting the family was solely to discover the identity of the real father. She denied knowing anything about the poison.

Bailey was in London, intending to sail for Spain, but, before he did so, he stayed briefly with his wife. She, however, anxious about the fate of the Clifton business, which must have been her main means of support, persuaded him to return there with her, to show her what needed to be done. He complied and, undeterred by placards announcing the arrest of Anne Barry, went to his shop and donned his apron. He was arrested soon afterwards. Bailey and Barry were tried at Gloucester Assizes on 23 December. Both were found guilty of murder and, on 12 January, they were hanged at Gloucester Gaol. Shortly afterwards, the chaplain told the press that they had both confessed their guilt.

A vermin killer was also the means by which Elizabeth Pearson of Gainford, near Darlington, poisoned her sick seventy-five-year-old grandfather in 1875, by disguising it as medicine; however, it was not only children and the elderly who could fall victim. In 1886, thirty-nine-year-old Mary Ann Britland, a factory worker of Ashton-under-Lyne, managed, in a way that was never fully explained, to overcome the problems of the colour, the flavour and the insolubility of vermin killers to poison three healthy adults and very nearly get away with murder.

The deaths of Mary Ann's nineteen-year-old daughter Elizabeth Hannah on 9 March and her own forty-four-year-old husband Thomas on 3 May were initially assumed to be due to natural causes. Elizabeth's medical attendant Charles Thompson had been told that the girl had had a fit, and he suspected an 'hysterical attack',[5] while the surgeon called to see Thomas was led to believe that the patient was suffering from delirium tremens.

Mary Ann was friends with a young married couple, near neighbours Thomas and Mary Dixon, who worked in the same factory. After her husband's death, Mary Ann stayed at the Dixons' house, but, on 13 May, Mrs Dixon was taken ill with spasmodic fits and died the next day, aged twenty-nine. Charles Thompson certified the death as due to abdominal spasms and exhaustion, but the neighbours were unconvinced and reported their suspicions to the police, who alerted the coroner.

A post-mortem examination and the sacrifice of two mice revealed the presence of both strychnine and arsenic in Mrs Dixon's stomach and intestines. One thousandth of a grain of the strychnine placed on the back of a frog caused tetanic convulsions: the frog recovered in twelve hours. Mary Ann was known to have purchased vermin killers, which she called 'mouse powder': a packet of Harrison's on 9 March, two packets of Hunter's on 28 April, and three more packets of Harrison's on 3 May.

The bodies of Elizabeth and Thomas Britland were exhumed, but the passage of time had made it difficult for analytical chemist Charles Estcourt to positively prove the presence of strychnine, although a solution prepared from Elizabeth's organs had a distinct bitter flavour. The Marsh test, however, revealed the presence of arsenic in both sets of remains. The vermin killers were also analysed. A threepenny packet of Hunter's contained two and half grains of strychnine, and a twopenny packet of Harrison's, two and half grains of strychnine and 3.52 grains of arsenic.

Mary Ann had received insurance money on the death of her daughter and husband and it was thought that she wanted to marry Thomas Dixon.

She and twenty-eight-year-old Dixon had often been seen in each other's company and, during Thomas Britland's lifetime, Dixon had visited Mary Ann when her husband was not at home. An 'unusual degree of intimacy'[6] was believed to have existed between Mary Ann and Dixon, who had been heard to say that he was tired of his wife and wished he had a wife like Mrs Britland. Dixon was arrested on suspicion of being an accessory to murder, but was eventually discharged due to lack of evidence.

How had Mary Ann managed to administer vermin killers to her victims without their suspecting they had been poisoned? The victims of strychnine remain conscious to the end, lucid and able to speak between convulsions. Mr Thompson had asked Elizabeth if she had taken anything for supper that had disagreed with her, and she had said no. Mary Ann had provided a clue, making anxious enquiries after the death of Mrs Dixon as to whether it was possible to tell if someone had been poisoned with 'mouse powder', asking, revealingly, 'Can they tell if she had it in tea?'[7]

Dr Thomas Harris, who had taken part in the post-mortem on Mr Britland, had experimented by adding a grain of strychnine to a quart of tea then sweetening it with sugar, but it still tasted of strychnine. He thought that half a grain, a fatal dose, in a pint might not be noticed if something were eaten with it. Strychnine was not, however, very soluble in water and, in the case of mouse powder, which included starch, a great deal of it would float on top, and any blue dye would be noticeable. It is not known if the Hunter's or Harrison's preparations were dyed blue. Mary Ann Britland was found guilty. She confessed to all three murders and was hanged on 9 August, the first woman to be executed at Strangeways Prison.

Although many of the problems of identifying poisons in human remains had been solved, juries were still left with the difficulty that, in most cases of homicidal poisoning, there was no eyewitness to the administration. Rarely did a poisoner commit murder in front of witnesses with no attempt being made at concealment. Rarely did he or she do it with such bare-faced nerve and panache as Dr George Henry Lamson. His chosen poison was unusual – bitter burning aconite. He would have been aware that, while the Stas test could identify alkaloid poisons, there was, as yet, no chemical test that would distinguish aconite. Toxicologists faced with looking for evidence of aconite had to fall back on some very old techniques, first tasting a sample, and then observing its effect when administered to animals. While aconite had been used with understandable caution as an internal medicine at the time of the Pritchard case in 1865, its dangers had led to this practice being largely abandoned in the UK, where aconite was mainly used in ointments although it was still an ingredient of some patent medicines available in France and Germany.

Lamson, born in New York in 1852, studied medicine in Paris and the United States, spending much of his career at military hospitals on the continent of Europe. In 1878, on the Isle of Wight he married Swansea-born Kate George John. Kate, the twenty-five-year-old orphaned daughter of a linen merchant, became entitled on marriage to claim her portion of her parents' legacy, which then became the property of her husband. Kate's sister had married and inherited her share in the previous year. Their two brothers were still minors, Hubert born in 1861 and Percy in 1862. In June 1879, Hubert died suddenly at the Lamsons' home, and his share of his parents' bequest passed to his two married sisters and brother. Percy's share of the legacy was £3,000. He was due to inherit either on marriage or achieving his majority, but if he died unmarried before reaching the age of twenty-one, his share would be distributed equally between his two sisters.

Percy suffered from a curvature of the spine and had lost the use of his legs, but was otherwise in good health. In 1881, he was boarding at Blenheim House School, St George's Road, Wimbledon, which had been his home for the Past three years. An intelligent and popular boy, he was well cared for, and both masters and pupils made sure that, as far as possible, he was involved in the activities of the school and had plenty to occupy his mind. It was only natural that Percy became depressed from time to time, being unable to take part in sports he loved to watch, but he generally bore his difficulties with cheerful acceptance.

In 1880, Lamson purchased a medical practice in Bournemouth, but, early in 1881, he was experiencing severe financial problems, for reasons that were not revealed until very much later, and his attempts to obtain funds were becoming increasingly desperate. Lamson pawned his watch and medical instruments and borrowed from acquaintances. He tried to get cash advances on cheques from an account that was already overdrawn, and, in one instance, on a bank with which he had no account. In March 1881, the Bournemouth Medical Society, after finding that Lamson was not entitled to a number of qualifications he had been claiming, removed him from its rolls and, in April, he sold his practice and left Bournemouth.

By then, even a half share in Percy's legacy must have been looking very attractive, but the boy showed every sign of living long enough to inherit. Lamson, under the guise of concern for his brother-in-law's health, began dosing him with a variety of medications, which invariably made him feel worse. Perhaps Lamson was trying to pave the way for an early death not to be regarded as suspicious. He sent a box of pills to William Bedbrook, the headmaster of Blenheim House School, saying that they would benefit Percy's condition. Bedbrook gave Percy one of the pills, but, the next morning, the boy complained of feeling unwell and said he would take no more. Lamson himself had given Percy a pill during a family visit in August 1881, and the boy had felt very ill afterwards. The persistent doctor also supplied powders, which he said were quinine, together with medicinal wafers made of flour paste intended to enclose the powders so that they could be swallowed without being tasted, but there is no evidence that Percy ever took one. No one appears to have suspected that Lamson was deliberately trying to harm Percy, but the boy may have learned to distrust medicines from that source.

On 1 December, with Percy approaching his nineteenth birthday, Lamson wrote to him saying that he would shortly be going abroad and would like to call to see him before he left. Lamson arrived at the school on the evening of 3 December and was shown into the ground-floor dining room while Percy was carried up the stairs from the basement rooms where he studied. It was some weeks since Lamson had last visited, and he looked so much thinner and paler than on the last occasion that Bedbrook did not at first recognise him and remarked on the change in his appearance. Bedbrook offered Lamson a glass of sherry, and Lamson accepted, saying that he always took sugar with it to counteract the effect of the alcohol. Bedbrook thought this odd, and said that he understood it would have the opposite effect, but he obligingly sent for the sugar, and a basin of powdered white sugar was brought and placed on the table. Lamson put a little sugar into his sherry glass and stirred it with his penknife.

Lamson was carrying a black bag, which he opened; he took out a Dundee cake and some sweets, which he said he had brought from New York. He then cut up the cake with his penknife, handed pieces to Percy and to Bedbrook and took one himself. Both Percy and Bedbrook helped themselves to sweets. Lamson next produced two boxes containing gelatine capsules, which he said he had brought back from a recent trip to America, suggesting that Bedbrook would find them useful in giving medicines to the boys. When he placed the boxes on the table, one was much nearer to Bedbrook. 'See how easily they can be swallowed,' said Lamson.[8] The master took a capsule from the nearer box, examined it, and checked that it was empty before swallowing it. Bedbrook later recollected that the capsules were of different sizes. He then saw that Lamson was spooning sugar from the basin into another capsule although, having been distracted when examining the capsule he had taken, he had not noticed from which box this capsule came. Lamson handed the filled capsule to Percy, saying, 'Here, Percy, you are a swell pill taker, take this.'[9] Percy obediently took the capsule and swallowed it. A few minutes later, Lamson announced that it was time for him to leave to catch his train. Bedbrook saw him to the door and, as he departed, Lamson said that he did not think Percy would live much longer, a comment the master thought quite unjustifiable from the boy's appearance and state of health.

Ten minutes later, Percy began to complain of heartburn, saying that he felt just the same as he had when Lamson had given him a quinine pill. His condition grew rapidly worse and he was carried up to bed where he suffered painful convulsions and vomited violently. The matron was called, and two doctors summoned. Both doctors, suspecting poison, made sure to preserve some vomit for testing. There was little they could do beyond injecting morphine to relieve the pain. Eventually, the tormented boy fell into a coma and, at 11.20, just four hours after Lamson's visit, Percy died. The police were notified and enquiries were made as to the whereabouts of Dr Lamson.

The post-mortem examination was carried out by Dr Thomas Bond, lecturer in forensic medicine at Westminster Hospital, who concluded that Percy had died from a vegetable alkaloid poison. The stomach contents were examined by a rising star of the criminal court, Dr Thomas Stevenson. Born in 1838, he was a former medical officer of health for St Pancras, and, by the time of the Lamson case, he was a Home Office analyst and lecturer in chemistry, and had recently succeeded Alfred Swaine Taylor, who had died in 1880, as lecturer in forensic medicine at Guy's Hospital. Stevenson was to give evidence at some of the most celebrated poisoning trials of the next twenty-four years and he was knighted in 1904.

Using a modified version of the Stas process, Stevenson obtained alkaloid extracts from portions of the remains and placed small amounts on his tongue. The extract obtained from the stomach contents in particular produced a burning and tingling, with a feeling of swelling at the back of the throat, 'followed by a peculiar seared sensation at the back of the tongue, as if a hot iron had been passed over it or some strong caustic applied'.[10] Other samples produced a similar effect – a 'burning sensation, extending down to the stomach . . . peculiar to aconitia'.[11] His symptoms lasted for over seven hours before they passed off. An extract of Percy's urine injected under the skin of a mouse caused death in thirty minutes. Some pills and powders supplied by Lamson were amongst the deceased's effects, and

Stevenson found that the pills and three of the powders produced the same burning sensation. No trace of poison was found in the cake, the sweets or in the gelatine capsules.

On 8 December, Lamson returned from Paris and went to Scotland Yard. He said he was unwell and upset, but that he had come home after reading about the death of Percy in the newspapers and had called to give his address, after which he planned to go to Chichester. He did not seem surprised when he was arrested, charged with the murder of Percy John and taken to Wandsworth police court. 'Do you think they would accept bail?'[12] he asked, hopefully. There was speculation that the body of Percy's brother Hubert might be exhumed, although in the event this was not done.

At Lamson's trial in March 1882, it was shown that he had bought two grains of aconite on 24 November. He had also bought some on the Isle of Wight on the day before Percy had been taken ill after taking a pill provided by Lamson. Thomas Stevenson was in no doubt that the cause of death was poisoning with aconite. The defence hoped to cast doubts on the cause of death, but Stevenson's experience was apparent to his listeners. He told the court, 'I have fifty to eighty vegetable preparations in my possession, and have tasted most of them.'[13]

The defence counsel, Mr Montagu Williams, who called no witnesses, did his best, and, in doing so, raised a question that had not seriously been tested in court before. In 1870, Francesco Selmi, professor of medical jurisprudence at the University of Bologna, used the Stas method to examine the viscera of a man thought to have been poisoned, and found a previously unknown alkaloid. He pursued the anomaly, examining bodies exhumed after several months' burial, and concluded that poisonous alkaloids were produced in the stomach of those who had died from natural causes by a process of putrefaction. It was several years before his conclusions were accepted and the existence of cadaveric alkaloids, some of which were very poisonous, was placed beyond doubt. Selmi called them ptomaines, from the Greek word for corpse. The implications of this discovery for criminal trials were alarming, to say the least. 'Amongst the various alkaloids of cadaveric origin, we find certain ones that are more easily confounded with poisonous vegetable alkaloids, and especial care must be taken not only to avoid condemnation of the innocent in judicial questions, but also to make the evidence so sure that poisoners may not escape the reward of their crimes,'[14] warned Dr Nelson Sizer in 1882.

One of the planks of Mr Montagu Williams's defence of Lamson was the concept of cadaveric alkaloids, which he hoped would create reasonable doubt in the minds of the jurors as to whether aconite had been the cause of death. Stevenson pointed out that the issue was 'still *sub judice* among experts';[15] however, while it had been asserted that some cadaveric alkaloids produced the same effects as vegetable alkaloids, he had seen none that mimicked aconite. Stevenson was re-examined by the prosecutor, Mr Poland, who established that the work on cadaveric alkaloids had been carried out on 'putrefied corpses'.[16] Williams did his best, telling the jurymen to pause in their fatal decision to consider that the question of cadaveric alkaloids was still undetermined. To decide if the deceased had died from the effects of aconite, they 'must be of the opinion that the matter is settled beyond any possibility of doubt';[17] Stevenson's evidence he dismissed as 'theoretical' and 'speculative'.[18]

Stevenson had been impressive in the witness box; clearly a master of his profession, calm and meticulous. In the face of this, Williams's defence, while vigorous, had an air of desperation. Mr Poland effectively destroyed the defence theory of cadaveric alkaloids in his final address when he pointed out that Percy's body had not yet started to decay when Dr Bond took his samples and placed them in spirit to preserve them. Lamson was found guilty, but his execution was delayed while his solicitor assembled information to demonstrate the condemned man's insanity. It was a fertile field to explore. The superintendent of Bloomingdale Asylum for the Insane, New York, revealed that Lamson's aunt, grandmother and great-uncle had all died there. Affidavits described Lamson's paranoid delusions, hallucinations and lack of self-control. It had been mentioned at the trial that Lamson had purchased morphine and atropine for his own use, but the extent of his addiction to those drugs was only now revealed. He had been suffering from pain, dizziness and insomnia, had developed a peculiar lurching walk, and his arms were scarred and abscessed from frequent injections. None of this evidence, however, suggested homicidal mania or an inability to tell right from wrong, and, as *The Times* pointed out, '[T]he immediate circumstances of the murder pointed to the exercise of a crafty deliberation . . . if Lamson could appreciate the benefit he would derive from Percy John's death – and why else should he have selected him as his victim? – he could realize the wickedness of his act.'[19]

There was to be no reprieve for Lamson, who, shortly before he died, admitted murdering Percy John, although he denied murdering Hubert. Lamson was executed at Wandsworth Gaol on 28 April 1882.

Only one question needs to be addressed: how did Lamson administer the poison under the nose of William Bedbrook? The simple and obvious answer is that it was in the capsule, but this has been disputed. Sir Henry Hawkins, the presiding judge, writing about the case in 1904,[20] believed that the poison was in the cake, the capsules being just a clever smoke-screen. He suggested that the poison was injected into one of the raisins in the cake, so that the skin would delay its effect, although he admitted that there was no evidence to support this theory. H. L. Adam, editor of the *Notable British Trials* volume, first published in 1912, stated, quoting the authority of an unnamed person present at the trial, that Lamson was able to apportion the poisoned slice to Percy because he brought the Dundee cake to the school already cut in slices. The essential evidence on this point is that of Bedbrook, recorded by the court stenographer and newspaper reporters at the trial, as well as in his statements at the inquest and police court and his original deposition. Bedbrook said he noticed nothing unusual about the cake except that it was wrapped in newspaper, and that Lamson sliced it into portions with his penknife. Bedbrook never said that the cake arrived pre-cut, and, if he had, the Solicitor General Sir Farrer Herschell, prosecuting, would surely have mentioned it. Adam also stated that the stomach contents included a raisin skin found to be impregnated with aconite. This is incorrect. Stevenson said that he had received a bottle containing some of the stomach contents, which were mainly fluid and included a raisin. He had found aconite not in the raisin but in the fluid.

Lamson was well aware that aconite has a burning taste. He had made use of it at a military hospital in Bucharest and, after his trial, two doctors signed affidavits about his reckless use of the drug when practising there. The best methods of supplying it as a poison were as a coated pill, in a powder meant to be swallowed enclosed in a wafer, or in a capsule. Lamson tried all three. The raisin idea would only work if Percy gulped it down whole and unchewed, hardly something on which a would-be poisoner could rely. Biting down on the raisin would cause immediate pain. Lamson's method shows careful planning and forethought. He first ensured that Bedbrook took a capsule from a box of empty ones pushed close within his reach. Lamson must have already primed one of the capsules in the other box with poison and had only to select the right one to fill with sugar while Bedbrook's attention was directed towards the capsule he was examining. If Bedbrook's observation – that the capsules he saw were of different sizes – was correct, then selecting the primed capsule should not have been difficult. Alternatively, Lamson might have had the fatal capsule in his pocket and only appeared to take it from the box, a piece of misdirection worthy of a conjuror. The cake did have a part to play. It was intended as an early birthday gift for Percy, who would have been nineteen on 18 December, and formed part of Lamson's excuse for his visit. Lamson might also have thought that the heavy fruit cake would delay the operation of the poison sufficiently for him not to be on the premises when it took effect.

CHAPTER EIGHTEEN

Disappearing Poison

In 1885, grocer's wife Adelaide Bartlett was perusing a copy of P. W. Squire's *Companion to the latest edition of the British Pharmacopoeia*, looking for the perfect poison with which to murder her husband, Edwin.

The marriage, which had taken place in 1875, when Adelaide was nineteen and Edwin thirty,[1] had got off to an unusual start, probably because there had been two serious obstacles to its happening at all. Edwin had an agreement with his business partner Edwards Baxter that neither would incur the expense of marriage and maintaining a family until their trade was well established, which, in 1875, was a position still some years in the future. Moreover, while Adelaide had charmed Edwin, she had failed to impress his father, Edwin senior, who made no secret of the fact that he distrusted her. Adelaide, who had no fortune to bring to the marriage, must have worried that her promising catch might escape her if there was a delay, so a compromise was arranged. Once the wedding had taken place, the bride was sent away to complete her schooling until such time as the couple could set up home. That home, two years later, was an apartment above the family grocer's shop under the critical eye of her father-in-law.

Edwin was not a bad husband, but he was devoted to his business and spent long hours at work. His main interest in life appears to have been grocery; his one hobby the breeding and showing of St Bernard dogs, which were housed at kennels. The neglected wife must have been bored and unhappy, and, in 1878, she ran away. Her father-in-law accused her of carrying on a romantic intrigue with Edwin's younger brother Frederick, but Adelaide denied this and claimed that she had been upset at Edwin senior's insults, and had gone to stay with an aunt.

Edwin, whose confidence and trust in his wife never wavered, managed to smooth over the situation and asked his father to sign a document admitting that his accusations were untrue. Edwin senior did so, but with extreme reluctance, and Frederick wisely left for America.

In 1881, after six years of marriage, Adelaide became pregnant. The couple had been trying to limit their family, probably for financial reasons, using condoms, but perhaps at long last they felt that the time was right. The pregnancy, as Adelaide was later to confide, was the result of unprotected intercourse. A child might have made a difference to the dull marriage, but the baby was stillborn and there were no further pregnancies. The sexual side of the marriage deteriorated, and Edwin, who carried condoms in his trouser pockets, may have sought solace outside the home.

One factor that must have contributed to the lack of marital tenderness was the unpleasant state of Edwin's mouth. Edwin's dentist had sawn off his patient's bad teeth at gum level with the intention of fitting him with a false set, but this had caused such terrible irritation that Edwin could tolerate neither dentures nor a toothbrush. The stumps rotted, and his gums were diseased, spongy and clogged with decaying food.

In 1884, the Bartletts took a cottage in the village of Merton Abbey. Edwin went to work every morning by train, leaving Adelaide to her reading, embroidery and light housework. A new friend who became a frequent visitor was a Methodist minister, George Dyson; three years younger than Adelaide, he was well educated and able to converse interestingly on a wide variety of subjects. He rapidly became smitten with the attractive young Frenchwoman, who hung on his every word. He gave her lessons in history, Latin and Greek; they took lunch together and went on long walks.

The bored wife blossomed, and Edwin, with complete confidence in the purity of purpose of a man of the cloth, was delighted at the change in his wife and did all he could to encourage the friendship. Edwin had originally made a will in which he left all his property to Adelaide on the condition that she did not remarry. She had long protested that this was unfair, but, in 1885, he made a new will, naming his good friend George Dyson as executor. Moved by this, George shamefacedly confessed his admiration of Adelaide to Edwin, but the trusting husband remained untroubled. When George was offered a new post in Putney in the autumn of 1885, the Bartletts gave up the Merton cottage and moved to a two-room apartment in Pimlico, which was convenient both for Edwin's work and George's continued visits.

Adelaide now set the stage for ridding herself of her dull smelly-mouthed husband with the assistance of her adoring dupe. Whether she ever intended to marry Dyson is uncertain. She was, as he later found to his cost, a convincing liar, able to make people believe the most unlikely tales if they were given an enticing gloss of scandal, tragedy and romance. The Victorian habit of 'delicacy' – shrouding true meaning in euphemisms and subtle allusions, allowing the listener to fill in the unmentionable blanks – was something she used to perfection. Adelaide told George that Edwin was suffering from a serious internal illness and had consulted a doctor, who had given him less than a year to live. George was distraught at this terrible news but she warned him not to discuss it with Edwin, who was very sensitive on the subject, and he promised faithfully that he would not. Adelaide used her reading of medical books to convince George not only that she was telling the truth, but also that she was knowledgeable enough about medicines to deal with her ailing husband's symptoms. George, always anxious to be useful, offered whatever help he could provide, but the idea must have occurred to him that, if he were patient, the enchanting Adelaide would one day be his.

That November, Edwin, who had previously been in good health, was suddenly taken ill with stomach pains, vomiting and diarrhoea. The dedicated man of business was obliged to take to his bed. Adelaide called in a nearby doctor, twenty-nine-year-old Alfred Leach, who noticed a blue line on the margins of Edwin's puffy gums, a symptom he recognised as caused by poisoning with a heavy metal such as mercury or lead. Leach asked Edwin if he had been taking mercury for venereal disease, but Edwin denied both the mercury and the disease. Leach's next theory was that Edwin had what was then termed an 'idiosyncrasy' (what we would now call a sensitivity or allergy) for mercury, which he had taken in a small amount in the one proprietary pill he had admitted to swallowing. Leach never seems to have suspected that Edwin was being deliberately poisoned.

* * *

If Adelaide was poisoning her husband, she might have been using lead acetate, known as sugar of lead. Easily obtainable from a chemist, it resembles sugar, has a sweetish taste and was used either internally or externally as an astringent. It is a far less active poison than arsenic and only occasionally featured in homicides. In 1883, Louisa Jane Taylor was executed for the murder of eighty-one-year-old Mrs Tregellis by poisoning over a period of more than six weeks with sugar of lead. Mrs Taylor, who was a widow, had had designs on eighty-five-year-old Mr Tregellis's annuity, and had unsuccessfully invited him to abandon his wife for her. The deceased's surgeon, Mr Smith, had not suspected foul play until Mrs Taylor was found to have stolen property from the Tregellises' house, and he recalled that she had purchased sugar of lead from his wife. Mrs Tregellis was still alive when Louisa was arrested, but the revelation came too late to save her.

The Times had this to say:

> [T]he consideration that Mrs Taylor's clumsy essay in the art of poisoning promised at one time to remain undetected is not comforting. She obtained her poison without difficulty; she continued to administer it while a qualified medical practitioner was visiting and prescribing for the patient. Sugar of lead is certainly not a poison the purchase of which is likely to occasion suspicion; and it is not one of those mentioned in the schedules of the Pharmacy Act which are to be sold under special restrictions. It is easy enough for poisoners to obtain their drugs as long as they are not too ambitious.

Recalling how Mr Smith had been deceived, *The Times* observed: 'From the numerous poisonings which have only been detected by an accident or an afterthought, the inference is only reasonable that there remains a margin of poisonings which are never detected at all.'[2]

* * *

Adelaide, if she was poisoning Edwin, had to rethink her methods in the light of Dr Leach's discovery. Edwin's condition improved, and Edwin senior, who had not been allowed to visit his son, was now invited to come and see him. Nevertheless, Adelaide retained total control over her husband and was constantly in attendance on him, even sleeping in a chair beside his bed, establishing his complete physical and emotional dependence on her. Just as he seemed to be rallying and was considering a therapeutic trip to the seaside, a worm was discovered in his faeces, which further puzzled Dr Leach and shook Edwin's nerves. This curious development was never explained, although it should be noted that Adelaide might have had access to worms at the dog kennels. Edwin, who felt that he drew strength from Adelaide's mere presence, was now reluctant to be apart from her for any length of time. Despite this, Leach was able to persuade his patient to go for a short drive and also to have some of his rotten tooth stumps removed. As the year drew to a close, Edwin began to feel stronger and more confident, and considered returning to work. He also talked about paying a regular allowance to his aged father. It was a recipe for murder.

Adelaide, worried by those giveaway blue lines, realised that what she required was a poison that would leave no trace in the body of the victim, and her studies of Squire's *Companion* suggested that she had found it. Her poison of choice was chloroform. She made a careful note on the page.

Soon after its first use as a surgical anaesthetic in 1847, chloroform began to acquire its legendary and wholly erroneous reputation as a harmless means of securing almost instantaneous and long-lasting unconsciousness.[3] Its supposed effects provided the perfect excuse for anyone caught in a potentially compromising position. The celebrated Victorian anaesthetist, Dr John Snow, recounted the case of a gentleman, discovered naked in a brothel, who attempted to cover his embarrassment and preserve his reputation by claiming that he had been rendered insensible by a chloroformed handkerchief being waved under his nose, and carried there against his will. No one was fooled. Unfortunately, the reputation persists to this day, kept alive in fictional crime dramas too numerous to mention. The actuality is that, in the hands of a skilled surgeon, it can take up to two minutes for a safe dilution of chloroform vapour in air to render a patient unconscious, a state that can only be maintained for any length of time by continuing to apply the chloroform while preserving the patient's airway. If the application ceases, the patient will regain consciousness in a few minutes. Chloroform is also far more dangerous than supposed. During the history of its use as an anaesthetic, approximately one patient in every 2,500 died during surgery from its effects, sometimes even before the first incision was made.

Since chloroform vapour produces unconsciousness only if it is actually inhaled, an able-bodied and unsecured victim of a criminal chloroform attack will have ample time to hold his or her breath, resist and escape. Weak or bound victims, however, forced to inhale undiluted chloroform by criminals unversed in anaesthesia, tend to die from cardiac arrest.

This public perception of the action of chloroform as producing fast, safe unconsciousness means that it is rare for someone to set out with the intention of using inhaled chloroform as a murder weapon, although he might unintentionally kill a victim when attempting to use it for robbery or rape. Squire's *Companion*, however, simply described how to administer chloroform internally as a sedative in doses of one to five minims, mixed with egg yolk, mucilage, syrup or brandy, a minim being 1⁄480 of a fluid ounce. Adelaide also learned that chloroform was sweet-tasting, dissolved well in alcohol, and 'evaporates speedily, and leaves no residue and no unpleasant odour'.[4] Squire's *Companion* gave no indication of the fatal dose, and did not mention chloroform's burning taste and potential for corrosive damage to the lining of the stomach.

Adelaide set the scene. Edwin had been suffering from insomnia, and she carefully dropped suggestions that chloroform was a pleasant thing to take. She got the obliging George to buy the chloroform for her, convincing him that she was skilled in its use. On 31 December, Edwin retired for the night, with Adelaide settling to sleep in a chair beside his bed. Next morning, he was dead.

Edwin's post-mortem examination revealed that his stomach contained a watery fluid smelling strongly of chloroform. Adelaide had found too late that the vaunted volatility of chloroform did not make it disappear without trace when it was inside the body. Edwin's stomach and its contents were sent for analysis to Dr Thomas Stevenson, who calculated that Edwin had swallowed about an ounce of liquid chloroform. There was nothing else in the remains to account for death.

George Dyson was terrified and appalled. When he tackled Adelaide about the bottle of chloroform he had supplied her with, there was a terrible row, which ended when she denied telling him that Edwin was a dying man. He demanded to know what had happened to the bottle, hoping that it would be found unopened, but Adelaide said she had thrown it away. At long last it dawned on George that he had been thoroughly duped by a clever woman, and he was ruined. In February 1886, Adelaide and George were arrested and charged with the murder of Edwin Bartlett.

* * *

It is unlikely that Adelaide had read about the death of Winifred Markland in 1876, or she might have chosen another poison. Twenty-seven-year-old Winifred had often been reprimanded by her father for her immoral behaviour, and, when he learned that she was pregnant, he turned her out of the family home in Stockport. The child was later stillborn. Winifred went to live in Manchester, where she made a meagre living with a sewing machine, supplementing her income with casual prostitution. She shared her lodgings with a respectable hat-maker, Fanny Stevenson, who knew about Winifred's sideline and preferred to go and stay with her sister when a male visitor was expected.

One of Winifred's more unusual men friends was the Reverend Father Joseph Daly, a thirty-seven-year-old Roman Catholic priest, formerly of

Stockport, who, in 1876, was officiating at a chapel in Runcorn. On 2 October, Winifred, who had just returned from a trip to Runcorn, told Fanny that she expected a gentleman visitor that night. He had already given her £1 and had promised her another two. Fanny departed, and the gentleman called at 10 p.m.; he was seen to arrive by the neighbours, one of whom knew Father Daly by sight and was sure it was he. When Fanny returned the next day, she found the front door locked and had to borrow a key from a neighbour. The bedroom door was open and she saw Winifred's arm hanging over the side of the bed. Assuming that Winifred was accompanied by a man, she decided not to disturb her. Fanny spent the next night at a friend's house but, on her return, found Winifred's arm in the same position, and investigated. The young woman was clearly dead and had been for some time.

Fanny called a doctor and a policeman. Bottles and glasses found by the bed contained nothing more suspicious than gin, water and sugar, and, at first, it was thought that Winifred had died from alcohol poisoning. Her stomach, however, contained, in addition to the gin, water and sugar, a fatal dose of liquid chloroform. No chloroform was found in the house.

Some three years previously, Winifred's father John had remonstrated with Father Daly about being over-familiar with his daughter. Following Winifred's death, rumours of an improper association started to circulate, even suggesting that the priest was the father of the stillborn child. Faced with these suspicions at the inquest, Daly boldly took to the witness stand and denied any impropriety with Winifred. He also claimed to have been in Runcorn at 10 p.m, the same hour at which the gentleman visitor called at Manchester, some twenty-five miles away, and, at his request, a number of his parishioners came forward to vouch for him. The jury was satisfied that Winifred had been murdered, but there was insufficient evidence for them to name a culprit. No mention was made of any perceived difficulty in getting the girl to drink chloroform, and it must have been assumed that she had been plied with it when drunk on gin.

Father Daly was not, however, the paragon of priestly virtue he liked to suggest. The following December, he was in the dock at Knutsford quarter sessions charged with the theft of £12 from the grocer's shop of his parishioner William Hailwood. Daly had seduced Hailwood's eighteen-year-old daughter, who supplied him with the key to the shop, and he had broken open the money drawer. When suspicion fell on the girl's brother, she confessed to the crime. Notes passed between the two lovers by servants proved the romantic association and, despite his denials, Daly was sentenced to eighteen months with hard labour. No one was ever convicted of the murder of Winifred Markland, but Daly, faced with her demands for money, and possible blackmail and ruin, is the obvious suspect.

* * *

In 1886, the distraught George Dyson was thought to be more valuable as a prosecution witness than a defendant. The charge against him was dropped, but his appearance to give evidence against Adelaide Bartlett when she was tried for the murder of her husband at the Central Criminal Court that April only served to excite sympathy for her. In the meantime, Adelaide had used her wiles and her imagination to effect an almost magical transformation from faithless wife to innocent victim of a bizarre marriage, claiming that her union with Edwin was almost sexless and that he had 'given' her to George Dyson. By this extraordinary piece of double-think, dull dead Edwin became the villain of the piece, while George was her 'true' husband, so it appeared that she had not been faithless after all. It is a tribute to the compelling nature of the story that many people, including the ingenuous Dr Leach, swallowed this ridiculous nonsense.

The crucial issues at the trial, however, were how Edwin had come to have an ounce of liquid chloroform in his stomach – was it a case of accident, suicide or murder? Here, the prosecution made a serious error. Attorney General Sir Charles Russell, whose parliamentary career was demanding a great deal of his time, had not devoted as much attention to the case as he should. He suggested that Adelaide had first administered chloroform vapour to Edwin while he slept and had then managed to trickle the liquid down the unconscious man's throat. Sir Edward Clarke for the defence had made a careful study of the use of chloroform, and was able to convince the jury that the prosecution theory was almost impossible. Attempts to chloroform a sleeping adult usually end in failure, as the pungent vapour tends to wake the sleeper before unconsciousness supervenes. Even if Adelaide had succeeded, she did not have the experience to recognise the precise moment when Edwin would have been sufficiently subdued not to feel pain but still be able to swallow. The final nail in the coffin of the prosecution theory was the evidence of Thomas Stevenson. Had Edwin been given liquid chloroform while in a recumbent position, it would have pooled in his mouth and at the top of the windpipe, leaving unmistakable marks of irritation. There had been no such irritation. If a seated person drank chloroform, however, the fluid would pass quickly over the sensitive interior of the throat and gullet without causing any damage, and the main patches of irritation would be where Stevenson had in fact found them: at the gullet end of the stomach. Edwin had, therefore, drunk the chloroform while sitting up, and had only lain down afterwards as he lost consciousness.

Throughout the trial, Adelaide, prohibited by law from speaking in her own defence or being questioned, stood silently in the dock, a drooping object of pity. Cross-examination by a skilled prosecutor might have shown her in a different light. Adelaide was acquitted, the main question remaining being not whether she had murdered her husband, but how. It is unlikely that Edwin, who was looking forward to returning to work, committed suicide, and, if he had taken the fatal overdose himself, either deliberately or by accident, his cries of pain as the chloroform attacked the lining of his stomach would have alerted Adelaide and, had she been innocent, would have sent her running for a doctor.

The error in the Bartlett trial, and in many subsequent theories about the case, was, as in the trial of Joseph Snaith Wooler (Chapter 12), to make it far too complicated. When an unskilled person uses poison, the answer is usually very simple. Adelaide had first primed Edwin to believe that chloroform was pleasant to take and the answer to his insomnia. Having no idea what constituted a fatal dose (although she knew that an ounce was almost a hundred times the maximum recommended), she might not even have been trying to kill him outright that evening. Perhaps she hoped to keep him gently drugged over a period of time until he quietly passed away. She probably gave her husband a glass of chloroform mixed with brandy and asked him to drink it down, which he obediently and trustingly did. She was known to have ordered some brandy from the grocer's shop on 30 December, and her landlord had noticed a sweetish smell emanating from a brandy glass found in the apartment just hours after Edwin's death – a glass that was soon afterwards washed. The abundance of fluid found in the stomach suggests that Edwin, on first feeling the searing pain as the chloroform attacked the lining of his stomach, asked for and was provided with a drink of water. He would have been unconscious a few minutes later.

George Dyson, his career as a minister over, left the country. Recent research[5] suggests that he changed his name and started a new life in America. Adelaide also appears to have left England and might have gone to America. It is doubtful that the two ever met again.

* * *

Chloroform used by someone without medical training is a highly unpredictable instrument at best, but, in the hands of a doctor or dentist bent on using it for crime, it is another matter. Not only does he have the skill to avoid accidentally killing his victim most of the time, but his professional status enables him to persuade a subject to submit willingly to the chloroform.

Etienne Deschamps was a French dentist born in 1830 who had settled in New Orleans in 1884, where he advertised his skills as a mesmerist, claiming that he could cure all maladies. In June 1887, a thirty-nine-year-old widowed carpenter Jules Dietsch arrived in America from Paris, together with his elderly mother and two daughters, ten-year-old Juliette and Laurence, seven.

The two Frenchmen became friends, and, in 1888, Dietsch confided in Deschamps that he was concerned for the virtue of his eldest daughter, who was now twelve and starting to attract the interest of the opposite sex. Deschamps also had a secret to confide to his good friend Dietsch. He had a plan to achieve great wealth by finding the legendary buried treasure of pirate Jean Lafitte by using his mesmeric powers. All he needed was a young virgin to act as a medium. Juliette was primed for her role and her father trustingly delivered her to Deschamps for the mesmeric sessions, often in the company of her younger sister.

Whether or not Deschamps actually had mesmeric skills is unknown. His technique involved the application of chloroform, which, when the vapour is carefully diluted with air, produces both analgesia and a sense of euphoria long before unconsciousness supervenes. When chloroform is applied directly to unprotected skin, it will cause burns and blisters, and the fact that no such marks were seen on Juliette's face over the ensuing weeks suggests that Deschamps did not allow his chloroformed handkerchief to touch her skin. By the 1880s, surgeons and dentists protected their patients' faces by use of a mask, but, even in the early days of anaesthesia, a handkerchief or towel, sometimes folded into a cone, was held a few inches from the patient's face. What is certain is that the only treasure Deschamps was interested in was Juliette's virtue.

Deschamps most probably did not intend to kill Juliette, but, on 30 January, something went horribly wrong. Laurence arrived home in hysterics and, when her father rushed to Deschamps's rooms and was unable to gain access, he alerted the police. Deschamps and Juliette were found lying on the bed; both were naked. Juliette was dead and Deschamps, who had stabbed himself in the chest with a dental implement, was only superficially wounded.

At the inquest on Juliette, Laurence revealed that the 'mesmeric' sessions usually involved Deschamps and Juliette removing their clothes and getting into bed. The application of chloroform had induced Juliette to experience religious visions. During the final fatal session, Deschamps had run out of chloroform, got out of bed, dressed and gone for further supplies. On his return, he had applied a further dose, stripped and got into bed again, but, carried away by excitement, he inadvertently killed the child. Even if Juliette had struggled, which surgical patients often did when chloroform was administered, she would have been unable to escape her fate. The post-mortem confirmed that the cause of death was an overdose of chloroform, and, on this occasion, it had left burn marks on the girl's lips. She had also been sexually abused over a long period.

While awaiting trial, Deschamps made wild allegations, claiming that Juliette had not been a virgin before the experiments and producing letters in evidence, which were shown to be forgeries. He also attempted suicide several times. The best defence that could be offered at his trials (there were two, as he was granted a retrial on the grounds that his defence had had insufficient time to prepare for the first) was that Juliette's death was accidental. Even if it was, however, Deschamps had applied the chloroform for the purposes of rape, knowing it to be a dangerous substance that could cause death. He had committed what is known as a 'felony murder' in which his undoubted malicious intent extended to the consequences of his actions even if they were unintended. Deschamps was found guilty and hanged on 12 May 1892.

Chloroforming an unbound victim who is more powerful than the attacker is a procedure fraught with difficulty, which, as shown by examples related by Dr John Snow, is almost always bound to fail, so the murder that took place in the United States in 1894 is more than usually mysterious.

On 3 September that year, the decomposing body of patent agent B. F. Perry was found lying on the floor of a bedroom in a run-down property in Philadelphia. Charring of the moustache, beard and shirt of the deceased, and the presence of a pipe, tobacco, matches and a bottle of cleaning solution, initially suggested accidental death by explosion, but Dr William Scott, called to examine the body, suspected that he was seeing a staged crime scene. The pipe and bottle were undamaged and the corpse was neatly laid out, positioned in such a way that the face had received the maximum sunlight possible, and was swollen and blackened.

The post-mortem, carried out by a Dr Mattern, showed that the cause of death was cardiac failure. He detected the odour of chloroform vapour in the lungs and there was liquid chloroform in the stomach. There were no marks to suggest that the deceased had been tied up or restrained. The condition of the kidneys and stomach showed that Perry had been a heavy drinker, but there was no alcohol in his system when he died. The charring had happened after death.

A colleague of Perry's, a Dr Holmes, now appeared to make a claim on an insurance policy, which necessitated his proving that the dead man's real name was not Perry, but Benjamin Freelon Pitezel. Holmes managed to overcome any doubts about the death and pocketed the payout of $10,000. He omitted, however, to pay $500 of the proceeds, which he had promised to a robber called Hedgepeth. Holmes had met Hedgepeth when in prison for fraud, and the robber had recommended a crooked lawyer, who agreed to assist Holmes and Pitezel in a scheme to defraud the insurance company. The plan was to insure Pitezel's life, and then, after a few months, Holmes would obtain a corpse, which would be identified as Pitezel, and the two would split the proceeds. Pitezel's wife Carrie knew about the scheme and received $500. Under stress, not knowing where her husband was, and having to care for five children, Carrie allowed Holmes to take charge of her fifteen-year-old daughter Alice and a younger brother and sister. Hedgepeth, meanwhile, having been cheated of his promised payout, wrote to the police, and Holmes was arrested.

Holmes admitted that the insurance scheme had been a fraud, and claimed that his partner was still alive, but, as his story unravelled, it became apparent to the horrified public that Holmes was not merely a scoundrel but a prolific serial murderer. The bodies of the three Pitezel children were discovered and, when a hotel built by Holmes in Chicago was searched, human remains were found. Holmes eventually admitted that the body found in Philadelphia was that of Benjamin Pitezel. He was tried for the murder in October 1895 and decided to conduct his own defence.

Holmes now claimed that Pitezel had committed suicide and that, on discovering this, he had faked the accident scenario in order not to invalidate the insurance policy. He said that he had found his associate lying on the floor with chloroform trickling into his mouth through a rubber tube connected to a gallon bottle of chloroform on a nearby chair. The room had been full of fumes, which presumably explained the vapour in Pitezel's lungs.

Holmes's case did not hold water. There was no physical evidence to substantiate his claims, and, had Pitezel attempted to drip chloroform into his mouth while lying down, it would have had to flow uphill to reach his stomach, something that Holmes, since he had a medical qualification, really ought to have known. In such a position, some liquid chloroform would have entered Pitezel's lungs, but there was no liquid chloroform there, only vapour. One aspect of the case that puzzled the medical witnesses was that the chloroform in Pitezel's stomach had caused very little irritation. Mattern even proposed that it had somehow been introduced after death.

As for the evidence that Pitezel had inhaled chloroform, the medical witnesses were in agreement that it would not have been possible for Holmes to forcibly chloroform a man who was sober, unbound, and taller, more muscular and heavier than him.

Despite these doubts, Holmes was found guilty of the murder of Benjamin Pitezel. He confessed to this and twenty-six other murders, then changed his story once more, saying that he had found Pitezel drunk and had tied him up, a claim wholly unsupported by the physical evidence. Holmes was hanged on 6 May 1896.

So, how did Holmes manage to administer chloroform to Benjamin Pitezel? An attempt to chloroform his victim while the man slept would almost certainly not have been successful, since the pungent vapour tends to wake a sleeper long before any anaesthetic effect can occur.[6] Even if he had succeeded, then, as in the Bartlett case, this does not explain how the liquid chloroform entered the stomach without pooling in the throat.

One scenario that was not explored at the time is that Pitezel, as a part of the fraud scheme, had been persuaded by Holmes to be voluntarily chloroformed, so that he would appear to be dead and could be identified at the initial examination for insurance purposes. He would have been led to believe that a body would be substituted for burial later. Holmes, of course, had always intended that Pitezel should die. He might not have known, however, that alcoholics are poor subjects for chloroform. If the stronger man had, after initial compliance, started to struggle against the anaesthetic, Holmes might have been able to induce his woozily uncooperative victim to swallow some chloroform, and then force further inhalation with a towel pressed over the face. This would have left telltale blisters, which Holmes tried to conceal first by burning and then by placing the body in the light where the face would decompose fastest. Pitezel's death from cardiac arrest on inhaling undiluted chloroform would have been rapid, the chloroform in his stomach introduced only very shortly before.

CHAPTER NINETEEN

Our American Cousins

In the 1890s, a small number of highly publicised poison trials in the United States, which formed only a small fraction of total murder cases, stimulated the fear that an epidemic of criminal poisoning was sweeping the land. The public was both chilled and fascinated by poison, which remained a popular device in crime fiction, while real murders made lurid headlines in what became known as the 'yellow press'. The chief exponents of this dramatic and deliberately shocking form of journalism were the *New York World*, bought by Joseph Pulitzer in 1883, and William Randolph Hearst's *New York Journal*, which he acquired 1895. These two publications vied with each other to attract a substantial readership with melodramatic and not always accurate accounts of crimes and disasters, liberally spiced with sex. The reality was that, from 1860 to 1891, there were only two trials in Manhattan for murder by poison, and, in the following ten years, there were six.[1]

Legislation on the sale of poisons in the nineteenth century had proceeded a little differently in the United States from the UK and continental Europe, since poison laws were enacted at state and not federal level. In 1829, New York passed a law requiring that, when poisons were sold, they should be labelled as such. Other states were to follow this example. In 1848, New Hampshire was the first state to require poison sales to be recorded. The American Pharmaceutical Association, founded in 1852, worked hard to combat the unrestricted sale of poisons, but reform came slowly, since, in isolated districts, there were often no drugstores for many miles, and the population relied on general storekeepers for their medicines. By the end of the nineteenth century, however, most US states had laws in place regarding the licensing of pharmacists and restrictions on the sale of poisons.

C. J. S. Thompson, in *Poison Mysteries Unsolved*, published in 1937, provided some undated United States statistics showing that five eighths of criminal poisonings were carried out by women who were housewives, housekeepers, nurses or domestic servants. He also stated that, in France between 1851 and 1871, 399 women and 304 men were accused of criminal poisoning. Thompson explained this differential by quoting a psychologist who said that 'every murder, excepting that by poison requires courage, the power to do [presumably he meant control and authority over the victim] and physical strength',[2] and that women made use of poison because they lacked these three qualities. So, poisoning was seen as a crime of cowardice, and women as deficient not only in strength, but also in the authority and courage required for more directly physical methods. In 1898, the *New York Times* theorised: 'Woman's nature is essentially subtle. From deeds of blood and violence she naturally shrinks because her nature steps in and prevents . . . She turns to poison as the easiest and surest method, because if handled deftly and cleverly, it insures less suspicion and less possibility of detection.'[3] These statements raise questions such as: if a woman chooses to poison her husband rather than strangle him with her bare hands, is this because she lacks the courage or because she is sensible enough to know that, while she might very much want to strangle him, she is not strong enough to succeed? But that is well outside the scope of this book. A German psychologist, who seems to have got carried away with this theme, said that, in a case of poisoning where the evidence did not point to a woman, one should look for 'an effeminate man who has feminine characteristics'.[4] The philandering male poisoners referred to in these pages so far – Castaing, Pommerais, Chantrelle, Pritchard, Bailey and de Bocarmé are amongst those who will spring to mind – are not, of course, on hand to express an opinion on this last theory.

Poison murders are, with a few exceptions, usually planned and rarely the result of an altercation or a sudden fit of temper, as is so often the case with crimes of violence. Unlike many other forms of murder, poisoning also offers the perpetrator hope that the death will be put down to a natural disease. The prerequisites are the ability to obtain poison and the opportunity to administer it, so, inevitably, there is a strong representation amongst poison murderers of people responsible for the preparation of food or the giving of medicine, who are in a close relationship with the victim and have easy access to noxious substances.

An analysis for the period 1750 to 1914 shows that women poisoners are most likely to be the mother, wife, other family member or servant of the victim. Men are most likely to be the husband, father, medical attendant, lover, son or friend.[5] Male murderers are not, as is often claimed, mainly members of the medical profession, although doctors are well placed to carry out such a crime. Their main difficulty is that, should poison be discovered, suspicion will point unerringly to the one person who has the trust of the victim, a motive for murder and the ability to acquire the means. When doctors murder, therefore, their activities often show not only careful planning of the crime, but also sophisticated thought as to how they are to avoid prosecution. In 1892, Dr Neill Cream got around the whole difficulty of obvious motive by randomly giving strychnine capsules to prostitutes he met in the street, taking 'advantage of the weakness that a large number of population have for taking pills'.[6] He was suspected only when he drew attention to himself with a letter-writing campaign accusing others of the crimes and by demonstrating his detailed knowledge of the murders.

In 1891, however, an American medical murderer with means, motive and opportunity thought he had all the answers. At 10 p.m. on 31 January 1891, nineteen-year-old Helen Potts[7] retired for the night in the room she shared with three fellow pupils at the Comstock Select Boarding School for Young Ladies in New York. She appeared to be perfectly well, but, later that night, she suddenly cried out that she felt ill, and her friends called the headmistress, Miss Day, who summoned Dr Edward Fowler. By the time he arrived, Helen was comatose with a bluish tinge to her skin and laboured breathing. The pupils of both her eyes were so contracted that they were hardly visible. Fowler was sure that Helen was suffering from morphine poisoning and he applied artificial respiration, as well as drugs and electricity, to stimulate the action of the heart.

Helen's room was searched for any medicines she might have taken, and an empty box was found, which had once contained some capsules prescribed by Carlyle Harris, a twenty-one-year-old medical student and friend of the Potts family.[8] Harris was sent for, and said that Helen had asked him for something to treat sick headaches, and a chemist had made up some capsules for him. He had ordered a total of twenty-five grains of quinine sulphate and one grain of morphine, to be divided between six capsules. The directions were that Helen should take one before retiring. Although one grain of morphine was not a dangerous amount, he had decided, for safety's sake, to supply her with only four of the capsules, retaining two. Harris confided that he was very attached to Helen and the two were due to marry when he graduated from medical school. Fowler, who believed from Helen's condition that she must have taken at least three or four grains of morphine, sent Harris to check with the chemist whether the prescription had been filled correctly and he returned with the assurance that there had been no mistake.

Despite strenuous efforts to save her, Helen died just before 11 a.m. the following day.

Saying that he could not find one of the remaining capsules, Harris passed the other to the coroner. It was shown to have been made exactly in accordance with the prescription. The cause of death remained mysterious until Helen's mother Cynthia revealed that her daughter had suffered from heart disease since childhood. An inquest was held, and, on 27 February, the coroner's jury returned the verdict that Helen had died from the effects of morphine in conjunction with her heart trouble. All concerns, as well as Helen's embalmed body, were laid to rest.

The press, however, was far from satisfied. On 21 March, after an investigation by the *New York Evening World*, the death of Helen Potts was sensational front-page news. Cynthia Potts and Helen's uncle Charles Treverton, a physician, had made statements strongly suggesting that Helen had been murdered. It was revealed that Carlyle Harris was more than just a friend or an admirer – he and Helen had been secretly married for almost a year, and Mrs Potts believed that Harris had killed Helen by some manipulation of the capsules. Her story of the heart disease, she now admitted, had been a lie, as she wanted to avoid a post-mortem examination that would have revealed that her daughter had undergone an abortion. Carlyle Harris was obliged to acknowledge the secret wedding, but denied performing an abortion or harming Helen in any way. He surrendered himself to the District Attorney.

The body of Helen Potts was exhumed. It showed no signs of disease, although there was a cyst on the uterus that could have been the result of surgery. There was, however, a marked congestion of the brain, a symptom of narcotic poisoning. Professor Rudolph A. Witthaus, lecturer on chemistry at City of New York University, undertook the analysis of the remains, having to overcome the fact that the body had been embalmed. The stomach and intestines of Helen Potts contained traces of morphine, but no quinine. Harris, it was surmised, must have emptied one of the capsules and substituted a fatal dose of morphine, which, as a medical student, he would have been able to obtain, before handing it and three others to Helen. The retention of the two innocuous capsules was now looking like part of a murder plan. If the poisoned capsule had been one of the first three she had taken, then evidence would have remained in her room to suggest a harmless dose, but, by keeping two, Harris was insuring against the possibility that the fatal capsule would be the last one. Carlyle Harris was charged with murder.

At the trial, which opened on 14 January 1892 and lasted three weeks, the history and behaviour of Carlyle Harris came under close scrutiny. Harris's maternal grandfather was a professor of medicine, and his mother an author, who wrote under the name Hope Ledyard. His father, a less impressive figure, was a stenographer. Harris had met Helen Potts, a sweet-natured, intelligent and attractive girl, at a dance in the summer of 1889. Helen's father was a railroad engineer and a man of means, and Helen must, therefore, have seemed an excellent catch for a less than prosperous medical student. Harris had shown an obvious interest in her, which she had reciprocated, but Mrs Potts had advised him that they were both far too young to think of an engagement. The issue had been discussed in a similar vein in January, but, on 8 February 1890, the couple slipped away and were secretly married. Within six weeks of this event, Mrs Potts started to notice that Harris's admiration of her daughter was waning. He called less often and, when he did, was not as courteous as he had once been, something that clearly upset Helen. He would also make appointments to meet her and then break them. In May, however, the young couple stayed away together for an afternoon and evening and, when Helen returned, she looked unwell.

In the latter part of May, a close friend of Helen's, Mae Scholefield, went to see her at her mother's home in Ocean Grove, NJ, and, a few days later, Harris arrived. Mae, Harris and Helen were due to go for a walk along the beach, but, after a whispered conversation between Helen and Harris, it was agreed that Harris and Miss Scholefield would walk on alone for a while. On the way, he told her that Helen had insisted that he reveal a secret – that the two of them had been married since February – but, when Mae suggested telling Mrs Potts about it, he flew into a temper and put her on her honour to tell no one. 'My prospects will be utterly ruined if this marriage is known. I would rather kill her and kill myself than have the marriage public. I wish that she were dead and I were out of it,'[9] he said. Mae, shaken by this venomous outburst, turned back to the house to find Helen, and saw her friend still on the steps. Helen asked Harris to come in, but he pleaded another engagement and left. There was no walk on the beach.

About a month later, Helen came to stay with her uncle, Dr Treverton. He noticed that she was pale and ill and asked if he might examine her, which, reluctantly, she allowed. Helen was pregnant, but the child had died in the womb. Her condition deteriorated and Treverton was obliged to induce labour in order to save her life, alleviating the pain with regular doses of a quarter of a grain of morphine. She eventually gave birth to a bruised and decomposed foetus of about four or five months' development. Mrs Potts was summoned to see her daughter, and it was only then that she learned about the secret wedding and that Helen's indisposition the previous May had been the result of a botched abortion. Carlyle Harris was sent for, and admitted to Dr Treverton that he had married Helen and performed two abortions on her. He added that he had had a 'connection'[10] with five young ladies, some of whom he had operated on, while one had had a child, and that he 'always got through with these things very nicely – rather slick',[11] and expected he would get through this one, too. Treverton remonstrated with him, but Harris said that he 'did not intend to drop the girl altogether'.[12] He always looked after the young ladies and 'never left them unprotected in that respect . . . never deserted them until they were all right'.[13]

Once Helen was on the way to recovery, Mrs Potts urged that her daughter and Harris should be married in a proper religious ceremony, but he had other worries on his mind. That August, he had been arrested when it was found that a restaurant business, of which he was secretary, had sold liquor without a licence and allowed gambling on the premises. In September, however, Harris, who appeared to have been drinking, approached Mrs Potts and suggested she go with him to the office of a lawyer, Mr Davison, to discuss the proposed marriage. Davison was obliged to send his clerk to get a copy of the marriage certificate and, while they were waiting, Harris told Mrs Potts he would not have his family know about the marriage for half a million dollars. She retorted that he should have thought of that sooner, to which he replied that, if she was so unhappy about it or the marriage grew to be a bore and they were tired of it, then it could be easily broken and no one would be any the wiser. Mrs Potts, clearly shocked by his uncaring attitude, said she would call that situation legalised prostitution. She demanded a ministerial marriage, but Harris said he did not want that at the present time, as Helen's name would be associated with the restaurant scandal.

When the clerk arrived with the certificate, Mrs Potts asked what had happened to the original one and Harris confessed that he had burned it. Mrs Potts could only stare at him in astonishment. The certificate, she now saw, showed that the couple had married under false names. Harris tried to dismiss the situation altogether by claiming that the ceremony had been performed by a saloon-keeper; however, Davison declared the marriage valid and got Harris to sign an affidavit to say that he was Helen's husband. Mrs Potts did not tell Helen of this hurtful conversation; neither did she mention the secret wedding to her husband, because of his temper. She decided, for the time being, not to press for a religious wedding. Soon afterwards, Harris arranged for Helen to board at Comstock. Mrs Potts saw very little of him thereafter, as he constantly made excuses not to see her.

There was, however, an exchange of letters between Harris and Mrs Potts, and, in January 1891, he suggested that nothing more should be done about the marriage for another two years, and Helen should stay at the school during the summer and prepare for college. Mrs Potts disagreed: she did not want her daughter to be an unacknowledged wife for another two years for no good reason. She demanded that Harris and her daughter go through a religious ceremony on the anniversary of the secret marriage. He replied to her on 20 January, 'I will concede to all your wishes. It shall be just as you say about this marriage *if no other means can be found of satisfying your scruples* [*sic*].'[14] On the same day, he went to the chemist and obtained the capsules.

Mrs Potts last saw her daughter the day before her death. Helen mentioned the capsules, saying that she had already taken three of them but they made her feel ill, and she had thought of throwing the last one away. Mrs Potts thought her daughter looked better and advised her to take the last capsule.

After Helen's death, Harris had begged Mrs Potts to say nothing about the secret wedding. Helen was not buried under her married name, and her husband did not attend the funeral.

At Harris's trial, Dr Treverton's testimony that he had given Helen doses of a quarter of a grain of morphine destroyed any defence theory that Helen might have had an undue sensitivity to the drug. It was also pointed out by the prosecution that, by retaining the capsules to protect himself, Harris had inadvertently thwarted another possible line of defence by proving that there had been no error by the chemist.

The initial questioning of Professor Witthaus occupies twenty-one pages of the trial transcript, the bulk of which describes in minute detail the tests he conducted. His testimony demonstrates the advances that had been made since the work of Stas, which enabled him not only to separate out alkaloids from organic material, but also to use the varying solubilities of alkaloids and their salts to separate them from each other. Witthaus used the Erdmann–Uslar process, which was a modification of the Dragendorff process, itself a modification of Stas–Otto. The residues he obtained were tested for the production of coloured precipitates. Some things had not changed, however, since he also tasted the material, and injected a sample into the lymph pouch of a frog. The eight-hour cross-examination of Witthaus by the defence, which attempted to raise the issue of ptomaines, did not shake his conclusions.

Harris's dishonesty over the restaurant was now shown to be a mere peccadillo compared with his dealings with women. It was revealed that, while Helen was at her uncle's home recuperating from the abortion, her husband was staying at a hotel under a false name with a young woman with whom he was on very affectionate terms. His status as a reprobate of the first water was complete when Helen's cousin Charles Oliver testified that Harris had boasted to him of his extensive experience with women, of whom he claimed to have made a study, saying that he had had sexual intercourse with a great many. His seduction technique was to spike their ginger ale with strong liquor. Once it had taken effect, 'he had never had any trouble in getting his desire',[15] except in two instances, in both of which he had been obliged to overcome female scruples with a secret marriage ceremony.

Seducer, date-rapist, abortionist and bigamist he might have been, but that did not necessarily make him a murderer; however, fellow medical students testified to conversations in which he had said 'it was easy enough to put any one out of the way without any one knowing of it'.[16]

Harris was found guilty of first-degree murder and, on 8 February 1892, the second anniversary of his marriage to Helen Potts, he was sentenced to die in the electric chair. Following a protracted legal battle, suggestions of reasonable doubt, allegations that Helen Potts was 'an impetuous willful girl'[17] with a morphine habit and the strenuous efforts of his mother to prove his innocence, he was executed on 8 May 1893. No one had seen him put morphine in a capsule, and the method of murder was all supposition, but, with his history of getting into trouble and then wriggling out of it, the threatened exposure of bigamy was motive enough.

Vain, manipulative and boastful, Harris believed that he could outwit medical men far more experienced than him. Morphine poisoning had been suspected from the start, but, although Helen's comatose state and laboured breathing could have been symptoms of other conditions, the significant feature that marked her death as due to a narcotic was the contraction of her pupils. It was only Mrs Potts's false information, given to avoid sullying her daughter's reputation, that had prevented a post-mortem being carried out initially.

* * *

Watching the trial of Carlyle Harris with more than academic interest was a Dr Robert Buchanan, who was planning to murder his second wife. Buchanan, a graduate of the Chicago College of Physicians and Surgeons, married Annie Brice Patterson in 1885. The couple had a daughter, Gertrude. He later practised medicine in New York, where he made the acquaintance of Mrs Anna B. Sutherland, a wealthy brothel owner, some twenty years his senior. Buchanan and his wife were divorced in November 1890 and, seventeen days later, he married Mrs Sutherland. The new Mrs Buchanan closed down her brothel, and the couple moved to Manhattan. She was taken ill on the morning of 23 April 1892, and physicians were summoned, one of whom diagnosed hysteria and prescribed medicine containing chloral, a sedative. Buchanan administered two teaspoons of the mixture. Soon afterwards, Mrs Buchanan became comatose and died the same night. Buchanan claimed that his father in-law had died from cerebral haemorrhage,[18] and, since Anna's symptoms matched that diagnosis, this was certified as the cause of death. There was no telltale contraction of the pupils; in fact, one was a little dilated.

Perhaps the physicians did not know Buchanan well, or they might have been more suspicious. He had openly admitted to his friends that he had married his second wife, whom he thought coarse, old and ugly, solely for her money. Soon after this marriage, he had started actively pursuing younger, more attractive women. Buchanan told a close friend, Richard Macomber, that he was going to 'dump the old woman',[19] and sent his daughter to board with Macomber because Anna used obscene language in front of the child. Macomber had never been told that the pair were married, although he suspected it. Other acquaintances were told she was a housekeeper. Although Macomber had assumed that the 'dumping' would be done legally, Buchanan had been unwise enough to tell a friend that: 'It is an easy matter for a doctor to get rid of anybody he does not want.'[20] He had also prepared the ground for Anna's death a little too thoroughly, by telling friends that she had kidney trouble and had threatened to poison herself.

Like Carlyle Harris, Buchanan thought he had outwitted everyone else, and he talked too much and unwisely. Within a few weeks of burying Anna, the unashamedly cheerful widower remarried his first wife. Suspicious acquaintances went to the District Attorney and filed affidavits alleging that Buchanan had poisoned Anna, while Richard Macomber recalled that Buchanan had called Carlyle Harris 'a damned young fool'[21] who didn't know how to poison his wife without leaving a trace. When Macomber asked Buchanan why, if Anna had died from morphine, her pupils were not contracted, Buchanan just couldn't resist showing off. He said that the effects of morphine on the eyes could be counteracted by belladonna, which had the opposite effect.

Buchanan was arrested and, the following year, he was tried for murder. The trial was a 'battle royal'[22] between the rival experts, all of whom arrived in court with impressive lists of credentials. The task of counsel was to cast doubt on expert testimony and, since they knew that a simple slip could invalidate a man's entire evidence, each side tried to wrong-foot the other's witnesses with unexpected questions. When Professor Witthaus said that he had confirmed the presence of morphine in Mrs Buchanan's viscera using colour reaction tests and by injecting suspect material into frogs, the defence attacked his methods as 'obsolete and discredited'[23] and brought its own witness, Victor C. Vaughan, who had published research on ptomaines. Vaughan arrived in court with an array of apparatus and chemicals, which he laid out on a table, creating a laboratory in miniature in the courtroom. He conducted colour reaction tests on both ptomaines and morphine mixed with ptomaines, after which he showed the results to the jury, claiming that the residues from Mrs Buchanan's body were more characteristic of ptomaines than morphine. Under cross-examination, however, he was forced to admit that individuals saw colours in different ways.

Sometimes simple demonstrations are the best. The prosecution wanted to prove that it was possible for a fatal dose of morphine to be dissolved in the chloral mixture Buchanan had been seen giving to his wife. A pharmacist passed a spoonful of chloral mixture to the jurors, for examination and cautious tasting, and, when it was handed back, he added the morphine. The spoon was then re-examined, and it was seen that the morphine had been dissolved. The spectacle of duelling experts was later to prompt a recommendation from a Professor Gray that opposing medical witnesses should meet before trials commenced to at least agree on the facts.[24]

Buchanan was convicted, but the *New York Times* may have been echoing public opinion in commenting that the complex medical testimony did not prove death by morphine beyond reasonable doubt; the jurors were convinced of the prisoner's guilt by 'his character and conduct and by the circumstances attending and following his wife's death'.[25] An interview conducted with a juror after the trial supported this view. The opposing medical experts had effectively cancelled each other out.

These two sensational poisoning cases cast a searching spotlight on scientific testimony in the courtroom. Where experts disagreed, how was a jury to decide? How could they judge the competence of analysts? It had always been assumed that experts were, at the very least, honest, although they could be mistaken, as Taylor had been over Smethurst, or that, as Herapath had done at the Palmer trial, they might say only what they were hired to say. The terrifying possibility of corruption for personal ends, however, was brought dramatically to the fore in 1896 when Mary Fleming was tried in New York for the murder of her mother with arsenic.

* * *

The evidence against Mary was sound. On 30 August 1895, her mother, Mrs Evelina Bliss, had died several hours after eating clam chowder and a piece of lemon meringue pie that had been sent to her by Mary. Evelina's symptoms as she lay dying, and the inflammation of her stomach seen at post-mortem, indicated that death was due to the administration of a corrosive poison. Her daughter, who was supporting three children and was pregnant with her fourth, was in financial need and stood to gain substantially from Mrs Bliss's death. Moreover, 'Fleming' was only an adopted surname. Mary was unmarried and, what was worse in the eyes of society, her children were the offspring of more than one lover.

Analytical chemist Dr Henry Mott discovered both arsenic and antimony in Evelina's stomach contents, vomit and the dregs of the clam chowder. No poison was found in the remains of the pie. Mary was arrested and achieved instant celebrity, newspaper headlines comparing her with the notorious female poisoners of history. Her trial, which opened on 11 May 1896, was a publicity feast of poison, matricide, slander and scandal. It also supplied a national stage on which to mount a very unusual defence.

Chief defence counsel Charles Brooke took an aggressive line from the start, making so many objections during Mott's testimony that he earned a rebuke from the judge for obstructing the proceedings. As a result, Mott's examination took so long that some of the jurors started to doze. The next witness was Walter T. Scheele, a German-born chemist who had immigrated to the US in 1865, aged twenty-five,[26] and owned a pharmacy in Manhattan. He was often called upon to give evidence at trials, and had worked with Brooke before. On those occasions, Scheele had assisted the defence, but this time he gave evidence for the prosecution, and his courtroom experience was to be very different. Scheele had conducted similar analyses to those of Mott and had obtained similar results. Brooke, however, was less interested in the findings than in planting the suggestion that Scheele had deliberately contaminated the samples in order to ensure that Mary was found guilty and thus enhance his own reputation. In cross-examination,

Brooke accused Scheele of saying: 'My reputation hinges on this case. With me, it is not a question of whether she is guilty or not. I have so fixed matters that she will be found guilty whether she did it or not. I will get a great deal of merit out of this case.'[27] Scheele, taken by surprise at this attack, turned pale with fury, leaped to his feet and denied the allegations.

The one thing the prosecution was unable to prove was how Mary had obtained the poison. If she read New York press reports of trials, she would have been aware that it was possible to trace the purchase of poison to an individual. In April 1895, Catherine W. Nolan and her sister Elizabeth were indicted for the murder of their brother John, and two druggists appeared at Catherine's trial to say they had sold her arsenic. The *New York Times* claimed that, in the eight months before John's death, his father, mother and an older sister had all died under mysterious circumstances, and all, like John, had been insured. It was also alleged that, before John was taken ill, Catherine had told a neighbour that her brother was diabetic and was going to die. It seemed that sugar had indeed been his death, since John had complained of burning pains in his stomach after a meal with his sisters, a meal in which only he had helped himself from the sugar bowl. His doctor suspected poison and refused to give a death certificate. The autopsy showed no disease that could have caused death, and all organs appeared normal apart from the stomach. An analysis of the viscera found almost five grains of arsenic. For reasons that remain unclear from the brief newspaper accounts of the trial, the jury acquitted Catherine after ten minutes' deliberation, and Elizabeth was discharged. There was an outburst of sympathy for the girls in court, since they had been held in gaol for many months and were very poor. At the suggestion of a juror, a hat was passed round and $60 collected. The *New York Times*, which had previously reported the finding of arsenic in John's remains, changed its tune and attacked the prosecutor who, it said, had 'tortured an innocent girl'[28] for no other reason than the fact that the brother's life had been insured.

Had Mary Fleming obtained arsenic or antimony from a druggist, some evidence would have emerged, but none did. Suspicion lighted on her lover and probable father of her third and fourth children, Ferdinand Wilckes, who had once worked in a shop supplying chemicals and had been known to obtain medicines without prescription.

In Brooke's energetic defence, he suggested just about every possible cause of Evelina's death – contamination in her doctor's medication, her age, suicide, arsenic eating, shock and food poisoning – apart from the obvious one, but his most violent attack was reserved for Walter Scheele, so much so that the *New York Times* ran the headline: 'Defense in Fleming Trial Using Chemist as a Target'.[29] Brooke brought a string of witnesses to attack Scheele's credibility, including newspaperman Max Mansfield, who had written an article claiming that all the arsenic found in the stomach, vomit and clam chowder had been placed there by Scheele. Mansfield, who had interests in the brewing industry, had his own reasons for attacking Scheele, who had recently written a report on impurities in beer. Only the day before this damning testimony, Scheele had angrily confronted Mansfield about his article, and Mansfield had punched him. Most of the witnesses who testified that they had heard Scheele state that he had fixed the evidence had connections to the beer industry. One of them was a friend of Wilckes who, with another defence witness, had been present at Mansfield's assault, and the prosecution, on establishing this information, suggested that Wilckes had been involved in arranging perjured testimony. By the time Brooke rose to make his closing statement, the hearing had been less a trial of Mary than a trial of Scheele. Alfred Swaine Taylor would have sympathised. After the Palmer trial, he had described 'personal attacks on witnesses whose evidence was of vital importance in the case' as 'the last refuge of a failing defence'.[30]

Mary Fleming was acquitted, and some jurors were later to say that the suggestion of corruption in Scheele's testimony had been the crucial factor in their verdict.[31] Swayed by the sheer number of men prepared to attack Scheele, the jurors had lost sight of evidence that clearly pointed to arsenic having been ingested during life. Walter Scheele continued as a druggist, but his days as an expert witness were over.

By 1900, there was, in the US, 'a growing restiveness on the part of judges, juries, and experts themselves' concerning the deficiencies in the way expert testimony was presented in court and 'a casting about for some better solution'.[32] The idea of establishing a class of official experts was widely discussed, but was felt to contravene the fundamental principles of the criminal trial: that the judge alone was the judge of the law, the jury was to decide on the facts, and the accused should be 'allowed to produce any relevant and competent evidence on his own behalf'.[33]

The editor of the *Journal of the American Medical Association* suggested humorously that the opposing experts, instead of testifying, should be 'stood up in line and cancelled in pairs so that by a sort of *reductio ad absurdum* it might be seen by the remainder which side to credit'.[34]

Some doctors approved of the concept of appointed experts, which Sir James Fitzjames Stephen described in 1883 as 'simply a protest made by medical men against cross-examination. They are not accustomed to it and they do not like it.'[35] Stephen suggested that discreditable scenes in court could be avoided if the medical men from both sides met in consultation before the trial, then 'they would be able to give a full and impartial account of the case which would not provoke cross-examination'.[36] In Leeds, where legal actions for injuries in railway accidents were common, physicians and surgeons had been following this practice for some years.

In France, however, where experts were appointed by the court and not engaged by counsel, concerns were expressed at the International Congress of Forensic Medicine in 1889 that this system was weighted against the accused, and a new system was proposed, akin to the American one.[37] The debate and proposals to reform the system were to continue, but the distinction between the inquisitorial trial system of France and the adversarial trials of the UK and USA remains to this day.

As the nineteenth century drew to a close, there was increasing concern, especially amongst medical men, that juries of laymen were not competent to make decisions based on scientific evidence, although Mr Justice Stephen, who believed that 'the great bulk of the working classes are altogether unfit to discharge judicial duties',[38] supported the jury system, as long as the twelve jurors were selected educated gentlemen. Many of these concerns have since been addressed by improved standards of literacy and education in the general population. The jury system is, of course, still alive, probably on the same principle as democracy – that it is the worst possible system, except for all the others.

As the nineteenth century drew to a close, many poisons such as laudanum were still freely sold to the general public. In London, George Chapman, a hairdresser who had once been apprenticed to a barber surgeon, was easily able to obtain tartar emetic from a chemist, and he used this to poison unwanted mistresses between 1897 and 1902. Desperate or angry people who committed unplanned murders, poisoning their victims with no thought of concealing what they had done, used whatever came to hand: oxalic acid, acetic acid, nitric acid, carbolic, turpentine, hydrochloric acid or phosphorus paste rat poison.

In 1895, there was a curious echo of the case of Eliza Fenning with which this book began: an attempted murder by a sixteen-year-old servant girl called Edith Fenn. Edith had been working as a kitchen maid in the South Kensington home of army officer Gordon Morris and his wife and three-and-half-year-old daughter Gwendolin. Edith did not prepare the child's food but was responsible for taking it upstairs to her. On 2 December, the nursemaid Florence Powell gave the child some milk, which was kept in a room adjoining the nursery. Gwendolin said it tasted nasty and spat it out, and, when Florence tried it, she agreed. Florence took it away and, meeting Edith on the landing, commented that the milk had gone sour. Mrs Morris was also shown the milk and tasted it. Shortly afterwards, Edith brought freshly cooked mince and potatoes for the little girl's dinner. On the way, she passed the housemaid, who noticed that the dish smelt strongly of ammonia, but, when she queried this, Edith said that the smell came from the milk. Edith took the dish into the nursery, put it on the table and went out. Florence, faced with her second suspect dish of the day, showed it to the housemaid then took it down to the kitchen and fetched Mrs Morris.

With the household by now in something of an uproar, Mrs Morris quietly took Edith aside and implored her to say what she had put in the child's milk. Eventually, Edith admitted that she had dosed the milk from a bottle that she had since thrown in the dustbin. When recovered, this was found to be a lotion for cleaning gold lace, clearly labelled 'POISON'. The principal ingredient was prussic acid. A doctor was sent for and antidotes administered; fortunately, none of those who had sampled the milk had taken enough to endanger life. The mince had been adulterated with ammonia from a bottle in the bathroom.

Edith was arrested on the same day. At her trial, which took place at the Central Criminal Court on 13 January 1896, she was charged with unlawfully administering poison to Gwendolin 'with intent to injure and annoy her'.[39] The only explanation Edith could offer for her actions was that she did not like taking food up to the child. She was found guilty, but, in view of her youth and previous good character, mercy was recommended. Mr Justice Hawkins, commenting that 'if this conduct was tolerated no household would be safe',[40] sentenced her to four months in prison with hard labour.

Epilogue
Dangerous Drugs and Cruel Poisons

The nineteenth century, the period with which this book is principally concerned, was a time of enormous advances in combating what was then widely feared to be a serious threat to society – homicidal poisoning. By the end of the century, as a result of the introduction of death certificates and advances in medicine leading to better diagnosis, more accurate statistics were available. The concept of poison murder could still send a chill of dread up the spine, but its incidence was far rarer than had once been believed. In 1895, M. W. and A. W. Blyth reported that, in the years 1883 to 1892, there had been 1,424 deaths from opium, laudanum and morphine, of which forty-five were attributable to soothing syrups and patent medicines, 876 were caused by accident or negligence, 497 were suicidal, and only six homicidal.[1] In the same period, strychnine and its preparations accounted for 325 deaths, of which eight were homicidal. In 1908, J. Dixon Mann published a comparison of the use of common poisons in suicide, accident and homicide. His table for murders shows that, in the years 1885 to 1894, cyanide (as hydrocyanic acid and potassium cyanide) headed the list with fourteen cases, followed by strychnine, seven, opium or morphine, five, carbolic acid and arsenic, two each, chloral hydrate, hydrochloric acid and ammonia, one each – a total of thirty-three cases. In the years 1895 to 1904, strychnine headed the list with six, then was carbolic acid, four, cyanide and opium or morphine, three each, arsenic, two, chloroform, sulphuric acid and oxalic acid, one each – a total of twenty-one. By comparison, in 1885–94, poison was used by 2,506 suicides and caused accidental death in 3,724 cases. In the years 1895 to 1904, there were 4,368 poison suicides and 4,632 accidental deaths due to poison.[2]

Toxicologists by then had a substantial battery of tests with which to identify individual alkaloid poisons, although in the case of aconite, Blyth stated that the most satisfactory test was still physiological: '[T]he minutest trace of an aconite-holding liquid, applied to the tongue or lips, causes a peculiar numbing, tingling sensation, which, once felt, can readily be remembered.'[3] There were chemical tests for aconite, but Blyth regarded these as supplementary to the old-fashioned way – tasting the poison. This advice remained the same in the 1920 edition.

The question of 'normal arsenic' might have seemed long dead and buried, but new work was being done by French analysts, who claimed that arsenic was normally present in some human organs. These results were refuted by German analysts, who stated that it was usually absent.[4] In 1902, Louis Danval, who had invited comparison with Eugène Chantrelle when convicted of the murder of his wife in 1878 (see Chapter 16) was still serving his life sentence in New Caledonia when a paper was published that gave him new hope. Gabriel Bertrand, head of the biological chemistry department at the Pasteur Institute in Paris, had carried out a research project on marine animals that had never been exposed to contamination with arsenic, and concluded that the body could contain 'normal arsenic', albeit in quantities so small they were unlikely to trouble forensic toxicologists. Doubts had been expressed at the time of Danval's trial regarding the thoroughness of the post-mortem examination of Mme Danval, and Bertrand's findings were enough to produce reasonable doubt. Danval was released and, in 1924, his conviction was quashed and he was awarded a pension. More recent studies have confirmed that arsenic is naturally present in rock, soil, water and all living organisms.[5] The human body contains between three and four milligrammes – a fatal dose being nearly 200mg.

Throughout the twentieth century, doors were gradually being closed to the would-be poisoner, especially to those who could not legitimately lay their hands on poisons for medical, industrial or agricultural use. In France, commented Dupré and Charpentier in 1909, most non-professionals did not have access to the alkaloid poisons and were not familiar with them, so turned to arsenic and phosphorus from matches when they committed poison murder. In Britain, the Poisons and Pharmacy Act of 1908 added cocaine and its preparations to the list of regulated drugs, and amended the restrictions on opium to all preparations containing 1 per cent or more of morphine.

Martin Wilbert, an American pharmacist who campaigned for the introduction of national legislation on poisons, was one of the few who recognised that even such laws had their limitations. He was mainly concerned with accidental or suicidal poisoning and saw that the choice of poison was influenced by the attention it was being given in the press. His article for *Public Health Reports* in 1914 was stimulated by the growing use of corrosive mercuric chloride by suicides, which was neither the most easily obtained nor the most potent poison, but was chosen 'because it was the most frequently mentioned and the most actively discussed of all the available toxic substances'.[6] While recognising that the main aim of controlling the sale and use of poisons was to protect the public, he regretted that 'it is practically impossible to make poison regulations foolproof, and it is also admittedly impracticable to dissuade the person bent on self-destruction from accomplishing his end. Easy access to poisons greatly increases their abuse, and it is difficult indeed to conceive of ways that will tend to prevent or even discourage the constantly growing abuse of poisonous substances.'[7]

Poison murders continued and made headlines; for example, the case of Frederick Seddon, who was hanged in 1912 for poisoning his tenant Miss Barrow with arsenic extracted from flypapers, but it was generally accepted that such cases were not numerous and there was no national epidemic of secret poisoning. Concerns expressed in the twentieth-century press about the sale of poisons mainly related to the availability of addictive drugs.

In 1913, Dr F. M. Sandwith, Gresham professor of physic, gave a lecture on drugs in which he stated that homicidal poisoning by arsenic 'was declining to a great extent owing to the restrictions on the sale of poisons, and the publicity given in the newspapers to cases of murder, undesirable as it was in many ways, inspired persons of criminal tendency with a wholesome fear of detection and a sense of insecurity in the face of the growing science of toxicology'.[8]

The Dangerous Drugs Act 1922 further controlled the availability of drugs such as cocaine and opium and its derivatives. It was still possible, however, for members of the public to buy arsenic and strychnine. In 1920, solicitor Harold Greenwood was acquitted of the murder of his wife with arsenical weedkiller. Another solicitor, Herbert Armstrong, may have been inspired by this and was able to buy substantial amounts of arsenic from his local chemist, which he said was for killing weeds. Most was coloured with charcoal, but one packet was pure white arsenic. He was hanged in 1922 for the murder of his wife. In 1924, inventor Jean-Pierre Vaquier openly purchased strychnine at a chemist's shop saying he wanted it for radio experiments, and then used it to poison the husband of his mistress.

It was not until the Births and Deaths Registration Act of 1926 that it became illegal to register a death without either a doctor certifying that death was due to natural causes, or a coroner's order. In the same year, a committee was appointed to consider the amendments needed to what was felt to be the largely outdated and even obsolete legislation on the sale of poisons. When, in 1930, concerns were expressed by L. Moreton Parry, the President of the Pharmaceutical Society, about children being able to get aspirin from vending machines, it was clear that things had moved on very substantially from the days when a child with a penny could legally be sold a quarter of a pound of arsenic by a grocer.[9]

The Pharmacy Act of 1933 and the Poisons Rules that came into operation in 1935 further restricted the availability of poisons to the public, notably strychnine, which could henceforward be purchased only by licensed pest controllers. Farmers, who had once used strychnine to destroy vermin, and salmon fishers, who had used it to kill the seals that damaged their nets, found that they could no longer obtain it, but the society felt that they would soon get used to the alternatives available.[10] The new legislation went hand in hand with the registration of every chemist's shop in Great Britain, of which there were some 14,000. All pharmacists had to become members of the Pharmaceutical Society – membership had previously been optional – and the society now had complete responsibility for the training, registration and overseeing of pharmacists in Great Britain.

Murderers did not, however, despair. There were still deadly poisons available for purchase.

In 1952, Alicia Millicent Roberts of Merioneth was charged with the murder of her forty-four-year-old husband, Jack, who had died on 6 March from acute arsenical poisoning after eating porridge she had prepared. The organs of the deceased were examined by Dr Gerald Roche Lynch, lecturer in forensic medicine and toxicology and senior official analyst for the Home Office, who had testified at most of the major British murder trials involving poison since 1920. Roche Lynch found almost seven grains of arsenic in the remains, which suggested, in view of the fact that Roberts had been vomiting, that a far greater amount had been consumed. Mrs Roberts had purchased weedkiller from two chemists by entering a false name and address in the register. When offered a non-poisonous product, she had insisted on the arsenical variety, which she said was more effective. The weedkiller she was sold was 50 per cent arsenic. Two tins of weedkiller with the chemists' labels still attached were later found where they had been thrown over the wall at the back of the Robertses' house.

At the trial, which opened in Swansea on 8 July, the Attorney General Sir Lionel Heald, QC, revealing that the attitude that had marked nineteenth-century cases was still very much in evidence, stated: 'There is no more deadly thing than poisoning, with its secrecy, deliberation and cruelty…'[11] The defence, conducted by Mr H. Edmund Davies, QC, suggested that Mr Roberts had taken the weedkiller himself, either to commit suicide or to make himself ill so that his wife would not leave him. It was threadbare but, astonishingly, it worked and Mrs Roberts was acquitted.

It had not, however, escaped police attention that Mrs Roberts had been married before and her previous husband John Hughes had also died in suspicious circumstances. Hughes had died aged fifty-six on 6 June 1949, just eleven days after his wife had purchased arsenical weedkiller. Hughes's body was exhumed, and arsenic was discovered. He had undoubtedly been ill with a chest complaint for some weeks before he died, so the question before the inquiry held at Holyhead on 31 July 1952 was whether arsenic was the cause of death. Roche Lynch told the inquiry that he had found an abnormal amount of arsenic in Hughes's remains, which could not be accounted for by seepage from an external source, and he believed that it had accelerated death. Pathologist Dr (later Professor) Francis Camps had another view. He thought that arsenic from the grave soil would have seeped into the coffin with rainwater, and death could have been due to natural causes.

At this point in the proceedings, Roche Lynch might well have experienced disturbing flashbacks to the 1931 Bodmin trial of Mrs Annie Hearn, accused of poisoning her neighbour and sister with arsenical weedkiller. The bodies had been exhumed and he had been certain in that case, too, that little, if any, contamination could have seeped into the coffins from the Cornish graveyard soil, but Mr Norman (later Baron) Birkett, in a masterful defence, had introduced sufficient reasonable doubt on the point to obtain an acquittal.

In 1952, offered a wide variety of possible conclusions, the inquest jury returned an open verdict on John Hughes, which meant that, while they accepted that his death was accelerated by arsenic, they were unable to decide how it had got into his body.

On 10 March 1953, elderly widow Sarah Ann Ricketts employed a new housekeeper, forty-six-year-old Louisa Merrifield, who moved into Mrs Ricketts's Blackpool bungalow two days later, together with her husband Alfred, who was seventy and suffered from ulcerated legs. On 24 March,

Mrs Ricketts changed her will; she had decided to leave her bungalow to the Merrifields. Louisa was so excited by this that she told her friends that a lady had died and left her a bungalow. At this point, the lady was still alive. The will was finalised on 31 March.

On 9 April, Louisa asked Mrs Ricketts's doctor, Burton Yule, to visit his patient, and, with the validity of the will in mind, to certify that the old lady was of sound mind. When Yule arrived, Mrs Ricketts said she felt perfectly well, and Yule thought her compos mentis. On 13 April, Louisa called another doctor, Albert Wood, saying that Mrs Ricketts was seriously ill and might die that night. Wood came to see Mrs Ricketts and was understandably surprised to find her in good health. Next morning, Louisa called Wood's partner, Dr Page, who saw that Mrs Ricketts was dying and sent for Yule. By the time Yule arrived, his patient was dead.

Perhaps Louisa thought that the death of a seventy-nine-year-old woman – even one who had just made a will leaving everything to her new housekeeper and her husband and had been well the previous day – would not arouse suspicion; and, even if it did, there was one precaution she could take. Two days after the death, she tried to arrange for the body to be cremated, but the funeral director told her that he would only take instructions from next of kin. Yule reported the death to the coroner, and the post-mortem revealed that Mrs Ricketts had been poisoned with phosphorus. The taste and smell of phosphorus is garlicky, but it can be masked by other things, such as rum or brandy. A wine merchant made regular deliveries of rum and Guinness to the bungalow and a bottle of brandy had been delivered on 12 April. During this visit, Mrs Ricketts had expressed her dissatisfaction with the Merrifields, who she thought were not feeding her properly and were taking her money, saying that she intended to change her will again and they would have to leave.

The Merrifields were arrested and tried for murder at Manchester Assizes in July. The probable source of the phosphorus was Rodine rat poison. A one-ounce tin of Rodine, costing a shilling, contained ten grains of phosphorus, enough to kill five people, and could be purchased at a chemist's shop without the need to sign a poison book. Bran, a main constituent of Rodine, had been found in Mrs Ricketts's stomach. A chemist and his assistant testified that, in March, they had sold a tin of Rodine to a man of Albert Merrifield's description, who had discussed treatments for ulcerated legs and had been accompanied by a woman.

Louisa had covered her tracks well, however. The Rodine tin was never found, and no trace of phosphorus was located in the bungalow, or in anything in the Merrifields' possession. The cause of death was in little doubt. Even the expert called for the defence, John Webster, professor of forensic medicine and toxicology at Birmingham University, while testifying that Mrs Ricketts's death had been due to natural causes, was obliged to say that she could have taken the phosphorus herself, adding, unconvincingly, that she might have died in the few minutes before it took effect.

The judge, in his final address to the jury, made a point that many armchair criminologists, including William Hone, would do well to take on board. Since murder by poison was 'rarely, if ever, committed in the presence of eye-witnesses who can give you direct evidence of the administration of the poison',[12] it followed that it had to be established by circumstantial evidence alone. 'You are not to suppose that circumstantial evidence is necessarily inferior to direct evidence. Indeed, it may be stronger.'[13] This same point had been made more eloquently many years before by David M. Stone, editor of the *New York Journal of Commerce*. 'Circumstantial testimony is sometimes the very strongest evidence of guilt, since the facts never commit perjury, and are not led astray by passion or prejudice.'[14]

Louisa was found guilty of murder, but the jury was unable to reach a verdict on her husband. The judge ordered a retrial of Alfred Merrifield, but this never proceeded. Louisa was hanged on 18 September.

Alfred, with the suspicion against him still unresolved, died in 1962, the same year in which the introduction of paraquat spelled the end of arsenical weedkillers. Further restrictions imposed by the Animals (Cruel Poisons) Act of 1962 meant that vermin were not to be destroyed by any poisons that involved suffering.

In 1957, Kenneth Barlow thought he had found the ultimate undetectable poison. His wife, thirty-two-year-old Elizabeth, who was eight weeks pregnant, was found drowned in the bath, but undisturbed water that had settled in the crook of her arm contradicted her husband's claim that he had made strenuous efforts to resuscitate her. It was suspected that she had somehow been rendered unconscious before she drowned, but the post-mortem showed no signs of any poison. Her dilated pupils and sweat-soaked pyjamas did, however, suggest a possible cause of unconsciousness – insulin. Elizabeth was not diabetic, but Barlow was a nurse and therefore able to obtain insulin and persuade his wife to allow him to inject her with something he represented as innocent. A magnifying glass identified two injection sites in her buttocks, and analysis of tissue samples confirmed the presence of insulin. It was later found that Barlow had, in the manner of so many poisoners, prepared the ground for his wife's early death by telling a colleague that he had once had to rescue her after he had found her collapsed in a hot bath. He had also inadvisedly boasted to others that it was easy to kill someone with insulin as it could not be detected in the body after death.[15] Barlow was found guilty of murder and sentenced to life imprisonment. He was released in 1984.

A murderer without such easy access to poison had to resort to more desperate methods of getting his or her supplies. On 31 December 1965, Frederick Whittle, a nightwatchman at Chelsea College of Science and Technology, encountered an intruder who coshed him with an iron bar and escaped. He was taken to hospital and discharged at 5 a.m., joining colleagues William Patrick Barnett and Herbert Weighell for the remainder of the night shift. They made a pot of tea but, soon afterwards, all were taken ill with violent vomiting. Once the early shift watchmen arrived, Barnett was able, despite his symptoms, to drive with the other sufferers to St Stephen's Hospital, but the casualty officer, who later claimed to have been ill himself, declined to see them and they were advised to consult their own doctors. Weighell and Whittle went home, but Barnett stayed in the porter's rest room, where he continued to be violently ill. No help was given to him, and, since it was New Year, there may well have been an assumption that his condition was caused by alcohol. Later that day, he was found in a state of collapse. He was taken to St Stephen's and declared dead on arrival.

Lethal amounts of arsenic were found in Barnett's body and it was traced to one source: the tea kettle. On 3 January, a laboratory technician noticed that the lock of a poisons cabinet had been forced and a bottle of arsenic was missing. The bottle, now empty, was found in a college dustbin. During the extensive police inquiry, 1,800 students were fingerprinted and 7,000 people questioned. The inquest returned a verdict of murder by a person or persons unknown. The murder remains unsolved. It emerged during enquiries that a similar incident, also unsolved, had occurred at the same college in 1938, when the contents of the tea kettle had been laced with atropine sulphate. The bitter taste of the tea ensured that very little was consumed and there were minimal ill effects. The incident was put down to a student prank.

Graham Young, Britain's most prolific, resourceful and imaginative serial poisoner of the twentieth century, was, even as a schoolboy unhampered by legal restrictions on the sale of deadly poisons. He started by looting bottles of acid and ether from the dustbins of local chemists' shops. He was widely read in both chemistry and the history of poison murder, and, when making his first purchase of antimony at the age of thirteen, was able to convince the chemist that he was an undersized seventeen, the legal age for the transaction. He obtained many other poisons in this way – arsenic, aconite, digitalin and, the one with which he was later to become most closely associated, thallium; he used his family as guinea pigs. When his father was diagnosed as suffering from poisoning, Graham was questioned and, as he was unable to resist showing off his knowledge, his obsession with poisons became apparent. Supplies of poison were found in his room, and, when he was searched, two bottles of antimony were found hidden under his shirt. He was still only fourteen when, in July 1962, he pleaded guilty to administering poison to his father, his sister and a school friend, causing them grievous bodily harm. His stepmother had died the previous April, and, at Graham's suggestion, she had been cremated. Her death was not investigated. He was committed to Broadmoor psychiatric hospital under the 1959 Mental Health Act, with the recommendation that he should not be released without the permission of the Home Secretary for fifteen years.

Less than a month later, a twenty-three-year-old Broadmoor patient, John Berridge, collapsed and died from cyanide poisoning. No supplies of cyanide were kept at Broadmoor, and Berridge had not received any visitors; neither could it be shown that the cyanide had been sent to him by post. However, laurel bushes, a source of cyanide, grew on adjoining farmland, and Graham had the skills to extract the poison. Whether the death was murder or suicide remains unresolved.

A knowledgeable poisoner like Graham Young can always find something to use. He hoarded cleaning chemicals and medicines, collected materials from the gardens, grew his own deadly nightshade plant, spiked the nurses' coffee with Harpic and emptied a packet of sugar soap into the tea urn. Eventually, he realised that this behaviour would not improve his chances of early release, and he was able to repress his instincts. In February 1971, aged twenty-three, he was released.

It does, on the face of it, seem extraordinary that he was able to obtain employment as a store man with John Hadland Ltd, a company based in the village of Bovingdon, manufacturing optical and photographic instruments and therefore providing a cornucopia of chemical opportunity to a poisoner, but, although the company made enquiries about him, they were never told of his record. While he waited for the results of his job interview, he purchased antimony from a chemist's shop, John Bell and Croyden, in London's Wigmore Street, under a false name, using a forged authorisation. His knowledge of chemistry gave him the air of a student.

Graham started work on 10 May, and lodged in a bedsit, where he began to amass a new collection of poisons. It was 3 June when the illnesses at Bovingdon began. Employees started to suffer from vomiting, diarrhoea and stomach pains. Although Young's poison of choice was thallium, a rare metal used in the manufacture of optical lenses, none was stocked at Hadlands, but, by then, he had established his bona fides with John Bell and Croyden.

The death of an employee, Bob Egle, on 7 July, and widespread sickness amongst other members of staff, led to talk of a local virus called 'the Bovingdon bug' being the cause, and, when another employee, Fred Biggs, died in November, the staff were in a state approaching panic. A meeting of all the employees and management, together with medical officer Dr Anderson, was held in the canteen. Anderson, not wanting to create further agitation, told the meeting that both radiation poisoning and heavy metal poisoning had been ruled out, the only possible conclusion being that there was some sort of virus. Graham could not let that go, and demanded to know why heavy metal poisoning had been ruled out. This naturally aroused suspicions of the young store man, and a train of enquiries revealed his record as a poisoner. Graham was arrested and, when his bedsit was explored, the police found pictures of Nazi leaders, drawings of people being poisoned, books on forensic medicine and an extraordinary array of chemicals. Graham had also kept a detailed diary of his poisonings. When interviewed, he boasted openly of his exploits, even admitting to the murder of his stepmother in 1962.

The body of Frederick Biggs was exhumed. Bob Egle had been cremated, but his ashes were exhumed. Both sets of remains showed the presence of thallium. This is believed to be the first time in the history of crime that the exhumation of cremains resulted in a charge of murder.[16]

Graham Young was found guilty on two counts of murder, two of attempted murder and two of administering poison. He was sentenced to life imprisonment. In August 1990, he died of a heart attack aged forty-two.

Poison murders, which, in figures published in 1999, form only about 2 or 3 per cent of homicides,[17] rarely make headlines nowadays and, when they do, they are very memorable, such as the 1978 murder of Bulgarian writer and dissident Georgi Markov with ricin, a highly toxic extract of the castor oil plant, delivered by a stab from an umbrella, or the 2006 murder of Russian journalist Alexander Litvinenko with radioactive polonium-210 administered in tea.

The internet is a whole new source of dangerous poisons. In 2014, Kuntal Patel was cleared of attempting to murder her mother with abrin, a plant toxin more powerful than ricin, extracted from the rosary pea, which she had purchased on the internet from an American black market supplier. There was no evidence that she had actually administered the poison, and she was convicted of obtaining a biological agent or toxin and jailed for three years. She was the first person to be sentenced under the 1974 Biological Weapons Act. Her supplier was nineteen-year-old Jesse William Korff, who manufactured and sold ricin and abrin to international purchasers, which he mailed hidden inside hollowed-out perfumed candles, including instructions on how much was required to kill people of different sizes and how the dose should be administered. He was arrested and pleaded guilty to the manufacture, sale and smuggling of toxins with conspiracy to murder. In February 2015, he was sentenced to 110 months in prison.

* * *

Of course, these high-profile cases reported in the press are only the ones we know about.

Glossary

Acids, organic – acids with origins in living matter e.g. oxalic.

Acids, inorganic – mineral acids e.g. sulphuric, which has numerous industrial uses.

Alkaloid – nitrogenous compounds, usually derived from plants.

Ammonia – NH_3, a gaseous compound of nitrogen and hydrogen. Domestically used in solution as a cleaner; medicinally, the pungent vapour released by ammonium carbonate is a feature of smelling salts.

Anodyne – medicinally, a painkiller.

Antimony – a white metallic element, Sb.

Arsenic trioxide – white arsenic, As_2O_3.

Arsine – AsH_3, a compound of arsenic, a toxic gas that, when oxidised, smells of garlic.

Battley's sedative solution – a preparation of opium used for pain relief.

Belladonna – the deadly nightshade, *atropa belladonna*, which yields the poisonous alkaloid, atropine. One symptom of atropine poisoning is dilation of the pupils.

Bismuth – an element, Bi, whose compounds have been used to treat gastric discomfort.

Black drop – a preparation made with opium, vinegar and spices, taken to relieve pain. There were many different formulations.

Blister – medicinally, a deliberate blistering of the skin thought to act as a counter-irritant.

Calomel – mercurous chloride, Hg_2Cl_2. Once used medicinally as a diuretic (promoting the production of urine) and laxative and in teething powders. Now known to be toxic.

Cantharidin or cantharides – a highly toxic irritant extracted from several species of beetle, colloquially known as 'Spanish Fly'.

Carbolic acid – or phenol, C_6H_5OH, a highly corrosive disinfectant.

Chloral – a sedative, chloral hydrate, $C_2H_3Cl_3O_2$.

Corrosive sublimate – mercuric chloride, $HgCl_2$, a toxic reagent once used as a treatment for syphilis.

Cream of tartar – $KC_4H_5O_6$, potassium bitartrate, has many culinary uses. In the nineteenth century, it was taken as a laxative.

Epsom salts – magnesium sulphate, $MgSO_4$, used domestically as bath salts or as a laxative.

Fomentation – a warm poultice.

Fowler's solution – A popular tonic containing potassium arsenite equivalent to 1 per cent arsenic trioxide.

Grey powder – a powder of mercury with chalk, prescribed as a laxative.

Gum arabic – a natural gum principally used as a stabiliser in manufactured foods and medicines.

Hydrogen sulphide – H_2S, also known as sulphuretted hydrogen; a gas produced by the decay of organic matter in the absence of oxygen, it has the characteristic odour of rotten eggs.

James's powders – Dr James's fever powders, an eighteenth-century remedy with a secret formula, later thought to be antimony and calcium phosphate.

Laudanum – a preparation of powdered opium and alcohol in which the opium was about one twelfth by weight.

Mercuric chloride – *see* corrosive sublimate.

Oil of vitriol – sulphuric acid, H_2SO_4.

Orpiment – arsenic trisulphide, As_2S_3, also known as sulphuret of arsenic, a bright yellow compound once used as a pigment.

Oxalic acid – an organic acid used in bleaching and cleaning, $H_2C_2O_4$.

Prussic acid – a solution of hydrogen cyanide, a compound first derived from the dye Prussian blue.

Ptomaines – alkaloids formed by the decay of organic matter.

Quinine – an alkaloid, employed to treat pain and fever, it was long used as a specific for malaria.

Realgar – arsenic sulphide, As_4S_4, a crystalline red compound once used as a pigment.

Steedman's soothing powders – a preparation intended to relieve the discomfort of teething in babies. They originally contained opium and mercury.

Sugar of lead – $Pb(CH_3COO)_2$ or lead acetate, a white crystalline compound with a sweet taste, which has many industrial uses.

Sugar soap – an alkaline cleaning product.

Tartar emetic – also known as tartarised antimony, antimony potassium tartrate or $K_2Sb_2(C_4H_2O_6)_2$, long known to have emetic properties, in the nineteenth century used as a treatment in poisoning and also for alcohol addiction.

White mercury – a popular name for arsenic.

Notes

Abbreviations

Affecting Case – Anon., *Affecting Case of Eliza Fenning Who Suffered the Sentence of the Law July 26, 1815,* eighth edition, including an account of the funeral, with other additional letters (John Fairburn, London, 1815)

Bell and Redwood – Bell, Jacob and Redwood, Theophilus, *Historical Sketch of the Progress of Pharmacy in Great Britain* (Pharmaceutical Society of Great Britain, London, 1880)

Borrow, *Celebrated Trials* – Borrow, George (ed.), *Celebrated Trials and Remarkable Cases of Criminal Jurisprudence from the earliest Records to the Year 1825,* six vols (Knight & Lacey, London, 1825)

Christison, *Treatise* 1829 – Christison, Robert, MD, *A Treatise On Poisons in Relation to Medical Jurisprudence, Physiology and the Practice of Physic* (Adam Black, Edinburgh, 1829)

Christison, *Treatise* 1836 – Christison, Robert, MD, *A Treatise On Poisons in Relation to Medical Jurisprudence, Physiology and the Practice of Physic* (Adam Black, Edinburgh, 1836)

Christison, *Treatise* 1845 – Christison, Robert, MD, FRSE, *A Treatise On Poisons in Relation to Medical Jurisprudence, Physiology and the Practice of Physic,* first American edition from fourth Edinburgh edition (ed. Barrington and Geo. D. Haswell, Philadelphia, 1845).

Circumstantial Evidence – Anon., *Circumstantial Evidence, the Extraordinary Case of Eliza Fenning* (Cowie & Strange, London, undated)

Full Report – Anon., *A full Report of the Evidence taken at the Thames Police Court, and the Coroner's Inquest, before Mr Baker, and a Respectable Jury, at Stepney, on the 10th of June, 1844, on the Alleged Poisoning Case, also the Trial of J. C. Belany [sic] etc.* (G. Pike, Alnwick, 1844)

Important Results – Watkins, John, LLD, *The Important Results of an Elaborate Investigation into the Mysterious Case of Elizabeth Fenning* (William Hone, London, 1815)

Irving – Irving H. B., *A Book of Remarkable Criminals* (Cassell & Co., London, 1918)

Marshall, *Five Cases* – Marshall, John, *Five Cases of Recovery from the Effects of Arsenic . . . To which are annexed many corroborating facts never before published, relative to the guilt of Eliza Fenni* (C. Chapple, bookseller to the Prince Regent, London, 1815)

Marshall, *Remarks* – Marshall, John, *Remarks on Arsenic* (J. Callow, London, 1817)

Orfila, *General System* – Orfila, Mathieu, *A General System of Toxicology, or a Treatise on Poisons, Found in the Mineral, Vegetable, and Animal Kingdoms, Considered in their*

Relations with Physiology, Pathology, and Medical Jurisprudence (M. Carey & Son, Philadelphia, 1817)

Orfila, *Directions* – Orfila, Mathieu, *Directions for the Treatment of Persons who have taken Poison and those in a state of Apparent Death* (Longman, Hurst, Rees, Orme & Brown, London, 1818)

Proof of Poison – Thorwald, Jürgen, *Proof of Poison* (Pan Books, London, 1969)

Rush, *Medical Inquiries* – Rush, Benjamin, *Medical Inquiries and Observations,* four vols, second edition (J. Conrad & Co., Philadelphia, 1805)

Queen v. Palmer – Anon., *The Queen v. Palmer. Verbatim Report of the Trial of William Palmer* (J. Allen, London, 1856)

Taylor, *On Poisons* 1848 – Taylor, Alfred Swaine, MD, FRS, *On Poisons in Relation to Medical Jurisprudence and Medicine* (Lea & Blanchard, Philadelphia, 1848)

Taylor's *Principles* 1894 – Stevenson, T., MD (ed.), *Taylor's Principles and Practice of Medical Jurisprudence,* two vols, fourth edition (J. & A. Churchill, London, 1894)

Taylor, *Strychnia* – Taylor, Alfred Swaine, MD, FRS, *On Poisoning by Strychnia, with Comments on the Medical Evidence Given at the Trial of William Palmer for the Murder of John Parsons Cook* (Longman, Brown, Green, Longmans & Roberts, London, 1856)

Trial of Carlyle W. Harris – Wellman, Francis L., *The Trial of Carlyle W. Harris for Poisoning His Wife, Helen Potts, at New York* (Yale Law Library, New York, 1892)

Trial of Mrs Burdock – Anon., *Trial of Mary Ann Burdock for the Wilful Murder of Mrs Clara Ann Smith by Administering Sulphuret of Arsenic to the Said Clara Ann Smith, Before the Recorder, Sir Charles Wetherell, at the Bristol Assizes, April the 10th, 11th and 13th, 1835* (W. H. Somerton, Bristol, 1835)

Trial of Pritchard – Roughhead, William (ed.), *Trial of Dr Pritchard* (W. Hodge & Co., Glasgow and Edinburgh, 1906)

Watson – Watson, Katherine, *Poisoned Lives: English Poisoners and their Victims* (Hambledon, London, 2004)

Epigraph

1. Attorney General Sir Lionel Heald addressing the jury at the trial of Louisa Merrifield, *Daily Mirror*, 31 July 1953 p. 16.
2. Cabanès, A. and Nass, L., *Poisons et Sortilèges* (Paris, Librarie Plon, 1903), p. 1.

Introduction: The Awful Sentence of Death

1. Anon., *Circumstantial Evidence, the Extraordinary Case of Eliza Fenning* (Cowie & Strange, London, undated), p. 7.
2. Born 10 June 1793. Watkins, John, LLD, *The Important Results of an Elaborate Investigation into the Mysterious Case of Elizabeth Fenning* (William Hone, London, 1815), p. 2.
3. It has been suggested that concern about the safety of the conviction led to an unusual delay in the date of Eliza's execution, but Oldfield and Adams – about whom there were no such concerns – were sentenced at the same sessions as Eliza. It was usual to hang prisoners not convicted of murder several weeks after the end of the sessions. Two other men, Harland and Box, who were sentenced to death that April, were executed on 27 July.
4. Newspaper reports that the Earl of Eldon attended those meetings were later denied. *Important Results*, p. 85.

5. Elected officials who attended the Justices.
6. It was later suggested by a spectator that, the usual time for execution being eight o'clock, the Sheriffs had been prevailed upon to delay the proceedings in case a reprieve should arrive. See *Birmingham Daily Post,* 8 June 1870, p. 5. A more prosaic explanation, that the executioner arrived late after officiating at another hanging in Ipswich, is given in Anon., *Affecting Case of Eliza Fenning Who Suffered the Sentence of the Law July 26, 1815,* eighth edition, including an account of the funeral, with other additional letters (John Fairburn, London, 1815), p. 34.
7. T. W. Wansbrough (1789–1859).
8. *The Times*, 27 July 1815, p. 3.
9. The home of a picture cleaner, Mr Millar, who had allowed the Fennings to lodge there after they had had to quit their home. Later affidavits from Mr Fenning bear that address.
10. The National Archives file on the case, which contains newspaper extracts, the official papers being missing, is marked: 'Supposed case of execution of an innocent woman.' HO 144/263/A56680.

Chapter 1: The Devilish Dumplings

1. Member of the Royal College of Surgeons, and apothecary to HRH the Duke of Gloucester's household.
2. Borrow, George (ed.), *Celebrated Trials and Remarkable Cases of Criminal Jurisprudence from the earliest Records to the Year 1825*, six vols (Knight & Lacey, London, 1825), vol. 6, p. 144.
3. *The Times*, 27 September 1815, p. 4.
4. I am indebted to John Marshall's very detailed account of the symptoms and treatment of the members of the Turner household in Marshall, John, *Five Cases of Recovery from the Effects of Arsenic . . . To which are annexed many corroborating facts never before published, relative to the guilt of Eliza Fenning* (C. Chapple, bookseller to the Prince Regent, London, 1815).
5. There is an excellent account of this in Christison, Robert, MD, *A Treatise On Poisons in Relation to Medical Jurisprudence, Physiology and the Practice of Physic* (Adam Black, Edinburgh, 1829), pp. 92–3.
6. The normal colour of faeces in the small intestine is green from bile secretions. It changes to brown in the large intestines. If excreted very rapidly, however, as in violent diarrhoea, it will still appear green.
7. Eliza was said to have challenged this by suggesting that she had told him not to eat them as she had had some herself and they had disagreed with her. *Important Results,* p. 119; *Circumstantial Evidence*, p. 5.
8. *Affecting Case*, p. 12.
9. Ibid.
10. Ibid.
11. *Important Results*, p. 4.
12. *Affecting Case*, p. 5.
13. Marshall, John, *Five Cases,* p. 35.
14. Eliza's champions made much of the fact that arsenic will not prevent dough from rising.
15. Marshall, *Five Cases*, p. 25.
16. Later experiments by one of Eliza's many champions suggested that if arsenic were sprinkled on dough left to rise, which was then made into dumplings by an unsuspecting cook, it might produce a similar appearance. *Important Results*, p. 77.
17. Thorwald, Jürgen, *Proof of Poison* (Pan Books, London, 1969), p. 16.
18. Ibid., p. 18.
19. Trial of William Henry Wyatt, proceedings of the Central Criminal Court, Sessions House, Old Bailey, 16 April 1806, fourth session, p. 266, case number 274, accessed via www.oldbaileyonline.org.

20. This was later to be a contentious point, and many of Eliza's defenders conducted experiments to prove that arsenic did not tarnish knives, although Marshall expressed the opinion at the trial that it did. He later commented that he did not regard the state of the knives to be a complete proof. His other tests more than adequately proved the presence of arsenic.
21. Borrow, *Celebrated Trials*, p. 146.
22. This point has been put forward as strongly suggesting Eliza's innocence; however, Mary Blandy – who undoubtedly gave poison to her father in 1752 – threw the packet on the fire, but missed the fact that undissolved powder had settled as a sediment in leftover gruel.
23. In apothecary's measure, a teaspoonful was one drachm or sixty grains.
24. This procedure is described in Rush, Benjamin, *Medical Inquiries and Observations*, four vols, second edition (J. Conrad & Co., Philadelphia, 1805), vol. 1, p. 240.
25. Marshall, John, *Remarks on Arsenic* (J. Callow, London, 1817), p. 54.
26. Rush, *Medical Inquiries*.
27. Scheele's Green, $AsCuHO_3$, a pigment discovered in 1775 by Carl Wilhelm Scheele.
28. A magistrates' court with six constables attached.
29. Marshall, *Five Cases*, p. 35.
30. Ibid.
31. *Affecting Case*, p. 6.
32. Trial of Eliza Fenning, proceedings of the Central Criminal Court, Sessions House, Old Bailey, 11 April 1815 fourth session, p. 220, case number 441, accessed via www.oldbaileyonline.org.
33. Petty treason ceased to be regarded as a distinct form of murder under the Offences Against the Person Act of 1828.
34. Ogilvy died on 6 March 1817, aged thirty-five.
35. *Important Results* p. 159.
36. Ibid., p. 78.
37. Ibid., p. 132.
38. *The Times*, 27 September 1815, p. 4.
39. Ibid.
40. *Important Results*, p. 84.
41. Trial of Mary Wittenback, proceedings of the Central Criminal Court, Sessions House, Old Bailey, 13 September 1827, seventh session, p. 608, case number 1597, accessed via www.oldbaileyonline.org.
42. *The Times*, 17 September 1827, p. 2.
43. Hackwood, Frederick William, *William Hone, His Life and Times* (T. Fisher Unwin, London, 1912), p. 99.
44. Paget, John, Esq., *Judicial Puzzles Gathered from the State Trials* (Sumner Whitney & Co., San Francisco, 1876), p. 97.
45. Trial of Isabella Hopes, proceedings of the Central Criminal Court, Sessions House, Old Bailey, fourth session, 2 February 1835, p. 509, case number 472, accessed via www.oldbaileyonline.org.
46. Ibid.
47. *Belfast News-Letter*, 14 July 1857, p. 1.
48. To date, it has not been possible to trace the date and location of Robert Gregson Turner's death. He and his father are listed in the Ipswich Poll Book of 1826. Orlibar (misspelled Olibar) was still listed there in 1831, but not Robert.

Chapter 2: Smothered in Onions

1. Watson, Katherine, *Poisoned Lives: English Poisoners and their Victims* (Hambledon Continuum, London, 2004), p. 33.
2. Often referred to in the nineteenth century in its Latin form as *strychnia*. In the text, 'strychnine' will be used unless in a quotation. Similarly with morphine and *morphia*.

3. I.e. those with origins in living matter, e.g. oxalic, used in cleaning and bleaching products.
4. I.e. mineral acids such as sulphuric, which has numerous industrial uses.
5. Nitrogenous compounds, usually derived from plants.
6. An Act for Better Regulating the Practice of Apothecaries throughout England and Wales, 55 Geo. III, c. 194.
7. See S. W. F. Holloway, 'The Apothecaries' Act, 1815: a Reinterpretation', *Medical History*, 10, no. 2 (April 1966), pp. 107–29.
8. Bell, Jacob and Redwood, Theophilus, *Historical Sketch of the Progress of Pharmacy in Great Britain* (Pharmaceutical Society of Great Britain, London, 1880), p. 70.
9. Ibid., p. 71.
10. Ibid.
11. *The Times*, 9 November 1825, p. 3.
12. Christison, Robert, MD, *A Treatise On Poisons in Relation to Medical Jurisprudence, Physiology and the Practice of Physic*, (Adam Black, Edinburgh, 1829), p. 515.
13. Ibid., p. 530.
14. Ibid., p. 531.
15. Bardsley, James Lomax, *Hospital Facts and Observations, Illustrative of the Efficacy of the New Remedies, Strychnia, Brucia, Acetate of Morphia, Veratria, Iodine, &c.* (Burgess & Hill, London, 1830), pp. 79–81.
16. Houlton, Joseph, *Formulary for the Preparation and Employment of Several New Remedies*, sixth edition (T. & G. Underwood, London, 1830), Appendix, p. 5.
17. A table of fatal cases from 1834 to 1855 may be found in Taylor, Alfred Swaine, MD, FRS, *On Poisoning by Strychnia, with Comments on the Medical Evidence Given at the Trial of William Palmer for the Murder of John Parsons Cook* (Longman, Brown, Green, Longmans & Roberts, London, 1856), pp. 78–85, with other cases listed pp. 117–46.
18. *The Times*, 16 August 1822, p. 3.
19. *The Times*, 1 March 1832, p. 6.
20. *Hampshire Advertiser*, 6 August 1831, p. 2.
21. *Salisbury and Winchester Journal*, 5 March 1832, p. 3.
22. Christison, *Treatise* 1845, p. 590.
23. *The Times*, 1 February 1833, p. 3.
24. Trial of Joseph Walters, proceedings of the Central Criminal Court, Sessions House, Old Bailey, 14 May 1838, seventh session, p. 147, case number 1351, accessed via www.oldbaileyonline.org.
25. Magnesium oxide, MgO, a common antacid.
26. *Morning Chronicle*, 23 March 1824, p. 4.
27. An account of this case is to be found in Cox, David J., *Foul Deeds and Suspicious Deaths in Shrewsbury and around Shropshire* (Wharncliffe Books, Barnsley, 2008).
28. Marshall, *Remarks*, p. 88.
29. *The Times*, 3 April 1817, p. 3.
30. Ibid.
31. A dish in which sautéed onions are layered with rabbit joints that have been dredged in seasoned flour and browned, the whole topped with sour cream and casseroled.
32. Donnall was baptised in 1789. Edwards took his Bachelor of Medicine and MD at Oxford in 1802.
33. *The Times*, 3 April 1817, p. 3.
34. Anon., *The Trial of Robert Sawle Donnall, Surgeon and Apothecary, late of Falmouth, in the County of Cornwall, for the Wilful Murder, by Poison, of Mrs. Elizabeth Downing, Widow, his Mother-in-law, at the Assize at Launceston, for the County aforesaid, On Monday, March 31, 1817, Before the Honourable Sir Charles Abbott, Knt.* (James Lake, Falmouth, 1817), pp. 1–2.
35. Ibid., p. 72.

36. Paris, J. A. and Fonblanque, J. S. M., *Medical Jurisprudence*, three vols (W. Phillips, London, 1823), vol. 2, p. 251.
37. Caudill, David S., 'Prefiguring the Arsenic Wars', *Chemical Heritage Magazine*, Chemical Heritage Foundation: http://www.chemheritage.org/discover/media/magazine/articles/27-1-prefiguring-the-arsenic-wars.aspx?page=1, accessed June 2015.

Chapter 3: The Greatest Refinement of Villainy

1. 1787–1853. Born Mateu Josep Bonaventura Orfila i Rotger, in Minorca, he became a French citizen in 1819.
2. Solution of calcium hydroxide $Ca(OH)_2$.
3. Orfila, Mathieu, *A General System of Toxicology, or a Treatise on Poisons* (M. Carey & Son, Philadelphia, 1817), Preface of the Author, p. vi.
4. Orfila, *General System*, p. 3.
5. Ibid., Preface of Joseph G. Nancrede, pp. xv–xvi.
6. Ibid., Preface of the Author, p. vi.
7. Ibid., p. 2.
8. Ibid., Report to the Institute of France, p. xlvii.
9. 1712–79.
10. Orfila, *General System* p. 62.
11. Ibid., p. 275.
12. 1777–1849.
13. 1742–1832.
14. Orfila, *General System*, p. 422.
15. Ibid., p. 444.
16. Ibid., p. 443.
17. Brodie, Benjamin, 'Further Experiments and Observations on the Action of Poisons on the Animal System', *Philosophical Transactions of the Royal Society of London*, 102 (1812), p. 209.
18. Ibid., pp. 205–6.
19. For a detailed discussion of this issue, see José Ramón Bertomeu-Sánchez, 'Animal Experiments, Vital Forces and Courtrooms: Mateu Orfila, François Magendie and the Study of Poisons in Nineteenth-century France', *Annals of Science*, 69, no. 1 (2012), pp. 1–26.
20. Translated as: Orfila, Mathieu, *A Popular Treatise on the Remedies to be Employed in Cases of Poisoning and Apparent Death* (Solomon W. Conrad, Philadelphia, 1818); and Orfila, Mathieu, *Directions for the Treatment of Persons who have taken Poison and those in a state of Apparent Death* (Longman, Hurst, Rees, Orme & Brown, London, 1818).
21. Bertomeu-Sánchez, José Ramón (ed.), 'Popularizing Controversial Science: A Popular Treatise on poisons by Mateu Orfila (1818),' *Medical History*, 53, no. 3 (July 2009), p. 366.
22. Ibid., pp. 371–2.
23. 1797–1882.
24. Christison (eds), *The Life of Sir Robert Christison, Bart, edited by his sons,* two vols (William Blackwood & Sons, Edinburgh and London, vol. 1, 1885; vol. 2, 1886), vol. 1, p. 288.
25. Ibid.,vol. 2, p. 247.
26. *Edinburgh Annual Register for 1823* (Constable & Co., Edinburgh, 1824), vol. 16, part III, p. 180.
27. According to Lebret in Anon., *Procès Complet d'Edme-Samuel Castaing, Docteur en Médecine*, (Pillet, Paris, 1823), p. 86, Auguste's three-quarters share of the estate was 195,000 francs. In 2015, 260,000 fr. is worth approx. £870,000.
28. Irving H. B., *A Book of Remarkable Criminals* (Cassell & Co., London, 1918), p. 172.
29. Anon., *Causes criminelles célèbres du dix-neuvième siècle rédigés par une Société d'Avocats*, four vols (H. Langlois Fils et Cie, Paris, 1828), vol. 4, p. 32.

30. *Proof of Poison*, p. 41.
31. *The Times*, 15 November 1823, p. 2.
32. *Journal des débats* quoted in *The Times,* 17 November 1823, p. 3.
33. Irving, p. 182.
34. Ibid., pp. 182–3.
35. *Edinburgh Annual Register for 1823*, vol. 16, part III, p. 188.
36. Taylor, *On Poisons* 1848, p. 496.

Chapter 4: Arsenic and Old Graves

1. Also sometimes spelt Lenergan. I have used the spelling that appears in his own statement.
2. A full report of the trial is in *Walker's Hibernian Magazine, or Compendium of Entertaining Knowledge* (Thomas Walker, Dublin, November 1781), pp. 593–600.
3. Ibid., p. 594.
4. Frank Thorpe Porter, in *Gleanings and Reminiscences* (Hodges, Foster & Co., Dublin, third edition, 1875), pp. 1–5, relates that Lonergan survived the hanging, was revived and escaped.
5. An account of this case is in Griffiths, Major Arthur, *The Fortresses of Romance and Crime* (Forgotten Books, London, 2013, reprint), pp. 82–93.
6. See Chapter 2, p. 28.
7. Mercuric chloride, $HgCl_2$.
8. Christison, *Treatise* 1829, p. 256.
9. Christison, Robert, MD, *A Treatise on Poisons in Relation to Medical Jurisprudence, Physiology and the Practice of Physic* (Adam Black, Edinburgh, 1836), p. 324.
10. See Stratmann, Linda, 'The Case of the Greek Gigolo', in Edwards, Martin (ed.), *Truly Criminal: A Crime Writers' Association Anthology of True Crime* (The History Press, Stroud, 2015).
11. *The Times*, 29 December 1829, p. 3.
12. Or Bouvier-Salazard following his marriage to Pierette Salazard.
13. *The Times*, 29 December 1829, p. 3.
14. Ibid.
15. Ibid.
16. The trial is reported in the *Gazette de Tribuneaux*, no. 1339, 26 November 1829, pp. 85–7.
17. *The Times*, 3 December 1829, p. 3.
18. The proceedings are reported in the *Journal de l'Ain*, no. 102, 27 August 1832, pp. 1–4, and no. 103, 29 August 1832, pp. 1–4.
19. According to a William Babbington, who gave evidence at the trial, Wade was a married man. He may have been the John Wade who was born in 1778 and died on 16 April 1834.
20. Anon., *Trial of Mary Ann Burdock for the Wilful Murder of Mrs Clara Ann Smith by Administering Sulphuret of Arsenic to the Said Clara Ann Smith, Before the Recorder, Sir Charles Wetherell, at the Bristol Assizes, April the 10th, 11th and 13th, 1835* (W. H. Somerton, Bristol, 1835), p. 12.
21. Arsenic trisulphide As_2S_3 produced by the reaction between arsenic and the hydrogen sulphide produced by decay.
22. Jennings, Rev. Thomas, *Two Sermons Occasioned by the Case of Mary Ann Burdock* (Seeleys, Bristol, 1835), p. 6.
23. In December 1834, she had a fifteen-year-old son and a seven-year-old daughter.
24. *The Times*, 13 April 1833, p. 5.
25. Ibid.
26. Ibid.
27. *Trial of Mrs Burdock*, p. 30.

28. Ibid., p. 38.
29. As_4S_4, a sulphide used as a red pigment.
30. It does seem extraordinary; however, at the trial, Mary Ann described following Mrs Burdock into her chamber, for what reason she did not state, and it is possible that Burdock was unaware of the maid standing behind her when she put the powder into the basin.
31. *The Times*, 27 April 1843, p. 5.
32. *The Times*, 1 May 1843, p. 8.
33. *Annals of Electricity, Magnetism, & Chemistry; and Guardian of Experimental Science*, (Sherwood Gilbert & Piper, London, 1837), conducted by William Sturgeon, vol. 1: October 1836 to October 1837 p. 25.
34. *The Times*, 25 July 1839, p. 7.

Chapter 5: The Suspicions of Mr Marsh

1. Born in Plumstead, Kent, 4 May 1810.
2. Plumstead burial records show that an Ann Bodle died in 1836, aged seventy-seven.
3. *The Times*, 11 November 1833, p. 3.
4. Ibid.
5. Ibid.
6. See Hempel, Sandra, *The Inheritor's Powder* (Phoenix, London, 2014), pp. 31–4, for a detailed analysis of the will.
7. *The Times*, 14 December 1833, p. 5.
8. Christison, *Treatise* 1829, pp. 187–8.
9. *Bucks Herald*, 16 November 1833, p. 2.
10. *The Times*, 14 December 1833, p. 6.
11. *Oxford Journal*, 21 December 1833, p. 3.
12. Marsh, J., 'Account of a Method of Separating Small Quantities of Arsenic From Substances With Which it May be Mixed', *Edinburgh New Philosophical Journal* (1836), vol. 21, p. 230.
13. A popular tonic containing potassium arsenite equivalent to 1 per cent AsO_3.
14. Barlow, George H. and Babington, P. James (eds), *Guy's Hospital Reports* (Samuel Highly, London, 1837), vol. 2, p. 75.
15. Rosenfeld, L., 'Alfred Swaine Taylor (1806–1880), Pioneer Toxicologist – and a Slight Case of Murder', *Clinical Chemistry,* vol. 31 (1985), p. 1235.
16. See Webster, Stewart H., 'The Development of the Marsh Test for Arsenic', *Journal of Chemical Education* (Division of the Chemical Education of the American Chemical Society, Easton, PA, October 1947), vol. 24, issue 10, pp. 487–99.
17. Thompson, L., 'On Antimoniuretted Hydrogen, with some Remarks on Mr. Marsh's Test for Arsenic', *London and Edinburgh Philosophical Magazine and Journal of Science* (R. and J. E. Taylor, London, 1837), vol. 10 (January–June), pp. 353–5.
18. *Bristol Mercury*, 18 May 1839, p. 4.
19. Ibid.
20. Marsh, J., 'On a New Method of Distinguishing Arsenic from Antimony, in Cases of Suspected Poisoning by the Former Substance', *London and Edinburgh Philosophical Magazine and Journal of Science* (R. & J. E. Taylor, London, 1839), vol. 15 (July–December), pp. 282–4.
21. *Lincolnshire Chronicle*, 13 April 1838, p. 3.
22. Ibid.
23. Thinus, F., 'Note sur l'emploi de la méthode de James MARSH, dans un cas de médecine légale', *Journal de Pharmacie et des Sciences Accessoires* (Louis Colas, Paris, 1838), vol. 24, pp. 500–2.
24. *Gazette des Tribuneaux*, no. 4043 (25 August 1838), pp. 1069–70 and no. 4044 (26 August 1838), pp. 1074–5.

25. Trial of John Clifton, proceedings of the Central Criminal Court, Sessions House, Old Bailey, tenth session, 12 August 1839, case number 2238, p. 633, accessed via www.oldbaileyonline.org.
26. Then used for its laxative properties, usually in combination with brimstone and treacle, it resembled arsenic and there were numerous fatal accidents as a result.
27. Trial of John Clifton, p. 635.
28. Ibid.
29. Orfila's work on dogs is reviewed in the *American Journal of the Medical Sciences* (Lea & Blanchard, Philadelphia, 1841), New Series, vol. 2, pp. 403–9.
30. Taylor, *On Poisons* 1848, p. 139.
31. 1805–67.
32. Orfila in *Am. J. Med. Sci.*, op. cit., pp. 409–12.
33. *Annales d'Hygiène Publique et de Médecine Légale* (Paris, Baillière), vol. 21 (January 1839), p. 465.
34. 1800–57.
35. *Am. J. Med. Sci.*, op. cit., p. 420, footnote.
36. Bertomeu-Sánchez, José Ramón, 'Sense and Sensitivity. Mateu Orfila, the Marsh Test and the Lafarge Affair', in Bertomeu-Sánchez, J. R. and Nieto-Galan, A. (eds), *Chemistry, Medicine and Crime, Mateu Orfila (1787–1853) and His Times* (Science History Publications, Sagamore Beach, MA, 2006), pp. 223–4.
37. Rognetta, M., *Nouvelle méthode de traitement de l'empoisonnement par l'arsenic et documents médico-légaux sur cet empoisonnement*, (Chez Gardenbas, Paris, 1840), pp. 99–104.
38. Ibid., p. 105 and Orfila, M. 'Empoisonnement par l'arsenic', *Annales d'Hygiène Publique et de Médecine Légale*, vol. 24 (July 1840), p. 313.
39. 1794–1878.

Chapter 6: French Porcelain

1. *Boston Medical Surgical Journal*, vol. 22, no. 8 (1840), p. 117.
2. 'C'est pris.' The trial is reported in the *Gazette des Tribunaux*, issue 4441 (2/3 December 1839), pp. 106–7 and issue 4442 (4 December 1839), p. 110.
3. *American Journal of the Medical Sciences* (n.s.), vol. 2 (1841), p. 412.
4. *Boston Medical Surgical Journal*, vol. 22, no. 8 (1840), p. 117.
5. *Am. J. Med. Sci.*
6. *Boston Medical Surgical Journal*, vol. 22, no. 8 (1840), p. 119.
7. About sixty-two grains.
8. Later Brive-la-Gaillarde, then a town with a population under 9,000.
9. *The Times*, 10 September 1840, p. 6.
10. Ibid.
11. Approx. 0.0077 of a grain.
12. *The Lancet*, vol. 1 (1840–1), p. 261.
13. She died of tuberculosis in 1852.
14. *The Lancet*, vol. 2 (1840–1), p. 335.
15. Rees, G. O., 'On the Existence of Arsenic as a Natural Constituent of Human Bones', *Guy's Hospital Reports*, 6 (1841), p. 163.
16. Anon., *The Year-book of Facts in Science and Art* (Charles Tilt, London, 1840), p. 172.
17. 'Some Difficulties in Forensic Medicine', *London Medical Gazette*, vol. 1 (1840–1), pp. 410–11.

Chapter 7: This Troublesome World

1. *The Times*, 4 August 1841, p. 6.

2. Bridget is listed in the 1841 census as a twenty-year-old factory hand living with her parents, Michael and Catharine, and three siblings in Stockport. Ages in that census were usually rounded to the nearest five years.
3. *London Standard*, 23 October 1840, p. 4.
4. *Manchester Courier and Lancashire General Advertiser*, 31 October 1840, p. 3.
5. Aged four years and five months at date of death, according to the *Manchester Guardian*, quoted in *Reading Mercury*, 31 October 1840, p. 4.
6. The ages of Jane and Edward are unknown, but they would have been between one and three years old.
7. *The Times*, 4 August 1841, p. 6.
8. Ibid.
9. Ibid.
10. *The Times*, 5 May 1837, p. 6.
11. Taylor, *Strychnia*, pp. 4–5.
12. *The Spectator*, 23 September 1848, p. 12.
13. 1841 census accessed via www.ancestry.com.
14. *The Times*, 4 August 1841, p. 6.
15. *The Times*, 5 August 1841, p. 7.
16. Chadwick, Sir Edwin, 'Supplementary Report on the Result of a Special Inquiry into the Practice of Interment in Towns', *British Parliamentary Papers*, 1843 [509], xii.
17. Martineau, H., *A History of the Thirty Years' Peace, AD 1816–1846,* four vols (George Bell & Sons, London, 1877), vol. 2, p. 315.
18. *The Times,* 18 January 1849, p. 4.
19. See Katherine Watson's excellent *Poisoned Lives.*
20. 4 & 5 William IV, c. 76.
21. Trial of Catherine Michael, proceedings of the Central Criminal Court, Sessions House, Old Bailey, 6 April 1840, sixth session, case number 1285, accessed via www.odbaileyonline.org
22. *London Standard* 16 April 1840 p. 1
23. *The Times,* 2 April 1844, p. 8.
24. *London Standard,* 8 September 1854, p. 4
25. Trial of Jane Harrington, proceedings of the Central Criminal Court, eleventh session, 18 September 1854, p. 1300, case 1068, accessed via www.odbaileyonline.org.
26. Ibid.
27. *London Standard,* 8 September 1854, p. 4.
28. *London Standard,* 22 September 1854, p. 1.
29. When he died in 1888, his estate was valued at over £1,400. Probate record accessed via www.ancestry.com.
30. *Lincolnshire Chronicle,* 9 August 1844, p. 3.
31. *The Times,* 24 July 1844, p. 7.
32. Ibid., p. 6.
33. Ibid., p. 6.
34. Ibid.

Chapter 8: Getting Away with Murder

1. Trial of Jane Bowler, proceedings of the Central Criminal Court, Sessions House, Old Bailey, 24 October 1842, twelfth session, p. 1363, case number 3062, accessed via www.oldbaileyonline.org.
2. *The Times,* 8 October 1842, p. 3.
3. Ibid.
4. Ibid.
5. Ibid.
6. Ibid.

7. Trial of Thomas Francis Dickman, proceedings of the Central Criminal Court, Sessions House, Old Bailey, 3 February 1845, fourth session, p. 482, case number 425, accessed via www.oldbaileyonline.org.
8. A feudal tenure of land.
9. Anon., *A Full Report of the Evidence Taken at the Thames Police Court, and the Coroner's Inquest, before Mr Baker, and a Respectable Jury, at Stepney, on the 10th of June, 1844, on the Alleged Poisoning Case, also, the Trial of J. C. Belany [sic] etc.* (G. Pike, Alnwick, 1844), p. 38.
10. Ibid., p. 22.
11. Ibid., p. 27.
12. Christison, *Treatise* 1829, p. 556.
13. *Full Report*, p. 7.
14. Ibid., p. 14.
15. Ibid.
16. Ibid.
17. Ibid.
18. Ibid.
19. Ibid., p. 15.
20. Ibid., p. 34.
21. Ibid.
22. Ibid., p. 35.
23. Ibid., p. 55.
24. *Full Report*, p. 59.
25. The scenario does call to mind the classic film *12 Angry Men*.
26. *The Times*, 24 August 1844, p. 6.
27. 'The Late Trial of Belaney for the Murder of His Wife', *London Medical Gazette*, (Longman, Brown Green & Longmans, London, 1843–4), New Series, vol. 2, 30 August 1844, pp. 745–9.
28. Christison, *Treatise* 1836, pp. 692–3.
29. Taylor, *On Poisons* 1848, p. 544.
30. *The Globe*, quoted in *the Newcastle Courant*, 13 September 1844, p. 3.
31. *Durham County Advertiser*, 30 August 1844, p. 2.
32. *Newcastle Journal*, 21 September 1844, p. 2.
33. *Leicester Mercury*, 19 October 1844, p. 4
34. *Fife Herald*, 28 November 1844, p. 3.
35. *Bristol Mercury*, 21 December 1844, p. 6.
36. See *Archie Belaney known as Grey Owl* by Barry Brown, which reveals that James was the great-uncle of the conservationist who claimed to be of First Nation descent, accessed http://wabbrown.co.uk/greyowl.pdf.
37. Trial of Louisa Susan Hartley, proceedings of the Central Criminal Court, Sessions House, Old Bailey, 6 May 1850, seventh session, p. 129, case number 1032, accessed via www.oldbaileyonline.org.
38. *The Times*, 13 May 1850, p. 7.
39. Trial of Martha Sharp, proceedings of the Central Criminal Court, 19 August 1850, tenth session, p. 538, case number 1503, accessed via www.oldbaileyonline.org.
40. The reputation of cantharides as a sexual stimulant persisted. In 1954, Arthur Kendrick Ford unintentionally killed two female co-workers after giving them coconut ice he had adulterated with cantharides. He was convicted of manslaughter.
41. Offences Against the Person Act 1861 (24 & 25 Vict, c. 100), para. 24.

Chapter 9: Alarm and Indignation

1. *The Times*, 27 December 1831, p. 4.
2. The Registration of Births and Deaths Act 1874 (1874 c. 88) required that death certificates should only be issued by 'registered medical practitioners', but this did not prevent the registration of uncertified deaths continuing.

3. *Morning Chronicle,* 8 December 1829, p. 1.
4. *Royal Cornwall Gazette,* Saturday 10 October 1829, p. 4.
5. *Bolton Chronicle,* quoted in *The Times*, 4 October 1842, p. 6.
6. *The Times*, 6 October 1842, p. 7.
7. Betty was later to deny ever having had a daughter called Nancy, and Henry was unable to recall the name of Betty's third daughter. A Nancy Haslam, daughter of Richard and Betty, was baptised in 1835.
8. The ages of the children as given in court sometimes differ from official records. Where there is any variance, the dates of baptism and burial have been used in preference.
9. *Bolton Chronicle,* quoted in *The Times*, 4 October 1842, p. 6.
10. *The Times*, 6 October 1842, p. 6.
11. James Heywood was the name of Betty's father so, even if this was not he, they may have been related.
12. *The Times*, 6 October 1842, p. 6.
13. *Bolton Chronicle,* quoted in *The Times*, 4 October 1842, p. 6.
14. *The Times,* 6 October 1842, p. 7.
15. *Bolton Chronicle,* quoted in *The Times*, 4 October 1842, p. 6.
16. *Bristol Times and Mirror*, 8 October 1842, p. 2.
17. *The Spectator*, 8 October 1842, vol. 15, p. 966.
18. *Bolton Chronicle,* quoted in *The Times*, 4 October 1842, p. 6.
19. *The Times*, 22 February 1844, p. 7.
20. *Stamford Mercury*, 26 July 1844, p. 2.
21. *The Times*, 13 January 1845, p. 3.
22. *The Times,* 15 January 1845, p. 3.
23. *The Times,* 16 January 1845, p. 6.
24. Ibid.
25. *The Times*, 11 September 1845, p. 8.
26. *Pharmaceutical Journal and Transactions,* vol. 4, no. 8 (1 February 1845), p. 342.
27. Ibid.
28. Ibid., p. 344.
29. 7 & 8 Vic. c.101.
30. Fathers were required to pay 2s. 6d. a week. This was increased to 5s. in 1872. Watson, p. 83.
31. *The Times*, 7 March 1851, p. 6.
32. *The Times*, 5 September 1846, p. 7.

Chapter 10: The Essex Factor

1. Eighty-seven women and seventy-seven men were tried either for murder or for attempted murder using poison in the years 1840 to 1850. Watson, p. 58.
2. *The Times*, 25 September 1847, p. 4.
3. Ibid.
4. 'Restriction on the Sale of Arsenic', *London Medical Gazette*, (Longman Green & Longmans, London, 1847), New Series, vol. 5, 29 October 1847, p. 780.
5. Trial of Hannah Leath, proceedings of the Central Criminal Court, Sessions House, Old Bailey, 21 August 1848, tenth session, p. 1363, case number 1849, accessed via www.oldbaileyonline.org.
6. The first was in 1832.
7. Letheby, H., 'On the Probability of Confounding Cases of Arsenical Poisoning with Those of Cholera', *Pharmaceutical Journal*, vol. 8 (1848–9), pp. 237–40.
8. *Morning Post*, 25 July 1848, p. 7.
9. *The Times*, 31 August 1848, p. 7.
10. *The Times*, 5 September 1848, p. 8.
11. *The Times*, 1 September 1848, p. 4

12. For a detailed account of the Essex poison cases illustrating that the scandal was largely a creation of the press, see Ainsley, Jill Louise, 'The Ordeal of Sarah Chesham', Master of Arts degree thesis 1997, Department of History, University of Victoria.
13. *The Times*, 20 October 1848, p. 4.
14. *The Times,* 10 March 1849, p. 6.
15. *Chelmsford Chronicle*, 14 December 1849, p. 3.
16. *The Times*, 10 August 1849, p. 4.
17. *Punch*, vol. XVII (8 September 1849), p. 97.
18. Taylor, *On Poisons* 1848, pp. 156–7.
19. Toogood, Jonathan, 'On the Crime of Secret Poisoning', *Provincial Medical and Surgical Journal*, vol. 13, no. 13 (27 June 1849), p. 362.
20. Ibid.
21. Ibid.
22. *Prov. Med. Surg. J.*, vol. 13 no. 17 (22 August 1849), pp. 437–8.
23. Ibid., p. 468.
24. Anon., *Report from the Select Committee of the House of Lords on the Sale of Poisons, &c. Bill (H.L.)* (House of Commons, London, 1857), pp. 96–7.
25. *The Times*, 8 March 1851, p. 4.
26. Ibid.
27. Ibid.
28. Ibid.
29. *Household Words* (London, Office 16 Wellington Street North, 1851), vol. 2, p. 155.
30. *Prov. Med. Surg. J.*, vol. 15, no. 7 (2 April 1851), p. 181.
31. *Hansard*, 13 March 1851, vol. 114, cols 1300–1.
32. *Hansard*, 24 March 1851, vol. 155, cols 422–4.
33. *Prov. Med. Surg. J.*, vol. 15, no. 17 (3 September 1851), p. 471.
34. Ibid., p. 456.
35. *Household Words*, vol. 4, p. 277.
36. Ibid.
37. Ibid.
38. From 1851 to 1860, there were fifty-six Old Bailey murder trials resulting in a guilty verdict, of which only five involved poison.
39. Spelt Rollinson in newspapers, but Rowlinson in the UK census.
40. Crowther, Anne and White, Brenda, *On Soul and Conscience: the Medical Expert and Crime. 150 Years of Forensic Medicine in Glasgow* (Aberdeen University Press, Aberdeen, 1988).

Chapter 11: Tobacco Kills

1. Taylor, *On Poisons* 1848, p. 114.
2. Ibid., p. 505.
3. *Arkansas Family Historian*, vol. 22, no. 3 (September 1984), p. 136; 1850 US census accessed via www.ancestry.com.
4. Perhaps about £8,500 today.
5. *Journal des débats*, 1 June 1851, p. 3.
6. *The Times*, 2 June 1851, p. 6.
7. Taylor, *On Poisons* 1848, pp. 630, 631.
8. *The Times,* 2 June 1851, p. 6.
9. Wennig, Robert, 'Back to the Roots of Modern Analytical Toxicology: Jean Servais Stas and the Bocarmé Murder Case', *Drug Test Analysis*, 1 (2009), pp. 153–5.
10. 'La Nicotine', by Félix Roubaud, in *L'illustration*, tome 17, numéro 434, vol. 17 (21 juin, 1851), p. 394
11. *The Times*, 2 June 1851, p. 6.
12. Ibid.

13. Ibid.
14. Ibid.
15. Ibid.
16. The account of the reading of the report at the trial is in the *Journal des débats*, 6 June 1851, p. 3.
17. *The Times*, 13 June 1851, p. 8.
18. *Manchester Guardian*, 18 June 1851, p. 3, reproduced from *Galignani's Messenger*.
19. 'The Bocarmé Tragedy', *Sharpe's London Magazine of Entertainment and Instruction for General Reading*, no. 3 (July 1853), pp. 41–8; also *Northern Whig*, 1 January 1853, p. 4.
20. Stevenson, T., MD (ed.), *Taylor's Principles and Practice of Medical Jurisprudence*, two vols, fourth edition (J. & A. Churchill, London, 1894), vol. 1, p. 386.
21. *Annales d'Hygiène publique et de médecine légale*, series 1, no. 48, vol. 2 (1852), pp. 359–69.

Chapter 12: Nil By Mouth

1. Burney, I., *Poison, Detection and the Victorian Imagination* (Manchester University Press, Manchester, 2006), pp. 19–20.
2. Christison, *Treatise* 1829, p. 37.
3. Ibid., p. 234.
4. Ibid., p. 235.
5. Ibid., pp. 235–6.
6. Ibid., p. 236.
7. *Provincial Medical & Surgical Journal*, vol. 1, no. 1 (3 October 1840), Introductory Address, p. 1.
8. Ibid.
9. Daly, Ann, MA, 'The Dublin Medical Press and Medical Authority in Ireland 1850–1890', thesis for the degree of PhD, Department of Modern History, National University of Ireland, Maynooth, January 2008, accessed 3 August 2015: http://eprints.maynoothuniversity.ie/5269/ p.4.
10. *Edinburgh Medical Journal* (Sutherland and Knox, Edinburgh, 1856), vol. 1, July 1855 to June 1856, p. 628.
11. *The Times*, 1 August 1855, p. 12.
12. A constant feeling of needing to pass stools, which may be accompanied by painful straining.
13. Anon., *Great Burdon Slow Poisoning Case, Report of the Investigation before the Magistrates at Darlington in the County of Durham . . . Together with the Evidence of Mr J. S. Wooler before the Coroner at Great Burdon* (Robert Swales, London, 1855), p. 9.
14. Ibid., p. 44.
15. Ibid., p. 9.
16. Ibid., p. 45.
17. Ibid., p. 11.
18. Ibid., p. 12.
19. Ibid., p. 15.
20. Ibid., p. 18.
21. About 39 mm.
22. Anon., *Great Burdon*, p. 20.
23. Ibid., p. 44.
24. Ibid.
25. Ibid.
26. Ibid.
27. Ibid.,
28. Ibid., p. 4.
29. Ibid.

30. Ibid.
31. *The Times*, 10 December 1855, p. 10.
32. Anon., *Great Burdon*, p. 13.
33. Ibid., p. 37.
34. Ibid., p. 43.
35. *Edinburgh Medical Journal*, op. cit., p. 628.
36. *The Times*, 27 December 1855, p. 4.
37. *Edinburgh Medical Journal*, op. cit., p. 709.
38. 'The German Poison Eaters', *The Chemist* (n. s.), vol. 1 (1854), pp. 762–4.
39. Parascandola, John, BS, MS, PhD, 'The Arsenic Eaters of Styria', *Hektoen International:* www.hektoeninternational.org accessed May 2015.
40. *The Times*, 12 December 1855, p. 8.
41. *Edinburgh Medical Journal*, op. cit., p. 628.
42. Ibid., p. 713.
43. *Association Medical Journal*, vol. 3, no. 154 (14 December 1855), p. 1114.
44. *Edinburgh Medical Journal*, op. cit., p. 761.
45. Ibid., p. 628.
46. *The Times*, 22 December 1855, p. 6.
47. Will of Joseph Snaith Wooler, obtained via https://probatesearch.service.gov.uk
48. 1881 census, accessed via www.ancestry.com.
49. 1891 census, accessed via www.ancestry.com.

Chapter 13: The Palmer Act

1. Nowadays, the words pill and tablet are used interchangeably, but the Victorian pill mixed medicinal substances with a pliable inert mass, which was then rolled into small balls. If recently made up, they would still have been soft enough to introduce small amounts of a poison.
2. Anon., *The Queen v. Palmer. Verbatim Report of the Trial of William Palmer* (J. Allen, London, 1856), p. 73.
3. *Queen v. Palmer*, p. 86.
4. Ibid., p. 87.
5. Ibid., p. 22.
6. *The Times,* 4 June 1856, p. 12.
7. According to Devonshire's testimony. Mr Serjeant Shee, counsel for Palmer at the trial, suggested that Frere was a solicitor of Shewsbury; surely an error.
8. *Queen v. Palmer,* p. 139.
9. Ibid.
10. *The Times*, 18 December 1855, p. 10.
11. Taylor, *Strychnia*, p. 69.
12. *Queen v. Palmer*, p. 99.
13. *The Times*, 18 December 1855, p. 10.
14. *Queen v. Palmer,* p. 11.
15. Ibid., p. 189.
16. *The Times*, 4 January 1856, p. 9.
17. Quoted in the *Staffordshire Advertiser*, 5 January 1856, p. 5.
18. Ibid.
19. *Daily News,* quoted in *Staffordshire Advertiser,* 12 January 1856, p. 8.
20. *The Times*, 23 January 1856, p. 11.
21. Taylor's *Principles* 1894, vol. 2, p. 639.
22. *Essex Standard*, 4 January 1856, p. 2. Ibid.
23. Ibid.
24. *Queen v. Palmer,* p. 147.
25. *Essex Standard,* 18 January 1856, p. 2.

26. At Palmer's trial, it was suggested incorrectly that it was Mayhew's brother Augustus who had called.
27. 'Our Interview with Dr Alfred Taylor', *Illustrated Times*, 2 February 1856, p. 27.
28. Ibid.
29. Ibid., p. 1.
30. See 'The Bank With No Scruples', in Stratmann, Linda, *Fraudsters and Charlatans* (The History Press, Stroud, 2010).
31. The Central Criminal Court Act (19 & 20 Vict., c. 16) received Royal Assent on 11 April 1856.
32. *Queen v. Palmer*, p. 3.
33. Ibid., p. 316.
34. Ibid., p. 120.
35. Ibid., p. 146.
36. Ibid., p. 147.
37. Ibid.
38. *The Times*, 21 May 1856, p. 9.
39. *Queen v. Palmer*, pp. 287–8.
40. *Morning Chronicle*, 12 June 1856, p. 6.
41. Bell and Redwood, p. 243.
42. Taylor, *Strychnia*.
43. Ibid. (italics in the original).
44. Ibid, p. 4.
45. 'Charges of Manslaughter against a Surgeon', *The Lancet*, vol. 1 (1856), pp. 692–3.

Chapter 14: Expert Witnesses

1. Bell and Redwood, p. 245.
2. Bell and Redwood, p. 246.
3. *The Times*, 14 November 1856, p. 12.
4. Spelling from the Lancaster criminal registers, 9 August 1856, p. 109, accessed via www.ancestry.com.
5. *The Times*, 23 August 1856, p. 10.
6. *The Times*, 25 August 1856, p. 9.
7. Ibid.
8. Anon., *Report from the Select Committee of the House of Lords on the Sale of Poisons, &c. Bill (H.L.)* (House of Commons, London, 1857), p. 3.
9. Ibid., p. 4.
10. Ibid.
11. Ibid., p. 21.
12. Ibid., p. 22.
13. Ibid., p. 39.
14. Ibid., p. 43.
15. Ibid., p. 64.
16. Ibid.
17. Ibid., p. 95.
18. Ibid., p. 96.
19. *Caledonian Mercury*, 9 July 1857, p. 4.
20. *Wells Journal*, 18 July 1857, p. 6.
21. *The Times*, 5 August 1857, p. 12.
22. *All the Year Round*, no. 327 (13 July 1865), p. 81.
23. Katherine Watson has supplied a table of English-based experts consulted in nineteenth-century poison trials. Taylor leads the field with thirty-one, Thomas Stevenson – who was active later in the century – twenty-four, Herapath, twenty-two, and Letheby, twelve. Seven other experts account for thirty-seven cases between them. Watson, p. 167.

24. Many nineteenth-century German universities offered medical degrees for a price. See M. Jeanne Peterson's *The Medical Profession in Mid-Victorian London* for more information.
25. Parry, Leonard A. (ed.), *Trial of Dr Smethurst* (W. Hodge & Co., Edinburgh and London, 1931), p. 241.
26. Her baptismal record shows that she was born on 15 September 1816.
27. Parry, *Trial of Dr Smethurst*, pp. 80–1.
28. Ibid., p. 3.
29. Ibid., p. 158.
30. *Morning Chronicle*, 9 May 1859, p. 7.
31. Parry, *Trial of Dr Smethurst*, p. 78.
32. Ibid., p. 11.
33. Ibid.
34. *British Medical Journal*, vol. 1, no. 139 (27 August 1859), p. 702.
35. Ibid., p. 703.
36. *Illustrated Times*, 3 September 1859, p. 11.
37. *The Times*, 24 August 1859, p. 12.
38. *Dublin Medical Press*, quoted in Parry, *Trial of Dr Smethurst*, p. 19.
39. *The Times*, 22 August 1859, p. 6.
40. *BMJ*, vol. 2, no. 140 (3 September 1859), p. 725.
41. Ibid., pp. 725–6.
42. *The Times*, 26 August 1859, p. 9.
43. Parry, *Trial of Dr Smethurst*, p. 22.
44. *The Times*, 17 November 1864, p. 5.
45. Parry, *Trial of Dr Smethurst*, p. 255.
46. Ibid., p. 257.
47. Ibid.

Chapter 15: Deadly Doctors

1. *The Times*, 30 August 1864, p. 6.
2. Ibid.
3. Ibid.
4. Ibid.
5. Ibid.
6. In 1864, he gave his age as thirty-four: *London Daily News*, 11 May 1864, p. 5.
7. Claude, Antoine, *Mémoires de M Claude, Chef de la Police de Sûreté*, two vols, eighteenth edition (Jules Rouff, Paris, 1881), vol. 2. p. 151.
8. Taylor's *Principles* 1894, vol. 1, pp. 478–9 and vol. 2, p. 640.
9. A legend soon arose and began to circulate in the newspapers that, after his execution, a doctor conducted some tests on Pommerais's head to discover how long signs of life remained. This was denied at the time, since the remains were taken to Montparnasse Cemetery direct from the scaffold, and no doctor had access to them. *London Daily News*, 25 June 1864, p. 5.
10. Conan Doyle, Sir Arthur, *The Penguin Complete Sherlock Holmes* (Penguin, London 1981), p. 270.
11. Roughhead, William (ed.), *Trial of Dr Pritchard* (W. Hodge & Co., Glasgow and Edinburgh, 1906), p. 15.
12. *Sheffield Telegraph* quoted in *Trial of Pritchard*, p. 18.
13. *Trial of Pritchard*, p. 19.
14. Based on her own evidence at the trial, she was born in October 1848.
15. In the 1860s, the basic tincture of aconite had not been standardised and preparations of varying strengths were available.
16. Geoghegan, T. G., 'Account of a Case of Poisoning by Monkshood which formed the Subject of a Criminal Trial', *Dublin Journal of Medical Science* (Hodges & Smith, Dublin, 1841), vol. xix, pp. 401–29.

17. Ibid., p. 406, footnote.
18. Ibid., p. 425.
19. *Trial of Pritchard,* p. 28.
20. Ibid., p. 140.
21. Ibid.
22. Ibid., p. 149.
23. Ibid., p. 94.
24. Ibid., p. 142.
25. Ibid., p. 33.
26. Ibid.
27. Ibid.
28. Ibid., p. 37.
29. Ibid., p. 38.
30. Ibid.
31. Ibid., p. 101.
32. Ibid., p. 102.
33. The former John Inglis, noted for his defence of Madeleine Smith in 1857.
34. *Trial of Pritchard,* p. 150.
35. Ibid., p. 183.
36. The bitter opiate taste of Battley's would have gone some way to disguise the aconite, and Mrs Taylor, taking her usual dose, could well have swallowed a fatal amount. Pritchard was probably following the same principle as Castaing and Palmer before him, using a powerful poison as the principal means of murder, with antimony as an emetic to remove the evidence.
37. *Trial of Pritchard,* p. 320, Appendix VIII, item III.

Chapter 16: Matrimonial Causes

1. *The Times*, 8 July 1865, p. 6.
2. Bell and Redwood, p. 377.
3. Registrar General's statistics, quoted in Berridge, Virginia and Edwards, Griffith, *Opium and the People* (Free Association Press, London, 1998), Part 5, Chapter 10, pp. 120–1.
4. *The Times*, 1 September 1871, p. 9.
5. *Brighton Examiner,* 27 June 1871, p. 4.
6. Text reproduced in *The Times*, 9 September 1871, p. 6.
7. Ibid.
8. Ibid.
9. Ibid.
10. Ibid., p. 9.
11. *Brighton Gazette*, 24 August 1871, p. 6.
12. *The Times,* 17 January 1872, p. 12.
13. One such petition was signed by many leading citizens of Brighton, including Dr and Mrs Beard, members of Mrs Beard's family, chemist Isaac Garrett, and Mr and Mrs Boys. See TNA papers HO 45/9297/9472.
14. *The Lancet*, vol. 99, issue 2526 (27 January 1872), p. 127.
15. Ibid.
16. In the 1861 census, he is listed as teaching French at a school in Hougham, Kent.
17. Born on 18 July 1851, according to parish records accessed via www.ancestry.com.
18. Later Sir Henry, emeritus professor of forensic medicine, University of Edinburgh.
19. Smith, A. Duncan, FSA (Scot.) (ed.), *Trial of Eugène Marie Chantrelle* (Canada Law Book Company, Toronto, 1906), p. 88.
20. Ibid.
21. Ibid., p. 86.
22. Ibid., p. 196.

23. *The Times,* 11 May 1878, p. 11.
24. Ibid.
25. Ibid.
26. Ibid.

Chapter 17: Sleight of Hand

1. *Western Daily Press*, 22 December 1873, p. 3.
2. Ibid.
3. Allen, A. H., 'Vermin Killers Containing Strychnine', *Pharmaceutical Journal and Transactions*, vol. 20 (12 October 1889), pp. 296–9.
4. Bailey's wife was not named in the newspapers, but she was probably Hannah Comer, who was married in Bristol on 28 May 1864, and whose brother Frederick was a London publican. Her subsequent history is unknown.
5. *Manchester Courier and Lancashire General Advertiser*, 30 June 1886, p. 6.
6. *Manchester Times*, 14 August 1886, p. 3.
7. *The Times,* 24 July 1886, p. 7.
8. Adam, Hargrave L. (ed.), *Trial of Dr Lamson*, second edition (William Hodge & Co., London, Notable British Trials series, 1951), p. 51.
9. Ibid.
10. Ibid., p. 97.
11. Ibid., p. 98.
12. Ibid., p. 21.
13. Ibid., p. 98.
14. Sizer, N. B., MD, 'The New Cadaveric Alkaloids and their Medico-Legal Interest to the Profession', *Proceedings of the Medical Society of the County of Kings*, vol. 7, no. 3 (May 1882), p. 52.
15. Hargrave, *Trial of Lamson,* p. 112.
16. Ibid.
17. Ibid., p. 136.
18. Ibid., p. 137.
19. *The Times*, 17 April 1882, p. 9.
20. Hawkins, Sir H.,*The Reminiscences of Sir Henry Hawkins* (Thomas Nelson & Sons, London, 1904).

Chapter 18: Disappearing Poison

1. Adelaide was born in December 1855, Edwin in October 1844.
2. *The Times,* 16 December 1882, p. 9.
3. Stratmann, Linda, *Chloroform: The Quest for Oblivion* (The History Press, Stroud, 2005).
4. Squire, P. W., *Companion to the latest edition of the British Pharmacopoeia*, twelfth edition (J. & A. Churchill, London, 1880), p. 99.
5. Vickers, John A., 'George Dyson alias John Bernard Walker', *Methodist History*, vol. 41, no. 1 (October 2002).
6. For a discussion of this, see Stratmann, *Chloroform.*

Chapter 19: Our American Cousins

1. Essig, Mark, 'Poison Murder and Expert Testimony: Doubting the Physician in Late Nineteenth Century America, *Yale Journal of Law and the Humanities*, vol. 14, issue 1 (2002), Article 4, p. 180.
2. Thompson, C. J. S., *Poison Mysteries Unsolved* (Hutchinson & Co., London, 1937), p. 19.

3. *New York Times*, 1 January 1898, p. 2.
4. Thompson, *Poison Mysteries Unsolved*, p. 19.
5. Watson, p. 47.
6. Blyth, A. W. and Blyth, M. W., *Poisons: Their Effects and Detection* (Charles Griffin & Co., London, 1895), p. 325.
7. Born on 3 May 1871; see Wellman, Francis L., *The Trial of Carlyle W. Harris for Poisoning His Wife, Helen Potts, at New York* (Yale Law Library, New York, 1892.
8. Born 23 September 1869; see Ledyard, Hope, *Articles Speeches and Poems of Carlyle W. Harris, edited by 'Carl's Mother' (Hope Ledyard)* (J. S. Ogilvie, New York, c. 1893), p. 15.
9. *Trial of Carlyle W. Harris*, p. 199
10. Ibid., p. 91.
11. Ibid.
12. Ibid.
13. Ibid.
14. Ibid., p.19.
15. Ibid., p. 119.
16. Ibid., p. 121.
17. Ledyard, *Articles, Speeches and Poems*, p. 11.
18. He had actually died from gangrene of the foot. *New York Times*, 28 March 1893, p. 9.
19. *New York Evening World*, 29 March 1893, p. 1.
20. *New York Times*, 15 April 1893, p. 12.
21. Trial of Buchanan, quoted in Essig, 'Poison Murder', p. 195.
22. *New York Times*, 11 April 1893, p. 3.
23. *New York Times*, 15 April 1893, p. 12.
24. *New York Times*, 11 April 1893, p. 4.
25. *New York Times*, 28 April 1893, p. 4.
26. The newspapers reporting the trial stated that he was a grandson of Carl Scheele, but this remains unproven.
27. *New York Times*, 6 June 1896, p. 1.
28. *New York Times*, 30 April 1895, p. 4.
29. *New York Times*, 18 June 1896, p. 8.
30. Taylor, *Strychnia*, p. 2.
31. Essig, 'Poison Murder', p. 8, footnote 31.
32. Hamilton, Allan McLane and Godkin, Lawrence, *A System of Legal Medicine* (E. B. Treat & Co., New York, 1900), Introduction p. 23.
33. Ibid., p. 24.
34. Editorial, *Medical Expert*, 48 JAMA 1355 (1907).
35. Stephen, Sir James Fitzjames, KCSI DCL, *A History of the Criminal Law of England*, three vols (Macmillan & Co., London, 1883), vol. 1, p. 575.
36. Ibid., vol. 1, p. 576.
37. Essig, 'Poison Murder', p. 183.
38. Stephen, *A History*, vol. 1, p. 572.
39. Trial of Edith Fenn, proceedings of the Central Criminal Court, Sessions House, Old Bailey, 13 January 1896, third session, p. 187, case 141, accessed via www.oldbaileyonline.org.
40. *Nottinghamshire Guardian*, 25 January 1896, p. 5.

Epilogue: Dangerous Drugs and Cruel Poisons

1. Blyth, A. W. and Blyth, M. W., *Poisons: Their Effects and Detection* (Charles Griffin & Co., London, 1895), p. 288.
2. Mann, J. Dixon, *Forensic Medicine and Toxicology*, fourth edition (Charles Griffin & Co., London, 1908), p. 407.
3. Ibid., p. 352.

4. Boos, William F. and Werby, A. Benjamin, 'Arsenic in Human Tissues and Food Animals. 1. So-called Normal Arsenic', *New England Journal of Medicine*, no. 213 (1935), pp. 520–7.
5. National Research Council Committee on Medical and Biological Effects of Environmental Pollutants, *Arsenic: Medical and Biologic Effects of Environmental Pollutants* (National Academies Press, Washington, DC, 1977).
6. Wilbert, M., 'Sale and Use of Poison', *Public Health Reports*, vol. 29, no. 46 (13 November 1914), pp. 3027–30.
7. Ibid., p. 3029.
8. *The Times*, 16 May 1913, p. 8.
9. *The Times,* 9 January 1930, p. 7.
10. *The Times,* 29 December 1936, p. 6.
11. *Aberdeen Evening Express,* 8 July 1952, p. 1.
12. *Yorkshire Evening Post,* 31 July 1953, p. 1.
13. Ibid.
14. Stone, David M., in *New York Journal of Commerce*, quoted in Ledyard, *Articles, Speeches and Poems*, p. 44.
15. Marks, Vincent and Richmond, Caroline, 'Kenneth Barlow: the First Documented Case of Murder by Insulin', *Journal of the Royal Society of Medicine*, vol. 101, no. 1 (January 2008), pp. 19–21.
16. Holden, Anthony, *The St Albans Poisoner* (Hodder & Stoughton, London, 1974), p. 121.
17. Watson, p. 209.

Bibliography

Archives

British Library

Letter to the chairman of Mr Wakley's Committee from T. W. Wansbrough 31 August 1830. British Library ref. 74/1880.c.1. Folio 180

The National Archives, Kew

ASSI 36/5 Assizes: Home, Norfolk and South-Eastern Circuit, 1846, Depositions. Sarah Chesham
ASSI 36/5 Assizes: Home, Norfolk and South-Eastern Circuit, 1850 Depositions. Sarah Chesham
CRIM 1/13/3 Central Criminal Court: Depositions. George Henry Lamson
CRIM 4/241 Central Criminal Court: Indictments. Felonies. (Includes John Bodle)
HO 12/122/37649 Home Office: Old Criminal Papers. Thomas Smethurst
HO 16/8 Home Office: Returns of committals for trial. (Includes John Bodle)
HO 27/25 p. 247 Criminal registers, England and Wales, A–L. (Includes John Bodle)
HO 45/9297/9472 Home Office: Registered Papers, 1872–1878. List of Criminal Cases, Christiana Edmunds
HO 144/90/A11385 Home Office: Registered Papers, Supplementary. George Henry Lamson
HO 144/263/A56680 Home Office Registered Papers, Supplementary. Eliza Fenning
MEPO 2/11423 Metropolitan Police: Office of the Commissioner: Correspondence and Papers. Death by arsenic poisoning of William Patrick Barnett. Enquiries and correspondence
MEPO 2/11424 Metropolitan Police: Office of the Commissioner: Death by arsenic poisoning of William Patrick Barnett. Divisional reports
MEPO 2/11425 Metropolitan Police: Office of the Commissioner: Divisional reports
MEPO 2/11426 Metropolitan Police: Office of the Commissioner: Correspondence and reports from provincial forces.
MEPO 2/11427 Metropolitan Police: Office of the Commissioner: Statements
MEPO 2/11428 Metropolitan Police: Office of the Commissioner: Appendices A–Z
PCOM 2/21 Millbank Prison, Middlesex: register of prisoners. (Includes John Bodle)
TS 25/513 Treasury Solicitor and HM Procurator General: Law Officers' and Counsel's Opinions. Suspected Poisoning of Richard Chesham by his wife Sarah Chesham

Published Sources

Adam, Hargrave L. (ed.), *Trial of Dr Lamson*, second edition (William Hodge & Co., Notable British Trials series, London, 1951)

Anon., *Affecting Case of Eliza Fenning Who Suffered the Sentence of the Law July 26, 1815*, eighth edition, including an account of the funeral, with other additional letters (John Fairburn, London, 1815)

Anon., *Causes criminelles célèbres du dix-neuvième siècle rédigés par une Société d'Avocats*, four vols (H. Langlois Fils et Cie, Paris, 1828)

Anon., *Circumstantial Evidence, the Extraordinary Case of Eliza Fenning* (Cowie & Strange, London, undated)

Anon., *The Edinburgh Annual Register for 1823*, vol. 16 (Constable & Co., Edinburgh, 1824)

Anon., *A Full Report of the Evidence Taken at the Thames Police Court, and the Coroner's Inquest, before Mr Baker, and a Respectable Jury, at Stepney, on the 10th of June, 1844, on the Alleged Poisoning Case, also, the Trial of J. C. Belany[sic] etc.* (G. Pike, Alnwick, 1844), http://www.forgottenbooks.com/readbook_text/A_Full_Report_of_the_Evidence_Taken_at_the_Thames_Police_Court_and_the_v1_1000853534/65

Anon., *The Hungry Forties, Life Under the Bread Tax* (T. Fisher Unwin, London, 1904)

Anon., *Great Burdon Slow Poisoning Case, Report of the Investigation before the Magistrates at Darlington in the County of Durham . . . Together with the Evidence of Mr J. S. Wooler before the Coroner at Great Burdon* (Robert Swales, London, 1855)

Anon., *The Life, Confessions and Execution of Mrs Burdock, who was Executed at the New Drop, Bristol Gaol, for the Murder of Mars Clara Ann Smith* (John Bonner, Bristol, 1835)

Anon., *Procès Complet d'Edme-Samuel Castaing, Docteur en Médecine* (Pillet, Paris, 1823)

Anon., *Procès de Madame Lafarge, empoisonnement* (Pagnerre, Paris, 1840)

Anon., *The Queen v. Palmer. Verbatim Report of the Trial of William Palmer* (J. Allen, London, 1856)

Anon., *Report from the Select Committee of the House of Lords on the Sale of Poisons, &c. Bill (H.L.)* (House of Commons, London, 1857)

Anon., *The Times Report of the Trial of William Palmer, for Poisoning John Parsons Cook, at Rugeley* (Ward & Lock, London, 1856)

Anon., *Trial of Mary Ann Burdock for the Wilful Murder of Mrs Clara Ann Smith by Administering Sulphuret of Arsenic to the Said Clara Ann Smith, Before the Recorder, Sir Charles Wetherell, at the Bristol Assizes, April the 10th, 11th and 13th, 1835* (W. H. Somerton, Bristol, 1835)

Anon., *The Trial of Robert Sawle Donnall, Surgeon and Apothecary, late of Falmouth, in the County of Cornwall, for the Wilful Murder, by Poison, of Mrs. Elizabeth Downing, Widow, his Mother-in-law, at the Assize at Launceston, for the County aforesaid, On Monday, March 31, 1817, Before the Honourable Sir Charles Abbott, Knt.* (James Lake, Falmouth, 1817)

Anon., *The Year-book of Facts in Science and Art* (Charles Tilt, London, 1840; Tilt & Bogue, London, 1842)

Bardsley, James Lomax, *Hospital Facts and Observations, Illustrative of the Efficacy of the New Remedies, Strychnia, Brucia, Acetate of Morphia, Veratria, Iodine, &c.* (Burgess & Hill, London, 1830)

Barlow, George H. and Babington, P. James (eds), *Guy's Hospital Reports*, vol. II (Samuel Highly, London, 1837)

Bartrip, Peter, 'A "Pennurth of Arsenic for Rat Poison": The Arsenic Act 1851 and the Prevention of Secret Poisoning', *Medical History*, 1992

Bates, Stephen, *The Poisoner: The Life and Crimes of Victorian England's Most Notorious Doctor* (Duckworth & Co., London, 2014)

Beal, E. (ed.), *The Trial of Adelaide Bartlett for Murder, complete and revised report* (Stevens & Haynes, London, 1886)

Bell, Jacob and Redwood, Theophilus, *Historical Sketch of the Progress of Pharmacy in Great Britain* (Pharmaceutical Society of Great Britain, London, 1880)

Berridge, Virginia and Edwards, Griffith, *Opium and the People* (Free Association Press, London, 1998)

Bertomeu-Sánchez J. R. and Nieto-Galan, A. (eds), *Chemistry, Medicine and Crime, Mateu Orfila (1787–1853) and his Times* (Science History Publications, Sagamore Beach, MA, 2006)

Blaker, Nathaniel Paine, MRCS, *Reminiscences* (The Southern Publishing Co., Brighton, 1900)

Blyth, A. W. and Blyth, M. W., *Poisons: Their Effects and Detection* (Charles Griffin & Co., London, 1895 and 1906)

Borrow, George (ed.), *Celebrated Trials and Remarkable Cases of Criminal Jurisprudence from the earliest Records to the Year 1825*, six vols (Knight & Lacey, London, 1825)

Bosell C. and Thompson, L., *The Girls in Nightmare House* (Gold Medal Books, London, 1959)

Buckingham, John, *Bitter Nemesis* (CRC Press, Abingdon, 2007)

Burney, I., *Poison, Detection and the Victorian Imagination* (Manchester University Press, Manchester, 2006)

Cabanès, A. and Nass, L., *Poisons et Sortilèges* (Librarie Plon, Paris, 1903)

Campbell, W. A., 'William Herapath, 1796–1868: a Pioneer of Toxicology', *Analytical Proceedings*, 1980

Chadwick, Sir Edwin, 'Supplementary Report on the Result of a Special Inquiry into the Practice of Interment in Towns', *British Parliamentary Papers*, 1843

Christison (eds), *The Life of Sir Robert Christison, Bart, edited by his sons,* two vols (William Blackwood & Sons, Edinburgh and London, vol. 1, 1885; vol. 2, 1886)

Christison, Robert, MD, *A Treatise On Poisons in Relation to Medical Jurisprudence, Physiology and the Practice of Physic* (Adam Black, Edinburgh, 1829)

Christison, Robert, MD, *A Treatise On Poisons in Relation to Medical Jurisprudence, Physiology and the Practice of Physic*, second edition (Adam Black, Edinburgh,1832)

Christison, Robert, MD, *A Treatise On Poisons in Relation to Medical Jurisprudence, Physiology and the Practice of Physic*, third edition (Adam Black, Edinburgh, 1836)

Christison, Robert, MD, FRSE, *A Treatise On Poisons in Relation to Medical Jurisprudence, Physiology and the Practice of Physic*, first American edition from fourth Edinburgh edition (ed. Barrington and Geo. D. Haswell, Philadelphia, 1845)

Clarke, E.G.C., 'Isolation and Identification of Alkaloids', in vol. 1 of Lundquist, Frank (ed.), *Methods of Forensic Science*, four vols (John Wiley & Sons, New York, 1962)

Clarke, Kate, *Bad Companions* (The History Press, Stroud, 2013)

Clarke, Kate, *Deadly Service* (Carrington Press, Hay-on-Wye, 2011)

Clarke, Kate, *In the Interests of Science: Adelaide Bartlett and the Pimlico Poisoning* (Mango Books, London, 2015)

Clarke, Kate, *Lethal Alliance* (Carrington Press, Hay-on-Wye, 2013)

Clarke, Kate, *The Pimlico Murder*, revised edition (Carrington Press, Hay-on-Wye, 2011)

Claude, Antoine, *Mémoires de M Claude, Chef de la Police de Sûreté,* two vols, eighteenth edition (Jules Rouff, Paris, 1881)

Collins, Wilkie, *Jezebel's Daughter* (Chatto & Windus, London, 1880)

Cox, David J., *Foul Deeds and Suspicious Deaths in Shrewsbury and Around Shropshire* (Wharncliffe Books, Barnsley, 2008)

Crowther, Anne and White, Brenda, *On Soul and Conscience: The Medical Expert and Crime. 150 Years of Forensic Medicine in Glasgow* (Aberdeen University Press, Aberdeen, 1988)

Dickens, Mamie and Hogarth, Georgina (eds), *The Letters of Charles Dickens*, three vols (Chapman and Hall, London, 1880)

Dupré, E. and Charpentier, R., *Les Empoisonneurs* (A. Rey & Cie, Lyon, 1909)

Edwards, Martin (ed.), *Truly Criminal: A Crime Writers' Association Anthology of True Crime* (The History Press, Stroud, 2015)

Fenning, Eliza, *Eliza Fenning's Own Narrative* (John Fairburn, London, 1815)

Goldsmith, Robert H., 'The Search for Arsenic', in Gerber, Samuel M. and Saferstein, Richard (eds), *More Chemistry and Crime: From Marsh Arsenic Test to DNA Profile* (American Chemical Society, Washington, DC, 1997)

Griffiths, Major Arthur, *The Fortresses of Romance and Crime* (Forgotten Books, London, 2013, reprint)

Hackwood, Frederick William, *William Hone, His Life and Times* (T. Fisher Unwin, London, 1912)

Hamilton, Allan McLane and Godkin, Lawrence, *A System of Legal Medicine* (E. B. Treat & Co., New York, 1900)

Hawkins, Sir H., *The Reminiscences of Sir Henry Hawkins* (Thomas Nelson & Sons, London, 1904)

Hempel, Sandra, *The Inheritor's Powder* (London, Phoenix, 2014)

Holden, Anthony, *The St Albans Poisoner* (Hodder & Stoughton, London, 1974)

Hone, W., *The Four Important Trials at Kingston Assizes, April 5, 181.6 With a preface containing thirteen questions to Isaac Espinasse Esq. . . . Respecting Mr Turner, the Prosecutor of Elizabeth Fenning*, second edition (W. Hone, London, c. 1816)

Houlton, Joseph, *Formulary for the Preparation and Employment of Several New Remedies*, sixth edition (T. & G. Underwood, London, 1830)

Irving H. B., *A Book of Remarkable Criminals* (Cassell & Co., London, 1918)

Knapp, Andrew and Baldwin, William, *The Newgate Calendar*, four vols (J. Robins & Co., London, 1828)

Jack, A. Fingland, MCom, *An Introduction to the History of Life Assurance* (P. S. King & Son, London, 1912)

Jennings, Rev. Thomas, *Two Sermons Occasioned by the Case of Mary Ann Burdock* (Seeleys, Bristol, 1835)

Lafarge, Marie, *Memoirs of Madame Lafarge, written by herself*, two vols (Henry Colburn, London, 1841)

Ledyard, Hope, *Articles, Speeches and Poems of Carlyle W. Harris, edited by 'Carl's Mother' (Hope Ledyard)* (J. S. Ogilvie, New York, c. 1893)

Livingston, James D., *Arsenic and Clam Chowder: Murder in Gilded Age New York* (State University of New York Press, Albany, 2010)

Lowndes, William Thomas, *The Bibliographer's Manual of English Literature* (George Bell & Sons, London, 1890)

Magendie, François, *Formulary for the Preparation and Employment of Many New Medicines*, trans. J. Baxter (New York, George T. Evans, 1828)

Mann, J. Dixon, *Forensic Medicine and Toxicology*, fourth edition (Charles Griffin & Co., London, 1908)

Marshall, John, *Five Cases of Recovery from the Effects of Arsenic . . . To which are annexed many corroborating facts never before published, relative to the guilt of Eliza Fenning* (C. Chapple, bookseller to the Prince Regent, London, 1815)

Marshall, John, *Remarks on Arsenic* (J. Callow, London, 1817)

Martineau, H., *A History of the Thirty Years' Peace, AD 1816–1846,* four vols (George Bell & Sons, London, 1877)

National Research Council Committee on Medical and Biological Effects of Environmental Pollutants, *Arsenic: Medical and Biologic Effects of Environmental Pollutants* (National Academies Press, Washington, DC, 1977)

Orfila, Mathieu, *Directions for the Treatment of Persons who have taken Poison and those in a state of Apparent Death* (Longman, Hurst, Rees, Orme & Brown, London, 1818)

Orfila, Mathieu, *A General System of Toxicology or a Treatise on Poisons, Found in the Mineral, Vegetable, and Animal Kingdoms, Considered in their Relations with Physiology, Pathology, and Medical Jurisprudence* (M. Carey & Son, Philadelphia, 1817)

Orfila, Mathieu, *Mémoire sur la nicotine et sur la conicine* (J-B. Baillière, Paris, 1854)

Orfila, Mathieu, *A Popular Treatise on the Remedies to be Employed in Cases of Poisoning and Apparent Death* (Solomon W. Conrad, Philadelphia, 1818)

Orfila, Mathieu, *A Practical Treatise on Poisons and Asphyxies Adapted to General Use* (Hilliard, Gray, Little & Wilkins, Boston, 1826)
Paget, John, esq., *Judicial Puzzles Gathered from the State Trials* (Sumner Whitney & Co., San Francisco, 1876)
Parascandola, John, *King of Poisons: A History of Arsenic* (Potomac Books, Washington, DC, 2012)
Paris, J. A. and Fonblanque, J. S. M., *Medical Jurisprudence*, three vols (W. Phillips, London, 1823)
Parker, C. E., *Some Micro-Chemical Tests for Alkaloids* (Charles Griffin & Co., London, 1921)
Parry, Leonard A. (ed.), *Trial of Dr Smethurst* (W. Hodge & Co., Edinburgh and London, 1931)
Peterson, M. Jeanne, *The Medical Profession in Mid-Victorian London* (University of California Press, Berkeley, 1978)
Porter, Frank Thorpe, *Gleanings and Reminiscences*, third edition (Hodges, Foster & Co., Dublin, 1875)
Rognetta, M, *Nouvelle méthode de traitement de l'empoisonnement par l'arsenic et documens médico-légaux sur cet empoisonnement* (Chez Gardenbas, Paris, 1840)
Roughhead, William (ed.), *Trial of Dr Pritchard* (W. Hodge & Co., Glasgow and Edinburgh, 1906)
Roughhead, William (ed.), *Trial of Mary Blandy* (W. Hodge & Co., Edinburgh and London, 1914)
Rush, Benjamin, *Medical Inquiries and Observations*, four vols, second edition (J. Conrad & Co., Philadelphia, 1805)
Ryan, Michael, *A Manual of Medical Jurisprudence* (Sherwood, Gilbert & Piper, London, 1836)
Saunders, Edith, *The Mystery of Marie Lafarge* (Clerke & Cockeran, London, 1951)
Schechter, H., *Depraved* (Pocket Books, New York, 1996)
Schechter, H., *The Devil's Gentleman* (Ballantine Books, New York, 2008)
Seleski, Patty, 'Domesticity is in the Streets: Eliza Fenning, Public Opinion and the Politics of Private Life', in Harris, Tim (ed.), *The Politics of the Excluded* (Palgrave, Hampshire and New York, 2001)
Sly, Nicola, *Murder by Poison: A Casebook of Historic British Murders* (The History Press, Stroud, 2009)
Smith, A. Duncan, FSA (Scot.), (ed.), *Trial of Eugène Marie Chantrelle* (Canada Law Book Company, Toronto, 1906)
Smith, G. Munro, *A History of the Bristol Royal Infirmary from 1750 to 1899* (J. Arrowsmith, Bristol, 1917)
Snow, John, MD, *On Chloroform and Other Anaesthetics: Their Action and Administration* (ed. B. W. Richardson, John Churchill, London, 1858)
Sonnedecker, Glenn and Kremers, Edward, *Kremers and Urdang's History of Pharmacy* (American Institute of the History of Pharmacy, USA, 1986)
Squire, P. W., *Companion to the latest edition of the British Pharmacopoeia*, twelfth edition (J. & A. Churchill, London, 1880)
Stephen, Sir James Fitzjames, KCSI DCL, *A History of the Criminal Law of England*, three vols (Macmillan & Co., London, 1883)
Stephen, Leslie (ed.), *Dictionary of National Biography*, twenty-one vols (Smith, Elder & Co., London, 1885–1900)
Stevenson, T., MD (ed.), *Taylor's Principles and Practice of Medical Jurisprudence*, two vols, fourth edition (J. & A. Churchill, London, 1894)
Stratmann, Linda, *Chloroform: The Quest for Oblivion* (The History Press, Stroud, 2005)
Stratmann, Linda, *Gloucestershire Murders* (The History Press, Stroud, 2005)
Taylor, Alfred Swaine, MD, FRS, *On Poisoning by Strychnia, with Comments on the Medical Evidence Given at the Trial of William Palmer for the Murder of John Parsons Cook* (Longman, Brown, Green, Longmans & Roberts, London, 1856)

Taylor, Alfred Swaine, MD, FRS, *On Poisons in Relation to Medical Jurisprudence and Medicine*, first American edition (Lea & Blanchard, Philadelphia, 1848)
Taylor, Alfred Swaine, MD, FRS, *On Poisons in Relation to Medical Jurisprudence and Medicine,* second American edition from second London edition (Blanchard & Lea, Philadelphia, 1859)
Taylor, Alfred Swaine, MD, FRS, *On Poisons in Relation to Medical Jurisprudence and Medicine,* third American edition from third London edition (Henry C. Lea, Philadelphia, 1875)
Taylor, Bernard and Clarke, Kate, *Murder at the Priory* (Grafton, London, 1988)
Thompson, C. J. S., *Poison Mysteries Unsolved* (Hutchinson & Co., London, 1937)
Thorwald, Jürgen, *Proof of Poison* (Pan Books, London, 1969)
Tilstone, William, Savage, Kathleen A. and Clark, Leigh A., *Forensic Science: An Encyclopedia of History, Methods and Techniques* (ABC-CLIO, Santa Barbara, 2006)
Wansbrough, T. W., *An Authentic Narrative of the conduct of Elizabeth Fenning from the time that the warrant arrived for her death till her execution: with copies of original letters, &c. By the gentleman who attended her during the whole period* (Ogles, Duncan & Cochran, London, 1815)
Watson, Katherine, *Poisoned Lives: English Poisoners and their Victims* (Hambledon Continuum, London, 2004)
Watkins, John, LLD, *The Important Results of an Elaborate Investigation into the Mysterious Case of Elizabeth Fenning* (William Hone, London, 1815)
Wellman, Francis L., *The Trial of Carlyle W. Harris for Poisoning His Wife, Helen Potts, at New York* (Yale Law Library, New York, 1892)
Whorton, James C., *The Arsenic Century: How Victorian Britain Was Poisoned at Home, Work, & Play* (Oxford University Press, New York, 2010)
Young, Winifred, *Obsessive Poisoner* (Robert Hale, London, 1973)

Periodicals

Newspapers

Aberdeen Evening Express
Berkshire Chronicle
Birmingham Daily Post
Bolton Chronicle
Brighton Examiner
Brighton Gazette
Brighton Guardian
Bristol Mercury
Bristol Times and Mirror
Bucks Herald
Bury and Norwich Post
Caledonian Mercury
The Champion
Chelmsford Chronicle
Daily Gazette for Middlesbrough
Daily Mirror
Durham County Advertiser
Essex Standard
The Examiner
Fife Herald
Galignani's Messenger
Gateshead Observer
Gazette des tribunaux, journal de jurisprudence et des débats judiciaires
Glens Falls NY Morning Star

The Globe
Gloucester Journal
Hampshire Advertiser
Illustrated Times
Journal de l'Ain
Journal des Débats Politique et Littéraires
Leamington Spa Courier
Leicester Chronicle
Leicester Mercury
Lincolnshire Chronicle
London Daily News
London Standard
Maidstone Journal
Manchester Courier and Lancashire General Advertiser
Manchester Mercury
Manchester Times
Methodist History
Morning Chronicle
Morning Post
Newcastle Chronicle
Newcastle Courant
Newcastle Journal
New York Evening World
New York Times
Northampton Mercury
Northern Whig
Nottinghamshire Guardian
Oxford Journal
Preston Chronicle
Punch
Reading Mercury
Royal Cornwall Gazette
Salisbury and Winchester Journal
Schuylerville Standard
Sharpe's London Magazine
The Spectator
Staffordshire Advertiser
Stamford Mercury
The Sun
The Times
Tyne Mercury
Walker's Hibernian Magazine, or, Compendium of Entertaining Knowledge
Weekly Despatch
Wells Journal
Western Daily Press
Yorkshire Evening Post

Journals and Magazines

All the Year Round
American Journal of the Medical Sciences
Annales d'Hygiène Publique et de Médecine Légale
Annals of Electricity, Magnetism, & Chemistry; and Guardian of Experimental Science
Annals of Science

Annual Register
Arkansas Family Historian
Association Medical Journal
Boston Medical and Surgical Journal
Boston University Journal of Science and Technology Law
British Medical Journal
Chemical Heritage Magazine
The Chemist
Clinical Chemistry
Dictionary of National Biography
Drug Test Analysis
Dublin Journal of Medical Science
Edinburgh Annual Register
Edinburgh Medical Journal
Edinburgh New Philosophical Journal
Guy's Hospital Reports
Hansard
Household Words
Journal of the American Medical Association
Journal of the American Medical Society
Journal of Chemical Education
Journal de Pharmacie et des Sciences Accessoires
Journal of the Royal Society of Medicine
The Lancet
London and Edinburgh Philosophical Magazine and Journal of Science
London Medical Gazette
Medical Expert
Medical Gazette
Medical History
New England Journal of Medicine
Pharmaceutical Journal and Transactions
Philosophical Transactions of the Royal Society of London
Proceedings of the Medical Society of the County of Kings
Provincial Medical and Surgical Journal
Public Health Reports
Sharpe's London Magazine of Entertainment and Instruction for General Reading
Yale Journal of Law and the Humanities
Yearbook of Facts

Acts of Parliament

An Act for the Amendment and Better Administration of the Laws relating to the Poor in England and Wales 1834 (4 & 5 William IV, c. 76)
An Act for Better Regulating the Practice of Apothecaries Throughout England and Wales 1815 (55 Geo. III, c. 194)
An Act for Registering Births, Deaths, and Marriages in England 1836 (6 & 7 William IV, c. 86)
Births and Deaths Registration Act 1926 (Regnal. 16 & 17 Geo. 5, c. 48)
The Central Criminal Court Act [Palmer Act] (19 & 20 Vict., c. 16)
Offences Against the Person Act 1861 (24 & 25 Vict., c. 100)
Pharmacy Act 1852 (15 & 16 Vict., c. 56)
Pharmacy Act 1868 (31 & 32 Vict., c. 121)
The Sale of Arsenic Regulation Act 1851 (14 & 15 Vict., c. 13)
The Registration of Births and Deaths Act 1874 (c. 88)

Genealogical Sources

Family history records accessed via www.ancestry.com
The will of Joseph Snaith Wooler supplied by the UK probate office

Online Sources

Ainsley, Jill Louise, 'The Ordeal of Sarah Chesham', Master of Arts degree thesis 1997, Department of History, University of Victoria, accessed June 2015: http://dspace.library.uvic.ca/bitstream/handle/1828/4351/Ainsley_Jill_MA_2012.pdf?sequence=1&isAllowed=y

Archie Belaney known as Grey Owl by Barry Brown: http://wabbrown.co.uk/greyowl.pdf

Hektoen International, Journal of Medical Humanities, accessed May 2015: http://www.hektoeninternational.org/

Old Bailey Online: http://www.oldbaileyonline.org

Caudill, David S., 'Prefiguring the Arsenic Wars', *Chemical Heritage Magazine*, Chemical Heritage Foundation, accessed June 2015: http://www.chemheritage.org/discover/media/magazine/articles/27-1-prefiguring-the-arsenic-wars.aspx?page=1

'Kuntal Patel jailed over 'dark web' poison purchase', BBC News, 7 November 2014: http://www.bbc.co.uk/news/uk-england-london-29958993

Daly, Ann, MA, 'The Dublin Medical Press and medical authority in Ireland 1850–1890', thesis for the degree of PhD, Department of Modern History, National University of Ireland, Maynooth, January 2008, accessed 3 August 2015: http://eprints.maynoothuniversity.ie/5269/

Index